Lecture Notes in Mathematics 1627

Editors:
A. Dold, Heidelberg
F. Takens, Groningen

Subseries: Fondazione C.I.M.E., Firenze

Advisor: Roberto Conti

Springer
Berlin
Heidelberg
New York
Barcelona
Budapest
Hong Kong
London
Milan
Paris
Santa Clara
Singapore
Tokyo

C. Graham Th. G. Kurtz S. Méléard
Ph. E. Protter M. Pulvirenti D. Talay

Probabilistic Models for Nonlinear Partial Differential Equations

Lectures given at the 1st Session of the
Centro Internazionale Matematico Estivo
(C.I.M.E.)
held in Montecatini Terme, Italy,
May 22-30, 1995

Editors: D. Talay, L. Tubaro

Fondazione
C.I.M.E.

Springer

Authors

Carl Graham
CMAP, URA CNRS 756
Ecole Polytechniqe
F-91128 Palaiseau, France
e-mail: carl @ cmapx. polytechique.fr

Thomas G. Kurtz
Dept. of Mathematics and Statistics
University of Wisconsin-Madison
Madison, WI 53706-1388, USA
e-mail: kurtz @ math.wisc.edu

Sylvie Méléard
MODAL'X, UFR SEGMI
Université Paris X
200, Av. de la République
F-92000 Nanterre, France

and

Laboratoire de Probabilités
Université Paris VI
4, Place Jussieu
F-75252 Paris Cedex 5, France
e-mail: sylm @ ccr.jussieu.fr

Philip E. Protter
Dept. of Mathematics and Statistics
Purdue University
West Lafayette, IN 47907-1395, USA
e-mail: protter @ math.purdue.edu

Mario Pulvirenti
Dipartimento di Matematica
Università di Roma "La Sapienza"
I-00185 Roma, Italy
e-mail: pulvirenti @ sci.uniroma1.it

Denis Talay
INRIA
2004 Route des Lucioles
BP 93
F-06902 Sophia-Antipolis, France
e-mail: talay @ sophia.inria.fr

Editors

Denis Talay
INRIA

Luciano Tubaro
Dipartimento die Matematica
Università di Trento
I-38050 Povo (Trento), Italy
e-mail: tubaro @ science.unitn.it

Cataloging-Data applied for

Die Deutsche Bibliothek - CIP-Einheitsaufnahme

Centro Internazionale Matematico Estivo <Firenze>: Lectures given at the ... session of the Centro Internazionale Matematico Estivo (CIME) ... - Berlin ; Heidelberg ; New York ; London ; Paris ; Tokyo ; Hong Kong : Springer
Früher Schriftenreihe. - Früher angezeigt u.d.T.: Centro Internazionale Matematico Estivo: Proceedings of the ... session of the Centro Internazionale Matematico Estivo (CIME)
NE: HST
1995,1. Probabilistic models for nonlinear partial differential equations. - 1996

Probabilistic models for nonlinear partial differential equations : held in Montecatini Terme, Italy, May 22 -30, 1995 / Carl Graham ... Ed.: D. Talay ; L. Tubaro. - Berlin ; Heidelberg ; New York ; Barcelona ; Budapest ; Hong Kong ; London ; Milan ; Paris ; Santa Clara ; Singapore ; Tokyo : Springer, 1996
(Lectures given at the ... session of the Centro Internazionale Matematico Estivo (CIME) ... ; 1995,1)
(Lecture notes in mathematics ; Vol. 1627 : Subseries: Fondazione CIME)
ISBN 3-540-61397-8
NE: Graham, Carl; Talay, Denis [Hrsg.]; 2. GT

Mathematics Subject Classification (1991): 60H10, 60H30, 60H15, 60K35, 60K30, 60G07, 65C20, 65M15, 65U05

ISSN 0075-8434
ISBN 3-540-61397-8 Springer-Verlag Berlin Heidelberg New York

Typesetting: Camera-ready TEX out put by the authors
SPIN: 10479756 46/3142-543210 - Printed on acid-free paper

Preface

These last years, there have been important developments in the probabilistic interpretation of nonlinear Partial Differential Equations, the theory of the convergence of law of stochastic processes and the numerical approximation of stochastic processes.

All these developments offer the appropriate theoretical background to analyse probabilistic algorithms used to solve equations as important in practice as the Navier-Stokes equation, the Boltzmann equation and certain Stochastic Partial Differential Equations. They also permit us to construct new methods.

For example, all the works around the propagation of chaos, particularly those of A-S. Sznitman, permit a quite new and fruitful point of view on the random vortex methods in Fluid Mechanics and on Monte–Carlo methods for Boltzmann-like equations. Likewise, the numerical analysis of stochastic differential equations has recently progressed in several interesting directions (variance reduction techniques, simulation of reflected diffusion processes, convergence in law of the normalized trajectorial error, asymptotic expansions of the discretization error).

Weak limit theorems for stochastic integrals naturally are among the main ingredients in the study of interacting particle systems, approximation procedures for solutions of stochastic differential equations, etc. A selection of such theorems in view of the analysis of applied problems should be useful. Besides, quite new weak limit theorems have just appeared for stochastic integrals with respect to infinite dimensional semimartingales.

We therefore enthusiastically answered Prof. V. Capasso's suggestion to submit to the CIME a proposal for a Summer School on the probabilistic models for nonlinear PDE's and their numerical applications with a three-fold emphasis: first, on the weak convergence of stochastic integrals; second, on the probabilistic interpretation and the particle approximation of equations coming from Physics (conservation laws, Boltzmann-like and Navier-Stokes equations); third, on the modelling of networks by interacting particle systems.

We thank all the participants to this Summer School which was held in Montecatini from May 22th to May 30th. The exchanges between the lecturers and the audience were very useful for everybody.

We thank all the lecturers (Carl Graham, Tom Kurtz, Sylvie Méléard, Philip Protter, Mario Pulvirenti) for having given fascinating lectures and for having written pedagogic and deep contributions to the present volume.

We hope that this book will be useful for our colleagues working on stochastic particle methods and on the approximation of SPDE's and in particular, for Ph.D. students and for young researchers.

We thank CIME for its generous financial support and for arranging the location in Montecatini offering us the combined delights of beautiful Tuscan surroundings and gastronomical excellence.

Sophia–Antipolis and Trento, December 1995

Denis Talay and Luciano Tubaro

Table of Contents

Weak Convergence of Stochastic Integrals and Differential Equations

Thomas G. Kurtz[1]* and Philip E. Protter[2] **

[1] Departments of Mathematics and Statistics University of Wisconsin - Madison, Madison, WI 53706-1388, USA
[2] Departments of Mathematics and Statistics, Purdue University, West Lafayette, IN 47907-1395, USA

1. Semimartingales

Let W denote a standard Wiener process with $W_0 = 0$. For a variety of reasons, it is desirable to have a notion of an integral $\int_0^1 H_s dW_s$, where H is a stochastic process; or more generally an indefinite integral $\int_0^t H_s dW_s$, $0 \le t < \infty$. If H is a process with continuous paths, an obvious way to define a stochastic integral is by a limit of sums: let $\pi^n[0,t]$ be a sequence of partitions of $[0,t]$, with $\text{mesh}(\pi^n) = \sup_i (t_{i+1} - t_i)$, where $0 = t_0 < t_1 < \ldots < t_n = t$ are the successive points of the partition. Then one could define

$$\int_0^t H_s dW_s = \lim_{n\to\infty} \sum_{t_i \in \pi^n[0,t]} H_{t_i}(W_{t_{i+1}} - W_{t_i}) \tag{1.1}$$

when $\lim_{n\to\infty} \text{mesh}(\pi^n) = 0$. If one wants the natural condition that (1.1) holds for all continuous processes H, then it is an elementary consequence of the Banach-Steinhaus theorem that W must have a.s. paths of finite variation on compacts. Of course this is precisely not the case for the Wiener process. The key insight of K. Itô in the 1940's was to ask for condition (1.1) to hold only for *adapted* continuous stochastic processes. We will both explain this idea and extend it to a large class of stochastic processes: exactly those for which both the integral exists as a limit of sums, and for which we also have a dominated convergence theorem.

We suppose given a filtered probability space $(\Omega, \mathcal{F}, P, \mathbf{F})$, where $\mathcal{F}$ is a P-complete σ-algebra and where $\mathbf{F} = (\mathcal{F}_t)_{0\le t<\infty}$ is a filtration of σ-algebras: i.e., $\mathcal{F}_s \subset \mathcal{F}_t$ if $s \le t$. We also assume that $\mathcal{F}_0$ contains all the P-null sets of $\mathcal{F}_0$ and that $\mathbf{F}$ is right continuous: that is, $\mathcal{F}_t = \mathcal{F}_{t+} = \cap_{u>t}\mathcal{F}_u$. (Note that if W is a standard Wiener process with its natural filtration $\mathbf{F}^0 = (\mathcal{F}_t^0)_{0\le t<\infty}$, where $\mathcal{F}_t = \sigma(W_s; s \le t)$, then if one adds the P-null sets of $\mathcal{F}_t^0$ to $\mathcal{F}_t^0$, all t, the resulting filtration $\mathbf{F}$ satisfies the preceding hypotheses, which are known

* Research supported in part by NSF grants DMS 92-04866 and DMS 95-04323
** Research supported in part by NSF grant INT 94-01109 and NSA grant MDR 904-94-H-2049

as the *usual hypotheses*. The same holds for Lévy processes and for most strong Markov processes.)

Let X be an *adapted* process with càdlàg paths: that is, X_t is $\mathcal{F}_t$-measurable, each $t > 0$, and a.s. has paths which are right continuous with left limits.[1]

Definition 1.1. *A process H is* simple predictable *if H has a representation*

$$H_t = H_0 1_{\{0\}}(t) + \sum_{i=1}^{n} H_i 1_{(T_i, T_{i+1}]}(t) \tag{1.2}$$

where $0 = T_1 \leq \ldots \leq T_{n+1} < \infty$ is a finite sequence of stopping times, $H_i \in \mathcal{F}_{T_i}$, $|H_i| < \infty$ a.s., $0 \leq i \leq n$. The collection of simple predictable processes is denoted $\mathbf{S}$.

Let $\mathbf{L}^0$ *denote all a.s. finite random variables. We topologize* $\mathbf{L}^0$ *with convergence in probability, and we topologize* $\mathbf{S}$ *with uniform convergence (in (t,ω)) and denote it* $\mathbf{S}_u$. *For a given X we define an operator I_X mapping* $\mathbf{S}$ *to* $\mathbf{L}^0$ *by (with H as in (1.2)):*

$$I_X(H) = H_0 X_0 + \sum_{i=1}^{n} H_i (X_{T_{i+1}} - X_{T_i}). \tag{1.3}$$

Definition 1.2. *A process X is a* semimartingale *if $I_X : \mathbf{S}_u \to \mathbf{L}^0$ is continuous on compact time sets.*

Definition (1.2) is not customary. We give the customary definition here, and to distinguish it from ours we call it a "classical" semimartingale.

Definition 1.3. *A process X is a* classical semimartingale *if it is adapted, càdlàg, and has a decomposition $X = M + A$, where M is a local martingale, and A (is adapted, càdlàg, and) has paths of finite variation on compacts.*

One of the deepest results in the theory of semimartingales is the following, proved around 1978, primarily by C. Dellacherie and K. Bichteler.

Theorem 1.4 (Bichteler-Dellacherie). *An adapted, càdlàg process X is a semimartingale if and only if it is a classical semimartingale.*

We remark that the deeper implication is the "only if".

Also note that the Bichteler-Dellacherie theorem gives us many *examples of semimartingales:*

(i) Any local martingale, such as the Wiener process, is a semimartingale.
(ii) Any finite variation process, such as the Poisson process, is a semimartingale.

---

[1] "càdlàg" is the French acronym for right continuous with left limits

(iii) The Doob-Meyer decomposition theorem states that any submartingale Y can be written $Y = M + A$, where M is a local martingale and A is an adapted, càdlàg process with nondecreasing paths. Thus, any submartingale (and hence any supermartingale) is a semimartingale.

(iv) If Z is a Lévy process (i.e., a càdlàg process with stationary and independent increments), then if $E\{|Z_t|\} < \infty$, each t, one has $E\{|Z_t|\} = \alpha t$ (assuming $Z_0 = 0$) and thus $Z_t = (Z_t - \alpha t) + \alpha t$ is a decomposition for Z,a nd Z is a semimartingale. More generally it can be shown that any Lévy process is a semimartingale.

(v) Most "reasonable" real valued strong Markov processes are semimartingales.

(vi) An illustrative example of a Lévy process that is a martingale is as follows: let N^i be a sequence of i.i.d. Poisson processes with arrival intensities $\alpha_i (\alpha_i > 0)$. Let $|\beta_i| \leq c$ and assume $\sum_{i=1}^{\infty} \beta_i^2 \alpha_i < \infty$. Then

$$M_t = \sum \beta_i (N_t^i - \alpha_i t)$$

is a Lévy process. Note that if, for example, $\alpha_i = 1$ (all i) and $\beta_i = \frac{1}{i}$, then if $\Delta M_s = M_s - M_{s-}$ (the jump at time s), we have $\sum_{0<s\leq t} |\Delta M_s| = \sum_{0<s\leq t} \Delta M_s = \sum_{i=1}^{\infty} \frac{1}{i} N_t^i = \infty$ a.s. This is an example of a martingale that cannot be used, path by path, as a classical differential because of behavior arising purely from the jumps; that is, M has paths of infinite variation on compacts and one cannot define a Lebesgue-Stieltjes pathwise integral for M.

Finally let us note some simple but important properties of semimartingales.

Theorem 1.5. *The set of semimartingales is a vector space.*

Theorem 1.6. *If Q is another probability absolutely continuous with respect to P, then every P-semimartingale is a Q-semimartingale.*

Theorem 1.7 (Stricker). *If X is a semimartingale for a filtration $\mathbf{F}$, and if $\mathbf{G}$ is a subfiltration such that X is adapted to $\mathbf{G}$, then X is a $\mathbf{G}$-semimartingale as well.*

Proof. Theorem 1.5 is immediate from the definition. For Theorem 1.6 it is enough to remark that if $Q \ll P$, then convergence in P-probability implies convergence in Q probability. For Theorem 1.7, let $\mathbf{S}(\mathbf{F})$ denote $\mathbf{S}$ for the filtration $\mathbf{F}$. Since $\mathbf{S}(\mathbf{G}) \subset \mathbf{S}(\mathbf{F})$, if I_X is continuous for $I_X : \mathbf{S}_u(\mathbf{F}) \to \mathbf{L}^0$, then it is *a fortiori* continuous for $\mathbf{S}_u(\mathbf{G})$. □

Stricker's theorem shows one can easily shrink the filtration since one is only shrinking the domain of a continuous operator. Expanding the filtration, on the other hand, is more delicate, since one is then asking a continuous operator to remain continuous for a larger domain. An elementary result in this direction is the following:

Theorem 1.8 (P. A. Meyer). *Let $\mathcal{A}$ be a countable collection of disjoint sets in $\mathcal{F}$. Let* **H** *be the filtration given by $\mathcal{H}_t = \sigma(\mathcal{F}_t, \mathcal{A})$. Then every* **F** *semimartingale is an* **H** *semimartingale.*

Proof. Without loss of generality assume $\mathcal{A}$ is a partition of Ω, and $P(A_n) > 0$, each $A_n \in \mathcal{A}$. Define $Q_n \ll P$ by $Q_n(\Lambda) = P(\Lambda | A_n)$. Then X is a Q_n-semimartingale by Theorem 1.6. Let $\mathbf{I}^n$ be the filtration generated by **F** and all Q_n null sets. Let X be a $(\mathbf{I}^n, Q_n)$-semimartingale, each n. Moreover $\mathbf{F} \subset \mathbf{H} \subset \mathbf{I}^n$. By Stricker's theorem, X is an **H** semimartingale under Q_n. Note that $dP = \sum_{n=1}^{\infty} P(A_n) dQ_n$. Suppose $H^n \in \mathbf{S}(\mathbf{H})$ converges to $H \in \mathbf{S}(\mathbf{H})$ uniformly. Then $I_X(H^n)$ converges to $I_X(H)$ in Q_n-probability for each n, and it follows that it converges in P-probability as well. Thus X is an $(\mathbf{H}, P)$-semimartingale. □

2. Stochastic Integration

We wish to define a stochastic integral of the form $\int_0^t H_{s-} dX_s$, where H is càdlàg, adapted, and H_{s-} represents its left continuous version; and X is a semimartingale. We recall **S** is the space of simple predictable processes and $\mathbf{L}^0$ is the space of finite valued random variables.

We also define:

D = the space of adapted processes with càdlàg paths
L = the space of adapted processes with càdlàg paths (left continuous with right limits)

Note that if $H \in \mathbf{D}$, then H_- (its left continuous version) is in **L**; and if $H \in \mathbf{L}$, then H_+ is in **D**. We next define a new topology, ucp, which will replace uniform convergence.

Definition 2.1. *A sequence of processes Y^n converges to a process Y* uniformly on compacts in probability *(denoted ucp) if for each $t > 0$, $\sup_{s \le t} |Y_s^n - Y_s| = (Y^n - Y)_t^*$ tends to 0 in probability as n tends to ∞.*

We note that this topology is metrizable.

Theorem 2.2. **S** *is dense in* **L** *under ucp.*

Proof. By stopping, $b\mathbf{L}$ is dense in **L**, where $b\mathbf{L}$ denotes the bounded processes in **L**. For $Y \in b\mathbf{L}$, let $Z = Y_+$, and for $\varepsilon > 0$, define $T_0^\varepsilon = 0$ and

$$T_{n+1}^\varepsilon = \inf\{t : t > T_n^\varepsilon \text{ and } |Z_t - Z_{T_n^\varepsilon}| > \varepsilon\}.$$

Then T_n^ε are stopping times and they are increasing since Z is càdlàg. Pose $Z_1^\varepsilon = Y_0 1_{\{0\}} + \sum_{i=1}^n Z_{T_i^\varepsilon} 1_{(T_i^\varepsilon \wedge n, T_{i+1}^\varepsilon \wedge n]}$. This can be made arbitrarily close to $Y \in b\mathbf{L}$ by taking ε small enough and n large enough. □

The operator I_X defined in (1.3) was, effectively, an operator giving a definite integral for processes $H \in \mathbf{S}$ and semimartingales X. We now wish to define an operator which will be an indefinite integral operator. Thus its range should be processes rather than random variables. Therefore for a given process X and a process $H \in \mathbf{S}$ as given in (1.2) we define the operator $J_X : \mathbf{S} \to \mathbf{D}$ by:

$$J_X(H) = H_0 X_0 + \sum_{i=1}^{n} H_i (X^{T_{i+1}} - X^{T_i}), \tag{2.1}$$

where the notation X^T, for a stopping time T, denotes the process $X_t^T = X_{t \wedge T} (t \geq 0)$.

Definition 2.3. *For an adapted, càdlàg process X and $H \in \mathbf{S}$, the process $J_X(H)$ is called the* stochastic integral *of H with respect to X.*

We will also use the notations $\int_0^t H_s dX_s$ and $H \cdot X$ or $H \cdot X_t$ to denote the stochastic integral. That is

$$\begin{aligned} J_X(H) &= \int H dX &&= H \cdot X \\ J_X(H)_t &= \int_0^t H_s dX_s &&= H \cdot X_t. \end{aligned}$$

Theorem 2.4. *Let X be a semimartingale. Then $J_X : \mathbf{S}_{\mathrm{ucp}} \to \mathbf{D}_{\mathrm{ucp}}$ is continuous.*

Proof. Suppose $H^k \in \mathbf{S}$ tends to H uniformly. By linearity, we can suppose without loss H^k tends to 0. Let $T^k = \inf\{t : |(H^k \cdot X)_t| \geq \delta\}$. Then $H^k 1_{[0,T^k]} \in \mathbf{S}$ tends to 0 uniformly as k tends to ∞. Thus for every t

$$\begin{aligned} P\{(H^k \cdot X)_t^* > \delta\} &\leq P\{|H^k \cdot X_{T^k \wedge t}| \geq \delta\} \\ &= P\{|(H^k 1_{[0,T^k]} \cdot X)_t| \geq \delta\} \\ &= P\{|I_X(H^k 1_{[0,T^k \wedge t]})| \geq \delta\} \end{aligned}$$

which tends to 0 by definition because X is a semimartingale. Therefore $J_X : \mathbf{S}_u \to \mathbf{D}_{\mathrm{ucp}}$ is continuous. We next show $J_X : \mathbf{S}_{\mathrm{ucp}} \to \mathbf{D}_{\mathrm{ucp}}$ is continuous. Let $\delta > 0$, $\varepsilon > 0$, $t > 0$. We now know there exists η such that $\|H\|_u \leq \eta$ implies $P(J_X(H)_t^* > \delta) < \varepsilon/2$. Let $R^k = \inf\{s : |H_s^k| > \eta\}$, and set $\widetilde{H}^k = H^k 1_{[0,R_k]} 1_{\{R_k > 0\}}$. Then $\widetilde{H}^k \in \mathbf{S}$ and $\|\widetilde{H}^k\|_u \leq \eta$ by left continuity. When $R^k \geq t$ we have $(\widetilde{H}^k \cdot X)_t^* = (H^k \cdot X)_t^*$, whence

$$\begin{aligned} P\left((H^k \cdot X)_t^* > \delta\right) &\leq P\left((\widetilde{H}^k \cdot X)_t^* \delta\right) + P(R^k < t) \\ &\leq \varepsilon/2 + P\left((H^k)_t^* > \eta\right) \\ &< \varepsilon, \end{aligned}$$

if k is large enough, since $\lim_{k \to \infty} P((H^k)_t^* > \eta) = 0$. □

Definition 2.5. *Let X be a semimartingale. The continuous linear mapping $J_X : \mathbf{L}_{\mathrm{ucp}} \to \mathbf{D}_{\mathrm{ucp}}$ obtained as the extension of $J_X : \mathbf{S} \to \mathbf{D}$ is called the* stochastic integral.

Suppose H is a process in $\mathbf{D}$. We can write the stochastic integral $H_{s-} \cdot X = (\int_0^t H_{s-} dX_s)_{t \geq 0}$ as defined above, as a limit of sums. Let σ denote a finite sequence of stopping times:

$$0 = T_0 \leq T_1 \leq \ldots \leq T_k < \infty \text{ a.s.} \tag{2.2}$$

Such a sequence is called a *random partition*. A sequence of random partitions σ_n

$$\sigma_n : T_0^n \leq T_1^n \leq \ldots \leq T_{k_n}^n$$

is said to *tend to the identity* if

(i) $\lim_n \sup_i T_i^n = \infty$ a.s.
(ii) $\|\sigma_n\| = \sup_i |T_{i+1}^n - T_i^n|$ converges to 0 a.s.

For a process H and a random partition σ as in (2.2) we define

$$H^\sigma = H_0 1_{\{0\}} + \sum_{i=1}^k H_{T_i} 1_{(T_i, T_{i+1}]}. \tag{2.3}$$

Thus if H is in $\mathbf{L}$ or $\mathbf{D}$, we have

$$\int_0^t H_s^\sigma dX_s = H_0 X_0 + \sum_{i=1}^k H_{T_i}(X^{T_{i+1}} - X^{T_i}). \tag{2.4}$$

Theorem 2.6. *Let X be a semimartingale and let $H \in \mathbf{D}$. Let $(\sigma_n)_{n \geq 1}$ be a sequence of random partitions tending to the identity. Then*

$$H_- \cdot X = \lim_{n \to \infty} \sum_i H_{T_i^n}(X^{T_{i+1}^n} - X^{T_i^n})$$

with convergence in ucp.

Proof. Let $H^k \in \mathbf{S}$ converge to H in ucp. Then

$$(H_- - H^{\sigma_n}) \cdot X = (H_- - H^k) \cdot X + (H^k - (H_+^k)^{\sigma_n}) \cdot X + ((H_+^k)^{\sigma_n} - H^{\sigma_n}) \cdot X. \tag{2.5}$$

The first term on the right side of (2.5) equals $J_X(H_- - H^k)$, which goes to 0 because J_X is continuous on $\mathbf{L}_{\mathrm{ucp}}$. The same applies to the third term for fixed k as n tends to ∞. Indeed, $(H_+^k)^{\sigma_n} - H^{\sigma_n}$ tends to 0 as $k \to \infty$ uniformly in n. As for the middle term on the right side of (2.5), for fixed k it tends to 0 as n tends to ∞. Thus one need only choose k so large that the first and third terms are small, and then choose n so large that the middle term is small. □

Theorem 2.6 gives an appealing intuitive description of the stochastic integral as a limit of Riemann-type sums. Of course one can only do this because of the path regularity of the integrands.

Let us next note some simple and quite nice properties of the stochastic integral. H will be assumed to be in $\mathbf{D}$, and X a semimartingale in Theorems 2.7 through 2.11.

Theorem 2.7. *If X has paths of finite variation a.s., then $H_- \cdot X$ agrees with the Lebesgue-Stieltjes integral, denoted $\int_{LS} H_{s-}dX_s$.*

Proof. The result is evident for $H \in \mathbf{S}$. For $H \in \mathbf{D}$, let $H^n \in \mathbf{S}$ converge to H_- in ucp. Then there exists a subsequence n_k such that H^{n_k} converges uniformly on compacts a.s. to $H_- \cdot X$. Since the convergence is uniform, $\int_{LS} H_s^{n_k} dX_s$ converges as well to $\int_{LS} H_{s-}dX_s$, whence the result. □

Recall that for a process $Y \in \mathbf{D}$, $\Delta Y_t = Y_t - Y_{t-}$, and ΔY denotes the process $(\Delta Y_t)_{0 \le t < \infty}$. An important feature of the stochastic integral is that the jumps behave "correctly" — that is, in the same manner as they do for the Lebesgue-Stieltjes integral. This is part of the reason we use $\mathbf{L}$, rather than, for example, $\mathbf{D}$, as our space of integrands. (See Pratelli [14] or Ahn-Protter [1] for more on this subject.)

Theorem 2.8. *The jump process $\Delta(H_- \cdot X)_s$ is indistinguishable[2] from the process $H_{s-}\Delta X_s$.*

Theorem 2.9. *Let $Q \ll P$. Then $H_{-Q} \cdot X$ is Q-indistinguishable from $H_{-P} \cdot X$.*

Theorem 2.10. *Let P and Q be any two probabilities and X a semimartingale for each. Then there exists $H_- \cdot X$ which is a version of both $H_{-P} \cdot X$ and $H_{-Q} \cdot X$.*

Theorem 2.11. *Let $\mathbf{G}$ be another filtration and suppose $H \in \mathbf{D}(\mathbf{G}) \cap \mathbf{D}(\mathbf{F})$, and that X is semimartingale for both $\mathbf{F}$ and $\mathbf{G}$. Then $H_{-\mathbf{G}} \cdot X = H_{-\mathbf{F}} \cdot X$.*

Proof. For Theorem 2.8 and 2.9, the result is clear for $H \in \mathbf{S}$ and follows for H_- with $H \in \mathbf{D}$ by taking limits in ucp (convergence in P-probability implies convergence in Q-probability). For Theorem 2.10, let $R = \frac{1}{2}(P+Q)$, and apply Theorem 2.9. For Theorem 2.11, we can use the construction in the proof of Theorem 2.2 to approximate $H \in \mathbf{D}$ constructively from H; thus the approximations $H^n \in \mathbf{S}$ are in $\mathbf{S}(\mathbf{F}) \cap \mathbf{S}(\mathbf{G})$; the result is clearly true for H in $\mathbf{S}$ and thus it follows by again taking limits. □

Theorem 2.9 can be used to show that many global results also hold locally.

We give an example.

---

[2] Two processes Y and Z are *indistinguishable* if $P\{\omega : t \to Y_t(\omega) \neq t \to Z_t(\omega)\} = 0$.

Theorem 2.12. *Let X, Y be two semimartingales and H, J be two processes in* $\mathbf{D}$. *Let*

$$A = \{\omega : H.(\omega) = J.(\omega) \text{ and } X.(\omega) = Y.(\omega)\}$$

where $H.(\omega)$ denotes the path of $H : t \to H_t(\omega)$. Let

$$B = \{\omega : X.(\omega) \text{ is finite variation on compacts}\}.$$

Then $H_- \cdot X = J_- \cdot X$ on A a.s., and $H_- \cdot X = \int_{LS} H_{s-} dX$ on B a.s.

Proof. Without loss of generality assume $P(A) > 0$. Define a new Q by $Q(\Lambda) = P(\Lambda|A)$. Then $H_- = J_-$ and $X = Y$ under Q. Note that X and Y are also semimartingales under Q. Thus $H_{-Q} \cdot X = H_{-P} \cdot X$, and one need only apply Theorem 2.9. The second assertion is a combination of the above idea with Theorems 2.7 and 2.9. □

The next result is quite important.

Theorem 2.13. *Let $H \in \mathbf{D}$ and X be a semimartingale. Then $Y = H_- \cdot X$ is again a semimartingale. Moreover if $G \in \mathbf{D}$ as well, then*

$$G_- \cdot Y = G_- \cdot (H_- \cdot X) = (GH)_- \cdot X.$$

Proof. If $G, H \in \mathbf{S}$, then clearly $Y = H_- \cdot X$ is a semimartingale, and $J_Y(G) = J_X(GH)$. The associativity property extends to H_-, G_- with $G, H \in \mathbf{D}$ by continuity. Therefore it remains only to show $Y = H_- \cdot X$ is a semimartingale. By taking subsequences if necessary, assume $H^n \in \mathbf{S}$ converges to H_- in ucp and also $H^n \cdot X$ converges a.s. to $H_- \cdot X$. For $G \in \mathbf{S}$, $J_Y(G)$ is defined for any process Y and hence makes sense *a priori*. Thus

$$\begin{aligned} J_Y(G) &= \lim_{n\to\infty} G \cdot Y^n = \lim_{n\to\infty} G \cdot (H^n \cdot X) \\ &= \lim_{n\to\infty} (GH^n) \cdot X = J_X(GH_-), \end{aligned}$$

since X is a semimartingale. Next let G^n converge to G in $\mathbf{S}_u$. We wish to show $I_Y(G^n)$ converges to $I_Y(G)$. But

$$\lim_{n\to\infty} J_Y(G^n) = \lim_{n\to\infty} J_X(G^n H_-) = J_X(GH_-)$$

since $G^n H_-$ converges to GH_- in ucp. Then since $J_X(GH_-) = J_Y(G)$ we have the result. □

3. Quadratic Variation

A process which plays a key role in the theory of stochastic integration is the quadratic variation process. We define it using stochastic integration:

Definition 3.1. *Let X be a semimartingale. The* quadratic variation process, *$[X,X]$, is defined to be*

$$[X,X]_t = X_t^2 - 2\int_0^t X_{s-}dX_s. \tag{3.1}$$

If X and Y are two semimartingales, the quadratic covariation process *is defined to be*

$$[X,Y]_t = X_tY_t - \int_0^t X_{s-}dY_s - \int_0^t Y_{s-}dX_s. \tag{3.2}$$

Note that if X is of finite variation, then (3.1) is simply integration by parts, and if X is also continuous then $[X,X]_t = X_0^2$, and in particular it is constant. Note further that the bracket $[\cdot,\cdot]$ satisfies a polarization identity:

$$[X,Y] = \frac{1}{2}\{[X+Y,X+Y]-[X,X]-[Y,Y]\}.$$

We make the convention that $X_{0-}=0$ a.s. always.

Theorem 3.2. *Let X be a semimartingale. Then $[X,X]$ is in* **D** *and has non-decreasing paths. Moreover $[X,X]_0 = X_0^2$ and*

(i) $\Delta[X,X] = (\Delta X)^2$;

(ii) If σ_n is a sequence of random partitions tending to the identity as defined in (2.2), then

$$\lim_{n\to\infty}\left\{X_0^2 + \sum_i (X^{T_{i+1}^n} - X^{T_i^n})^2\right\} = [X,X]$$

with convergence in ucp;

(iii) for a stopping time T, $[X^T,X] = [X,X^T] = [X^T,X^T] = [X,X]^T$.

Proof. $[X,X]$ is in **D** since the right side of (3.1) is in **D**. It is nondecreasing a consequence of (ii) above. That (i) holds follows from (3.1) and Theorem 2.8. Property (ii) is an elementary consequence of Theorem 2.6. Finally (iii) follows easily from (ii). □

Theorem 3.2 gives a method of extending the notion of quadratic variation to a wider class of processes than semimartingales; namely, those for which a limit of sums exists in ucp, as given in Theorem 3.2 (ii). This would include, for example, the Dirichlet processes.

It is worthwhile to calculate the quadratic variation of some basic processes. Theorem 3.2 (ii) allows one to deduce that $[W,W]_t = t$ a.s., where W is standard Wiener process. If A is of finite variation, again Theorem 3.2 (ii) allows one to conclude that $[A,A]_t = \sum_{0<s\le t}(\Delta A_s)^2$. In particular if N is the Poisson process then $[N,N]_t = N_t$. If A is continuous and of finite variation, $[A,A]_t = A_0^2$, and thus if $A_0 = 0$ then $[A,A]\equiv 0$.

The quadratic variation process has a particularly nice property with respect to stochastic integrals:

Theorem 3.3. *Let X and Y be two semimartingales, and let $H, K \in \mathbf{D}$. Then*

$$[H_- \cdot X, K_- \cdot Y]_t = \int_0^t H_{s-}K_{s-}d[X,Y]_s. \tag{3.3}$$

In particular,

$$[H_- \cdot X, H_- \cdot X]_t = \int_0^t (H_{s-})^2 d[X,X]_s.$$

Proof. Without loss we assume $X_0 = Y_0 = 0$. By symmetry it suffices to prove

$$[H_- \cdot X, Y]_t = \int_0^t H_{s-}d[X,Y]_s. \tag{3.4}$$

First assume that $H = 1_{[0,T]}$. Then (3.4) follows from Theorem 3.2 (iii). Next let $H = V1_{(S,T]}$ where X and T are stopping times and $V \in \mathcal{F}_S$. Then $H \cdot X = V(X^T - X^S)$, and by Theorem 3.2

$$\begin{aligned}[H \cdot X, Y] &= V\{[X^T, Y] - [X^S, Y]\} \\ &= V\{[X,Y]^T - [X,Y]^S\} = \int H_s d[X,Y]_s.\end{aligned}$$

The result now holds for $H \in \mathbf{S}$ by linearity. For $H \in \mathbf{D}$, let $H^n \in \mathbf{S}$ approximate H_- in ucp. Let $Z^n = H^n \cdot X$. Then $[Z^n, Y] = \int H_s^n d[X,Y]_s$, and since $H^n \in \mathbf{S}$ we have:

$$\begin{aligned}[Z^n, Y] &= YZ^n - \int Y_- dZ^n - \int Z_-^n dY \\ &= YZ^n - \int Y_- H^n dX - \int Z_-^n dY\end{aligned}$$

which converges to

$$YZ - \int Y_- H_- dX - \int Z_- dY = YZ - \int Y_- dZ - \int Z_- dY = [Z, Y].$$

Thus,

$$[Z,Y] = \lim_{n\to\infty} [Z^n, Y] = \lim_{n\to\infty} \int H_s^n d[X,Y]_s = \int H_{s-}d[X,Y]_s.$$

But $Z = \lim_{n\to\infty} Z^n = \lim_{n\to\infty} H^n \cdot X = H_- \cdot X$, and we have the result. □

An important special case is that of martingales. If M is a martingale and $E\{\sup_{s\le t}|M_s|^2\} < \infty$, then $E\{M_t^2\} = E\{[M,M]_t\}$. Therefore Doob's maximal quadratic inequality for martingales can be expressed as follows (see [15]):

Theorem 3.4. *Let M be a local martingale. Then*

$$E\{\sup_{s\le t}(M_s)^2\} \le 4E\{[M,M]_t\}.$$

In particular if $E\{[M,M]_t\} < \infty$, then M is a square integrable martingale on $[0,t]$.

Let W be a standard Wiener process, and let $H \in \mathbf{D}$. Then as we previously remarked $[W,W]_t = t$. Hence

$$[H_- \cdot W, H_- \cdot W]_t = \int_0^t H_{s-}^2 ds = \int_0^t H_s^2 ds.$$

Therefore,

$$E\left\{\left(\int_0^t H_{s-} dW_s\right)^2\right\} = \int_0^t E\{H_s^2\} ds,$$

by Fubini's theorem. It is this isometry that K. Itô used when he originally defined the stochastic integral for the Wiener process.

4. Change of Variables

The change of variables formula in the general case (that is, the case for semimartingales with jumps) often looks strange, but actually it is close to the formula for Lebesgue-Stieltjes integration. The problem is that the latter formula is not well known. Of course it is a corollary of the general formula, but we nevertheless state it first.

Theorem 4.1. *Let V be a process with càdlàg paths of finite variation on compacts, and let f be C^1. Then*

$$f(V_t) - f(V_0) = \int_{0+}^t f'(V_{s-}) dV_s + \sum_{0<s\leq t} \{f(V_s) - f(V_{s-}) - f'(V_{s-})\Delta V_s\}$$

and in particular $f(V)$ is again a process with paths of finite variation on compacts.

Note that it is not *a priori* obvious that the infinite sum above converges. Before we state the general theorem recall that for a semimartingale X the process $[X,X]$ is in $\mathbf{D}$ and is non-decreasing. Therefore ω by ω the paths $t \to [X,X]_t(\omega)$ have a Lebesgue decomposition into a continuous part and a pure jump part. Indeed, in light of Theorem 3.2 we can write

$$[X,X]_t = [X,X]_t^c + \sum_{0<s\leq t} (\Delta X_s)^2,$$

and we call $[X,X]^c$ the *continuous part of the quadratic variation*. Note that we also have for semimartingales X, Y:

$$[X,Y]_t = [X,Y]_t^c + \sum_{0<s\leq t} \Delta X_s \Delta Y_s,$$

and in particular we deduce $\sum_{0<s\leq t} (\Delta X_s)^2 < \infty$ a.s. for any semimartingale X. (It is of course not true in general that $\sum_{0<s\leq t} |\Delta X_s|$ is finite a.s.; see example (vi) where such a term is ∞ a.s., each $t > 0$.)

Theorem 4.2 (Change of Variables). *Let $X = (X^1, \dots, X^d)$ be a d-dimensional semimartingale, and let $f : \mathbf{R}^d \to \mathbf{R}$ be C^2. Then $f(X)$ is a semimartingale and moreover*

$$\begin{aligned} f(X_t) - f(X_0) &= \sum_{i=1}^{d} \int_{0+}^{t} \frac{\partial f}{\partial x_i}(X_{s-})dX_s^i \qquad (4.1)\\ &+ \frac{1}{2} \sum_{1\le i,j\le d} \int_{-+}^{t} \frac{\partial^2 f}{\partial x_i \partial x_j}(X_{s-})d[X^i, X^j]_s^c \\ &+ \sum_{0<s\le t} \left\{ f(X_s) - f(X_{s-}) - \sum_{i=1}^{d} \frac{\partial f}{\partial x_i}(X_{s-})\Delta X_s^i \right\}. \end{aligned}$$

Proof. We give the proof for $d = 1$; the case for $d > 1$ is analogous but messier. Thus we want to establish:

$$\begin{aligned} f(X_t) - f(X_0) &= \int_{0+}^{t} f'(X_{s-})dX_s + \frac{1}{2}\int_{0}^{t} f''(X_{s-})d[X, X]_s^c \\ &+ \sum_{0<s\le t} \{f(X_s) - f(X_{s-}) - f'(X_{s-})\Delta X_s\}. \end{aligned} \qquad (4.2)$$

We further assume $X_0 = 0$, to eliminate the plus symbols. First suppose f is a polynomial on $\mathbf{R}$. Obviously (4.2) holds for f a constant function. We will use induction and thus it suffices to prove the following: let g be such that $g(X)$ is a semimartingale and that (4.2) holds; then if $f(x) = xg(x)$, also $f(X)$ is a semimartingale and (4.2) holds for f.

Note that the product of two semimartingales is a semimartingale by integration by parts (formula (3.2) and Theorem 3.2), thus $Xg(X)$ is a semimartingale. Again using integration by parts (formula (3.2)) we have

$$f(X_t) = f(X_0) + \int_0^t X_{s-}dg(X_s) + \int_0^t g(X_{s-})dX_s + [X, g(X)]_t.$$

By hypothesis $g'(X)$ satisfies (4.2), hence by Theorems 2.8 and 2.13 we obtain:

$$\begin{aligned} f(X_t) &= f(X_0) + \int_0^t X_{s-}g'(X_{s-})dX_s + \frac{1}{2}\int_0^t X_{s-}g''(X_{s-})d[X, X]_s^c \\ &+ \sum_{0<s\le t} X_{s-}\{g(X_s) - g(X_{s-}) - g'(X_{s-})\Delta X_s\} \\ &+ \int_0^t g(X_{s-})dX_s + [X, g(X)]_t. \end{aligned} \qquad (4.3)$$

Next, using Theorem 3.3 we see that

$$\begin{aligned}[X,g(X)]_t &= \int_0^t g(X_{s-})d[X,X]_s \\ &\quad + \sum_{0<s\leq t} (\Delta X_s)\{g(X_s) - g(X_{s-}) - g'(X_{s-})\Delta X_s\} \\ &= \int_0^t g(X_{s-})d[X,X]_s^c + \sum_{0<s\leq t} \Delta X_s\{g(X_s) - g(X_{s-})\}.\end{aligned} \tag{4.4}$$

Combining (4.3) and (4.4) we see that f satisfies (4.2).

Now we consider the case where f is not a polynomial. Let

$$T_n = \inf\{t > 0 : |X_t| > n\}.$$

Then for each fixed n we can find a sequence $(g_{nm})_{m\geq 1}$ of polynomials that converge, together with their first and second derivatives, respectively to f and its first two derivatives, uniformly on $\{x : |x| \leq n\}$. There exists a constant K_n such that for $|x|, |y| \leq n$,

$$|f(x) - f(y) - f'(y)(x-y)| \leq K_n|x-y|^2 \tag{4.5}$$

and

$$|g_{nm}(x) - g_{nm}(y) - g'_{nm}(y)(x-y)| \leq K_n|x-y|^2. \tag{4.6}$$

Recall we remarked just before stating this theorem that $\sum_{s\leq t}(\Delta X_s)^2 < \infty$ a.s. for any semimartingale X; therefore using (4.5) and (4.6) and taking limits as m increases to ∞ we deduce that for $t < T_n$:

$$\sum_{0<s\leq t} |f(X_s) - f(X_{s-}) - f'(X_{s-})\Delta X_s| < \infty \text{ a.s.}$$

and moreover

$$\begin{aligned}\lim_{m\to\infty} &\sum_{0<s\leq t} \{g_{nm}(X_s) - g_{nm}(X_{s-}) - g'_{nm}(X_{s-})\Delta X_s\} \\ &= \sum_{0<s\leq t} \{f(X_s) - f(X_{s-}) - f'(X_{s-})\Delta X_s\}.\end{aligned}$$

Furthermore $\lim_{m\to\infty} g_{nm}(X_t) = f(X_t)$, for $t < T_n$. Note further that $\int_0^t g'_{nm}(X_{s-})dX_s$ tends to $\int_0^t f'(X_{s-})dX_s$ since J_X is continuous in ucp on L, and also $\int_0^t g''_{nm}(X_{s-})d[X,X]_s^c$ converges in ucp to $\int_0^t f''(X_{s-})d[X,X]_s^c$.

Since T_n increases to ∞ a.s., the process $\sum_{0<s\leq t}\{f(X_s) - f(X_{s-}) - f'(X_{s-})\Delta X_s\}$ is of finite variation (and thus absolutely convergent as a series a.s. for $t > 0$) on compacts. Since the other terms on the right side of (4.2) are all well defined semimartingales, we conclude that $f(X)$ is a semimartingale and that (4.2) indeed holds. □

We remark that the preceding proof, while quick, simple, and elegant, is not particularly intuitive. A more intuitive proof, using Taylor expansions, can be found for example in [15, pp. 71ff].

If $X = (W^1, \ldots, W^d)$ is a d-dimensional Wiener process, then W^i is independent of W^j for $i \neq j$, and one can check that $[W^i, W^j] \equiv 0$ for $i \neq j$ (we assume $W_0 = 0$). In this case for $f \in C^2$ we can write the change of variables formula (known here as *Itô's formula*) in the form

$$f(W_t) - f(W_0) = \int_0^t \nabla f(W_s) \cdot dW_s + \frac{1}{2} \int_0^t \Delta f(W_s) ds.$$

In particular if $\Delta f = 0$, then $f(W)$ is a local martingale.

Also note that if $X = (X^1, \ldots, X^d)$ is such that some of the components of X are finite variation processes, then f need only to be C^1 for the corresponding coordinates.

5. Stochastic Differential Equations

We consider here fairly general stochastic differential equations which are sufficient for most applications. Let $\mathbf{D}^d$ denote d-dimensional vectors of processes in $\mathbf{D}$.

Definition 5.1. *An operator $F : \mathbf{D}^d \to \mathbf{D}$ is said to be* functional Lipschitz *if for any processes X, Y in $\mathbf{D}^d$ we have*

(i) for any stopping time T, $X^{T-} = Y^{T-}$ implies $F(X)^{T-} = F(Y)^{T-}$.[3]
(ii) $|F(X)_t - F(Y)_t| \leq M \sup_{s \leq t} |X_s - Y_s|$.

The following theorem is a special case of a more general result to be proved in Section 7 of Part II of these notes (see also [15]).

Theorem 5.2. *Let $Y = (Y^1, \ldots, Y^d)$ be a vector of semimartingales and let J^i, $1 \leq i \leq k$, be processes in $\mathbf{D}$. Let F^i_j, $1 \leq i \leq k$, $1 \leq j \leq d$ be functional Lipschitz operators. Then the system of equations*

$$X^i_t = J^i_t + \sum_{i=1}^{d} \int_0^t F^i_j(X)_{s-} dZ^j_s \tag{5.1}$$

has a solution in $\mathbf{D}$, and it is unique. Moreover if the processes J^i are semimartingales, then X^i are semimartingales as well.

The reader may wonder if the condition $F(X)_{s-}$, instead of $F(X)_s$ is merely a technicality to ensure that the integrand of the stochastic integral is in $\mathbf{L}$. It is not, but rather is essential if one considers driving terms with jumps, and it corresponds to one's physical intuition: a jump at time t "kicks" the process according to where it was just before t. Indeed, if one takes the

---

[3] The notation X^{T-} denotes $X_t 1_{\{t<T\}} + X_{T-} 1_{\{t \geq T\}}$.

non-random example where $Z_t = 1_{\{t\geq 2\}}$ and $X_t = 1 + \int_0^t X_s dZ_s$, then one has $X_t = 1$ for $0 \leq t < 2$, and $X_2 = 1 + X_2$, which gives $0 = 1$.

A particularly important special case of (5.1) is the exponential equation:

$$X_t = X_0 + \int_0^t \alpha X_{s-} dY_s. \tag{5.2}$$

One can use the change of variables formula to give an explicit solution of (5.2), called the ***stochastic exponential*** and denoted $\mathcal{E}(Y)$:

$$\mathcal{E}(Y)_t = \exp\left(\alpha Y_t - \frac{\alpha^2}{2}[Y,Y]_t^c\right) \prod_{0<s\leq t} (1+\alpha\Delta Y_s)\exp(-\alpha\Delta Y_s). \tag{5.3}$$

The stochastic exponential has behavior similar to a true exponential, but of course slightly different due to the semimartingale calculus; for example we have the following pretty result:

Theorem 5.3 (Yor). *Let X and Y be semimartingales with $X_0 = Y_0 = 0$. Then $\mathcal{E}(X)\mathcal{E}(Y) = \mathcal{E}(X+Y+[X,Y])$.*

Proof. Let $U_t = \mathcal{E}(X)_t$ and $V_t = \mathcal{E}(Y)_t$. By integration by parts $U_t V_t - 1 = \int_{0_+}^t U_{s-} dV_s + \int_{0_+}^t V_{s-} dU_s + [U,V]_t$. Using that U and V are exponentials and letting $W = UV$ this becomes

$$W_t = 1 + \int_0^t W_{s-} d(X+Y+[X,Y])_s,$$

whence the result. □

Using a variation of constants technique we can generalize the stochastic exponential results.

Theorem 5.4. *Let H and Z be semimartingales and assume $P\{\Delta Z_t \neq -1, t \geq 0\} = 1$. Let X be the unique solution of*

$$X_t = H_t + \int_0^t X_{s-} dZ_s.$$

Then $X_t = \mathcal{E}_H(Z)_t$ has the form:

$$\mathcal{E}_H(Z)_t = \mathcal{E}(Z)_t\left\{H_0 + \int_{0_+}^t \mathcal{E}(Z)_{s-}^{-1} d\left\{H_s - [H,Z]_s - \sum_{0<u\leq s}\left(\frac{\Delta H_u (\Delta Z_u)^2}{1+\Delta Z_u}\right)\right\}\right\}.$$

Proof. Let us assume the solution is of the form $C_t\mathcal{E}(Z)_t$. Let $U_t = \mathcal{E}(Z)_t$, and we wish to determine C. Note that

$$\Delta X_t = \Delta H_t + X_{t-}\Delta Z_t. \tag{5.4}$$

Integration by parts yields:

$$\begin{aligned} dX_t &= C_t - dU_t + U_{t-} dC_t + d[C,U]_t \\ &= C_{t-} U_{t-} dZ_t + U_{t-} d\{C_t + [C,Z]_t\} \end{aligned} \tag{5.5}$$

from which we deduce

$$\Delta X_t = X_{t-} \Delta Z_t + U_{t-} \Delta C_t + U_{t-} \Delta C_t \Delta Z_t. \tag{5.6}$$

Combining (5.4) and (5.5) yields

$$\Delta C_t = \frac{\Delta H_t}{U_{t-}(1+\Delta Z_t)} = \frac{\Delta H_t}{\mathcal{E}(Z)_{t-}(1+\Delta Z_t)}. \tag{5.7}$$

¿From (5.6) we have

$$dH_t = U_{t-} d\{C_t + [C,Z]_t\}$$

which implies

$$\begin{aligned} \frac{1}{U_-} \cdot H &= C + [C,Z], \\ \left[\frac{1}{U_-} \cdot H, Z\right] &= [C,Z]_t + [\,[C,Z],Z]_t\,, \end{aligned} \tag{5.8}$$

and since $[\,[C,Z],Z] = \sum \Delta C (\Delta Z)^2$, and since we know ΔC by (5.7), we obtain

$$[C,Z] = \frac{1}{U_-} \cdot [H,Z] - \sum \frac{\Delta H (\Delta Z)^2}{U_-(1+\Delta Z)}. \tag{5.9}$$

Using (5.8) and (5.9) we get:

$$C_t = \int_0^t \frac{1}{U_{s-}} dH_s - d[C,Z]_t = \int_0^t \frac{1}{U_{s-}} d\Big\{ H_s - [H,Z]_s - \sum_{0<u\leq s} \frac{\Delta H_u (\Delta Z_u)^2}{(1+\Delta Z_u)} \Big\},$$

and the result follows. □

6. The Skorohod Topology and Weak Convergence

In this section we recall the essentials of the Skorohod topology and weak convergence. Since this material is by now classic, we omit the proofs except for Theorem 6.5 which is recent and the reader can consult any of several expository treatments both in books and research articles.

Recall that **D** has been used to denote adapted, càdlàg stochastic processes.

We now let $D = D_{\mathbf{R}^d} = D_{\mathbf{R}^d}[0,\infty)$ denote the space of càdlàg *functions* from $[0,\infty)$ to $\mathbf{R}^d$. A process in **D** has almost all of its sample paths in D. We wish to endow D with a topology for which it is a complete separable metric space. A natural candidate would be *uc* (uniform convergence on compacts), and the corresponding metric would be:

$$d_{uc}(x,y)=\sum_{n=1}^{\infty}\frac{1}{2^n}\left(\min\left(1,\sup_{s\le n}|x(s)-y(s)|\right)\right).$$

Such a topology, however, is not separable: for example, the family of functions $x_s(t)=1_{[s,\infty)}(t)$, $0\le s<1$, is uncountable, while $d_{uc}(x_s,x_u)=1$ for $s\ne u$.

Let $\mathbf{R}_+$ denote $[0,\infty)$.

Definition 6.1. *A time change function* λ *is an increasing, bijective function from* $\mathbf{R}_+$ *to* $\mathbf{R}_+$. *We let* Γ *denote the class of these functions.*

Definition 6.2. *A sequence of functions* $x_n\in D$ *converges in the* Skorohod topology *to* $x\in D$ *if there exists* $\lambda_n\in\Gamma$ *such that* $\lambda_n(t)$ *converges to* $\lambda(t)=t$ *uniformly and* $x_n(\lambda_n(t))$ *converges to* $x(t)$ *uniformly on compacts.*

In the *uc* topology, if x_n converges to x, then for large enough n the jumps of s_n must occur at the same time as those of x, and of course the sizes must also converge.

With the Skorohod topology the sizes of the jumps must still converge, but the jumps need not occur at the same time. The Skorohod topology also allows the *times of occurrence* of the jumps to converge. Note that if the limit process x is continuous, then $x_n\to x$ in the Skorohod topology if and only if it converges in *uc*.

Theorem 6.3. *The Skorohod topology is metrizable, and the resulting metric space is separable and complete.*

To prove Theorem 6.3 one can construct a compatible metric. A metric analogous to the *uc* metric is:

$$d^1(x,y)=\inf_{\lambda\in\Gamma}d^1_\lambda(x,y)$$

where

$$d^1_\lambda(x,y)=\sup_{t\ge0}|\lambda(t)-t|+\sum_{n=1}^{\infty}\frac{1}{2^n}\left(1\wedge\sup_{t\ge0}|x(n\wedge\lambda(t))-y(n\wedge t)|\right).\quad(6.1)$$

The metric d^1 is a compatible metric, but it is not complete.

We give two other compatible metrics which are in fact complete. We define Γ' to be the set of Lipschitz continuous functions $\lambda\in\Gamma$ such that

$$\gamma(\lambda)=\operatorname*{ess\,sup}_{t\ge0}|\log\lambda'(t)|=\sup_{0\le s<t}\left|\log\frac{\lambda(t)-\lambda(s)}{t-s}\right|<\infty.$$

Next define $d^2(x,y,\lambda,u)=\sup_{t\ge0}|x(\lambda(t)\wedge u)-y(t\wedge u)|$. Finally we can define

$$d^2(x,y)=\inf_{\lambda\in\Gamma'}\left\{\gamma(\lambda)\vee\int_0^\infty e^{-u}d(x,y,\lambda,u)du\right\}.$$

For the third metric, we define

$$k_n(t) = \begin{cases} 1 & \text{if } t \le n, \\ n+1-t & \text{if } n < t < n+1, \\ 0 & \text{if } t \ge n+1 \end{cases},$$

$$d^3(x,y,n) = \inf_{\lambda \in \Gamma'} (\gamma(\lambda) + \|(k_n x) \circ \lambda - k_n y\|_{L^\infty}),$$

$$d^3(x,y) = \sum_{n \ge 0} 2^{-n}(1 \wedge d^3(x,y,n)).$$

The first distance d^1 is close to that of the original distance proposed by Skorohod. The second distance d^2 is taken from Ethier and Kurtz (1986) and is a modernized variation of Prokhorov's distance. The third distance d^3 is actually that of Prokhorov. Note that if x_n converges to x in the Skorohod topology and if the limit function x is continuous, then convergence in the Skorohod topology is equivalent to convergence in the *uc* topology.

While the Skorohod topology seems to be quite nice from the above description, it has a few traits which can create problems.

Example 6.4. Let $z_n(s) = x(s)1_{\{r_n \le s\}} + y(s)1_{\{t_n \le s\}}$, with $r_n < t_n$, and x and y continuous, not zero. Then z_n converges to a limit z in the Skorohod topology if and only if:

(i) $\lim_{n\to\infty} r_n = \infty$; whence $z = 0$;
(ii) $\lim_{n\to\infty} r_n = r < \infty$; $\lim_{n\to\infty} t_n = \infty$; then $z(s) = x(s)1_{\{t \le s\}}$;
(iii) $\lim_{n\to\infty} r_n = r < \infty$; $\lim_{n\to\infty} t_n = t < \infty$; and $r < t$, then $z(s) = x(s)1_{\{t \le s\}} + y(s)1_{\{t \le s\}}$.

From Example 6.4 two important properties are clear:

It can happen that $\lim_{n\to\infty} x_n = x$ and $\lim_{n\to\infty} y_n = y$, but $\lim_{n\to\infty}(x_n + y_n) \ne x + y$. (Note that if y is continuous, then the above does hold.) Thus D with the Skorohod topology *is not a topological vector space*.

$D(\mathbf{R}^d) \ne \prod_{i=1}^d D(\mathbf{R})$ in the sense of Cartesian products as topological spaces. Indeed, the topology of $D(\mathbf{R}^d)$ *is finer* than the product topology $D(\mathbf{R})^d$.

We will say that a subset A of D is *relatively compact* if it has compact closure. Note that if A is relatively compact then every sequence has a convergent subsequence (in the Skorohod topology, of course).

We now wish to pass from convergence in the space of functions to convergence of stochastic processes. There is a minor problem to make this procedure measurable. We have the following result. (See [7] for more results of this type.)

Theorem 6.5. *Let X_n and X be E-valued stochastic processes, where E is a Polish space, and suppose $\lim_{n\to\infty} X_n = X$ a.s. in the Skorohod topology. Then there exists a sequence $(\Lambda_n)_{n\geq 1}$ of measurable processes with paths in Γ such that $\lim_{n\to\infty} d_{\Lambda_n}(X_n, X) = 0$ a.s.*

Proof. Recall $d^1_\lambda(x, y)$ is defined in (6.1).

$$U_n = \{(\omega, \lambda) \in \Omega \times \Gamma : d^1_\lambda(X_n(\omega), X(\omega)) \leq d^1(X_n(\omega), X(\omega)) + 2^{-n}\}.$$

Denote by $\mathcal{D}$ and $\mathcal{G}$ the Borel fields of D and Γ respectively, where Γ is endowed with the uniform topology. Then $(x, \lambda) \to x \circ \lambda$ is Borel from $D \times \Gamma$ into D, and $(x, y) \to d^1(x(t), y(t); t \leq n)$ is Borel from $D \times D$ into $\mathbf{R}$. Since X_n and X are measurable from $(\Omega, \mathcal{F})$ into $(D, \mathcal{D})$, by composition we have $U_n \in \mathcal{F} \otimes \mathcal{G}$.

The projection $\pi_n(U_n) = \{\omega : \exists \lambda \in \Gamma \text{ with } (\omega, \lambda) \in U_n\}$ is equal to all of Ω. By the measurable section theorem (cf, e.g., [2, p. 18], there exists a random variable Λ_n with values in $(\Gamma, \mathcal{G})$ such that $P\{\omega : (\omega, \Lambda_n(\omega)) \in U_n\} = 1$. Since $d^1(x, y) = \inf_{\lambda \in \Gamma} d^1_\lambda(x, y)$, we have $d^1_{\Lambda_n}(X_n, X) \leq d^1(X_n, X) + 2^{-n}$ a.s., whence the result. □

We remark that one can improve upon this result to obtain a sure result (instead of "almost sure"); the proof is complicated and uses a measurable selection theorem (see, e.g., [7]).

Let us now turn to *weak convergence*. For a given Polish space E (for our purposes one can think of E as a complete, separable metric space), let $\mathcal{P}(E)$ denote the space of all probability measures on $(E, \mathcal{E})$. We endow $\mathcal{P}(E)$ with the *weak topology:* this is the smallest topology making all the mappings

$$f \to \int f \, d\mu$$

continuous for all *bounded continuous functions* f defined on E. We have that $\mathcal{P}(E)$ is also a Polish space for this topology. We have the following elementary properties:

Theorem 6.6. *Let E, E' be Polish spaces. Suppose μ_n, $\mu \in \mathcal{P}(E)$ and μ_n converges to μ weakly. Then*

(i) if F is a closed subset of E then $\limsup_{n\to\infty} \mu_n(F) \leq \mu(F)$;

(ii) if f is a bounded function on E that is $\mu-$ a.s. continuous then $\lim_{n\to\infty} \mu_n(f) = \mu(f)$;

(iii) if $h : E \to E'$ then $\mu \to \mu \circ h^{-1}$ is continuous from $\mathcal{P}(E)$ to $\mathcal{P}(E')$ at each point μ such that h is $\mu-$ a.s. continuous.

Definition 6.7. *A subset A of $\mathcal{P}(E)$ is called* "tight" *if for every $\varepsilon > 0$ there exists a compact subset K of E such that $\mu(K^c) \leq \varepsilon$ for all $\mu \in A$.*

Perhaps the most important result in weak convergence is the following:

Theorem 6.8 (Prokhorov). *A subset A of $\mathcal{P}(E)$ is relatively compact for the weak topology if and only if it is tight.*

We now wish to consider the *weak convergence of stochastic processes.* Let X_n, X be $\mathbf{R}^d$-valued stochastic processes with paths in D. That is, X_n and X have càdlàg paths. (One could replace $\mathbf{R}^d$ with a Polish space E if desired.) Two obvious ways X_n could converge to X are:

$$X^n(\omega) \to X(\omega) \text{ in } D \text{ for the Skorohod topology for all } \omega; \quad (6.2)$$

$$X^n \to X \text{ in } D \text{ for the Skorohod topology almost surely.} \quad (6.3)$$

We wish to consider a third way, namely,

$$E\{f(X^n)\} \to E\{f(X)\} \text{ for all bounded Skorohod continuous functions } f. \quad (6.4)$$

Note that if we let μ_n, μ be the distributions respectively of X^n, X, then (6.4) is the same as

$$\int f\,d\mu_n \to \int f\,d\mu \text{ for all bounded Skorohod continuous } f. \quad (6.5)$$

The third type ((6.4) above) will be called *the convergence in distribution of* X_n to X and it will be implicitly understood that *we are always using the Skorohod topology.* We denote $X^n \Rightarrow X$ to mean X^n converges in distribution to X.

Observe that for convergence types (6.2) and (6.3), X^n and X must all be defined on the same probability space, whereas for convergence in distribution (6.4) each X^n and X can be defined on a different space. Such a nuance is important for limit theorems, since the limit X may of necessity "live" on a strictly bigger space than the converging sequence X^n. On the other hand, combining Theorem 6.5 with the classical Skorohod representation theorem, one can prove the following:

Theorem 6.9. *Let $X^n \Rightarrow X$. Then there exists a probability space $(\widehat{\Omega}, \widehat{\mathcal{F}}, \widehat{P})$ such that there exists processes $(\widehat{X}^n)_{n\geq 1}$, $\widehat{X}$ defined on $\widehat{\Omega}$ with $\mathcal{L}(\widehat{X}^n) = \mathcal{L}(X^n)$; $\mathcal{L}(\widehat{X}) = \mathcal{L}(X)$; and furthermore there exists a sequence of measurable processes Λ_n with paths in Γ such that $\lim_{n\to\infty} d_{\Lambda_n}(X^n, \widehat{X}) = 0$, $\widehat{P}$ a.s.*[4]

(Recall that the metric $d^1_\lambda(x, y)$ is given in (6.1)).

The definition (6.4) of convergence in distribution is stated in terms of functions which are "continuous for the Skorohod topology". We can relate (6.4) to the more familiar continuous functions (for the uniform topology) using the ideas of Theorem 6.5 or Theorem 6.9:

Theorem 6.10. *$X^n \Rightarrow X$ if and only if there exists a sequence of measurable processes Λ^n with paths in Γ such that $\lim_{n\to\infty} E\{f(X^n_{\Lambda^n})\} = E\{f(X)\}$ for all bounded, continuous f (f continuous in the uniform topology).*

---

[4] $\mathcal{L}(X)$ denotes the law of X; that is, the distribution of X

Since we will be concerned with the convergence in distribution of càdlàg processes (and not probability measures), it is useful to reformulate Prokhorov's theorem in terms of them:

Definition 6.11. *A sequence $(X^n)_{n\geq 1}$ of stochastic processes with paths in D is said to be* relatively compact in distribution *if the sequence $\mathcal{L}(X^n)$ of its distribution measures is relatively compact.*

Note that Definition 6.11 says essentially that (X^n) is tight if there exists a compact subset K of D such that $P^n(X^n \notin K) \leq \varepsilon$ for all n.

Theorem 6.12 (Prokhorov). *Let $(X^n)_{n\geq 1}$ be a sequence of stochastic processes with paths in D. The sequence $(\mathcal{L}(X^n))_{n\geq 1}$ is relatively compact in $\mathcal{P}(D_E)$ if and only if the collection of distribution measures of $(X^n)_{n\geq 1}$ is tight.*

7. Weak Convergence of Stochastic Integrals

Let (H^n, X^n) be a sequence of processes in $\mathbf{D}$. If we assume X^n are semimartingales, each n, a natural question to pose — which is useful in many applications — is when do the stochastic integrals $\int H^n_{s-} dX^n_s$ converge, and to what do they converge? If $(H^n, X^n) \Rightarrow (H, X)$, it would be desirable to have sufficient conditions such that X is a semimartingale too and $\int H^n_{s-} dX^n_s \Rightarrow \int H_{s-} dX_s$. We will see that we have a surprisingly nice answer to this question.

Since we are dealing with weak convergence we may assume that each (H^n, X^n) is defined on its own space. Let $\Theta^n = (\Omega^n, \mathcal{F}^n, P^n, \mathbf{F}^n)$ where $\mathbf{F}^n = (\mathcal{F}^n_t)_{t\geq 0}$ is a filtration satisfying the usual hypothesis, each $n \geq 1$.

Before we make the next definition, that of goodness, we must clear up an important ambiguity. If (H^n, X^n) is a sequence of processes in $\mathbf{D}^2$ converging to (H, X) in $\mathbf{D}^2$ in the Skorohod topology, then they could be considered to converge either in $D_{\mathbf{R}^2}[0,\infty)$ or in $D_{\mathbf{R}}[0,\infty) \times D_{\mathbf{R}}[0,\infty)$. The former convergence is stronger: for $D_{\mathbf{R}^2}[0,\infty)$ we assume there is *one* sequence Λ_n of changes of time such that $(H^n_{\Lambda_n(t)}, X^n_{\Lambda_n(t)})$ converges uniformly to (H_t, X_t); in $D_{\mathbf{R}}[0,\infty) \times D_{\mathbf{R}}[0,\infty)$ there are *two* changes of time, Λ^1_n and Λ^2_n, such that $(H^n_{\Lambda^1_n(t)}, X^n_{\Lambda^2_n(t)})$ converges uniformly to (H_t, X_t). We will always use the stronger topology $D_{\mathbf{R}^2}[0,\infty)$. That is, *if we write $(H^n, X^n) \Rightarrow (H, X)$ it will be understood that convergence is in the topology $D_{\mathbf{R}^2}[0,\infty)$*, and thus one change of time Λ_n applies to both H^n and X^n. It turns out that this is the natural convergence to use for most applications (eg to stochastic differential equations), since often the jumps of H^n will be intimately related to those of X^n. In addition, the following example shows that the fundamental theorem of weak convergence of stochastic integrals (Theorem 7.10) fails if one takes convergence in $D_{\mathbf{R}}[0,\infty) \times D_{\mathbf{R}}[0,\infty)$.

Example 7.1. Let $X_t^n = X_t = 1_{\{t\geq 1\}}$ for all n, and let $H_t^n = 1_{\{t\geq 1+\frac{1}{n}\}}$. Then $\int_0^t H_{s-}^n dX_s^n = 1$ for $t > 1+\frac{1}{n}$, but the limiting integral $\int_0^t H_{s-}dX_s = 0$ for all t.

Caveat 7.2. To keep our notation simple we make the convention that when we say, for example, that (H^n, X^n), (H, X) are vector processes in $\mathbf{D}$ and $(H^n, X^n) \Rightarrow (H, X)$, we mean that X^n, X are d dimensional vectors of processes with each component a process in $\mathbf{D}$, and H^n is a $k \times d$ matrix of processes with each component in $\mathbf{D}$. The convergence is of course weak convergence in the Skorohod topology $D_{M^{kd}\times \mathbf{R}^d}[0,\infty)$, where M^{kd} denotes $k \times d$ real valued matrices.

Definition 7.3. *Let X^n be a sequence of $\mathbf{R}^d$-valued semimartingales on Θ^n, $n \geq 1$ and assume $X^n \Rightarrow X$. The sequence X^n is* good *if for any sequence $(H^n)_{n\geq 1}$ of $d \times k$ matrix processes in* $\mathbf{D}$ *defined on Θ^n such that $(H^n, X^n) \Rightarrow (H, X)$, then X is semimartingale and $\int H_{s-}^n dX_s^n \Rightarrow \int H_{s-}dX_s$.*

Observe that in Definition 7.3 we are implicitly assuming that the limit process X is a semimartingale on a space $(\Omega, \mathcal{F}, P, \mathbf{F})$ relative to which H is an *adapted*, càdlàg process. Thus $\mathbf{F}$ may be required to be a bigger filtration than the minimal one generated by X that satisfies the usual hypotheses.

Recall that X was defined to be a semimartingale if for H in $\mathbf{S}$, satisfying (1.2), and I_X defined by (1.3), we have $I_X : \mathbf{S}_u \to \mathbf{L}^0$ were continuous on compact time sets. In other words, if $H^n \in \mathbf{S}$ converged uniformly to $H \in \mathbf{S}$, then $\int H_s^n dX_s$ would converge in probability to $\int H_s dX_s$. The following analogous property for sequences was proposed by Jakubowski, Mémin, and Pagès [9]:

Definition 7.4. *A sequence of semimartingales $(X^n)_{n\geq 1}$, with X^n defined on Θ^n, is said to be* uniformly tight, *denoted* UT, *if for each $t > 0$, the set $\{\int_0^t H_{s-}^n dX_s^n, H^n \in \mathbf{S}^n, |H^n| \leq 1, n \geq 1\}$ is stochastically bounded (uniformly in n).*

In the above definition $\mathbf{S}^n$ denotes the simple predictable processes on Θ^n.

Definition 7.4 gives a theoretically compelling criterion, but it is perhaps not easy to verify in practice. We will give another criterion that is indeed easy to verify in practice and which turns out to be equivalent. A first step is to modify a semimartingale in such a way as to work with processes with *bounded jumps.* A standard procedure in the theory of stochastic integration is simply to subtract away the jumps bigger than a certain size: that is, if X is a given semimartingale, and $\delta > 0$ is given, let

$$X_t = \left\{ X_t - \sum_{0<s\leq t} \Delta X_s 1_{\{|\Delta X_s|>\delta\}} \right\} + \sum_{0<s\leq t} \Delta X_s 1_{\{|\Delta X_s|>\delta\}} \qquad (7.1)$$

where of course $\Delta X_s = X_s - X_{s-}$. The sums converge since ω by ω they have only a finite number of terms before each $t > 0$, since X has càdlàg paths. The problem with this approach is that it is not a continuous operation for the Skorohod topology! We will instead propose a similar procedure which — while it is a bit more complicated — is indeed a Skorohod continuous procedure! Instead of removing the large jumps, we shrink them to be no larger than a specified $\delta > 0$. We define $h_\delta : \mathbf{R}_+ \to \mathbf{R}_+$ by $h_\delta(r) = (1-\delta/r)^+$, and $J_\delta : D(\mathbf{R}^d) \to D(\mathbf{R}^d)$ by

$$J_\delta(x)t = \sum_{0<s\leq t} h_\delta(|\Delta x_s|)\Delta x_s. \tag{7.2}$$

For a semimartingale X set $X^\delta = X - J_\delta(X)$, and analogously for a sequence $X^n : X^{n,\delta} = X^n - J_\delta(X^n)$. Then X^δ will have all of its jumps bounded by δ. A semimartingale with bounded jumps has many nice properties. The most important ones for us will be as follows. Let Y *be a semimartingale with jumps bounded by* $\delta > 0$; then we have:

Y is locally bounded; that is, there exist stopping times $(T^k)_{k\geq 1}$ increasing to ∞ a.s. such that $Y^{T^k} = (Y_{t\wedge T^k})_{t\geq 0}$ is bounded a.s.; (7.3)

$[Y,Y]$ is locally bounded; (7.4)

Y has a decomposition $Y = M + A$ where M is a local martingale and $A \in \mathbf{D}$ has paths of finite variation on compacts, and M and A both have bounded jumps (by, e.g., 2δ); (7.5)

The process A in (7.5) can be taken to be "natural" (see [15]), or equivalently, predictably measurable.[5] (7.6)

The process M and A in (7.5) are each locally bounded. Moreover the total variation process of A, denoted $\int|dA_s|$, is also locally bounded. (7.7)

Suppose X^n is a sequence of semimartingales and $\delta > 0$. We can form $X^{n,\delta}$ for each n and we then obtain decompositions such as (7.5) for each n : $X^{n,\delta} = M^{n,\delta} + A^{n,\delta}$. As in (7.4) and (7.7) there will exist stopping times $T^{n,k}$ increasing to ∞ a.s. in k such that $[M^{n,\delta}, M^{n,\delta}]$ and $\int|dA_s|$ are locally bounded; the next definition makes the dependence of each $T^{n,k}$ on k uniform in n; note that this is a little subtle, since each sequence $(T^{n,k})_{k\geq 1}$ is *a priori* defined on a different space Θ^n.

Definition 7.5. *A sequence of semimartingales* $(X^n)_{n\geq 1}$ *is said to have* uniformly controlled variations (UCV) *if there exists* $\delta > 0$, *and for each* $\alpha > 0$, $n \geq 1$, *there exist decompositions* $X^{n,\delta} = M^{n,\delta} + A^{n,\delta}$ *and stopping times* $T^{n,\alpha}$ *such that* $P(\{T^{n,\alpha} \leq \alpha\}) \leq \frac{1}{\alpha}$ *and furthermore*

[5] The predictable σ-algebra on $\Omega \times \mathbf{R}_+$ is $\mathcal{R} = \sigma(\mathbf{L})$.

$$\sup_n E^n\left\{[M^{n,\delta},M^{n,\delta}]_{t\wedge T^{n,\alpha}}+\int_0^{t\wedge T^{n,\alpha}}|dA_s^{n,\delta}|\right\}<\infty. \tag{7.8}$$

Note that it is implicit in Definition 7.5 that each semimartingale X^n (and hence also $X^{n,\delta}$, $M^{n,\delta}$, $A^{n,\delta}$ and $T^{n,\alpha}$) can be defined on a different probability space Θ^n. Definition 7.5 is taken from [10].

Theorem 7.6. *Let $(X^n)_{n\geq}$ be a sequence of semimartingales, $X\in\mathbf{D}$, and suppose $X^n\Rightarrow X$. Then X^n satisfies UT if and only if it satisfies UCV.*

Proof. Suppose first $(X^n)_{n\geq 1}$ satisfies UT. By considering stopping times of the form $T=\inf\{t>0:|X_t^n|\geq c\}$, and then H of the form $H=1_{[0,T]}(t)$ (which is in $\mathbf{S}^n$), it follows that $\{\sup_{s\leq t}|X_s^n|;n\geq 1\}$ is stochastically bounded. Using this and Theorem 3.2 (ii) which approximates $[X,X]$ as a limit of sums of squared increments of X in ucp), we see that $[X^n,X^n]$ is also stochastically bounded. Therefore, the number of jumps of X^n bigger than $\delta(n\geq 1)$ is stochastically bounded, whence $(X^{n,\delta})$ is stochastically bounded too. Apply the preceding again to deduce that $[X^{n,\delta},X^{n,\delta}]$ is also stochastically bounded.

Let $X^{n,\delta}=M^{n,\delta}+A^{n,\delta}$ be the decomposition of $X^{n,\delta}$ where $A^{n,\delta}$ is taken to be natural (as mentioned in (7.6)). Given $\varepsilon>0$, one can find K such that $P^n([X^{n,\delta},X^{n,\delta}]>K)<\varepsilon$. Let $T^{n,k}=\inf\{t>0:[X^{n,\delta},X^{n,\delta}]_t>K\}\wedge t$. Then $P(T^{n,k}<t)<\varepsilon$, and moreover

$$E\{[X^{n,\delta},X^{n,\delta}]_{T^{n,k}}\}\leq K+4\delta^2,$$

using $(a+b)^2\leq 2a^2+2b^2$ and also (7.5). Since $A^{n,\delta}$ is natural one has that $E^n\{[M^{n,\delta},A^{n,\delta}]_R\}=0$ for stopping times R such that $M^{n,\delta}$ is bounded. Since $[X^{n,\delta},X^{n,\delta}]=[M^{n,\delta},M^{n,\delta}]+2[M^{n,\delta},A^{n,\delta}]+[A^{n,\delta},A^{n,\delta}]$, we deduce

$$E^n\{[M^{n,\delta},M^{n,\delta}]_{T^{n,k}}\}\leq E^n\{[X^{n,\delta},X^{n,\delta}]_{T^{n,k}}\}\leq K+4\delta^2. \tag{7.9}$$

If $H^n\in\mathbf{S}^n$, $|H^n|\leq 1$, then by Doob's maximal quadratic inequality we have

$$\begin{aligned}E\left\{\left(\sup_{r\leq T^{n,k}}\int_0^r H_s^n dM_s^{n,\delta}\right)^2\right\}&\leq 4E\left\{\int_0^{T^{n,k}}(H_s^n)^2 d[M^{n,\delta},M^{n,\delta}]_s\right\}\\&\leq 4E\{[M^{n,\delta},M^{n,\delta}]_{T^{n,k}}\}\leq K+4\delta^2\end{aligned}$$

by (7.9), and combining this with the UT property of $X^{n,\delta}$, and since

$$P([M^{n,\delta},M^{n,\delta}]_{T^{n,k}}>K)\leq\varepsilon+\frac{K+4\delta^2}{K^2},$$

we have that

$$\left\{\int_0^r H_s^n dA_s^{n,\delta};H^n\in\mathbf{S}^n,|H^n|\leq 1,n\geq 1\right\}$$

is stochastically bounded. Note that

$$\frac{|dA_s^{n,\delta}|}{dA_s^{n,\delta}} = H_s^{n,\delta} \in \mathbf{D} \text{ has } |H_s^{n,\delta}| = 1,$$

and since $A^{n,\delta}$ is natural we can take $H^{n,\delta} \in \mathbf{L}$ without loss. Therefore we have that

$$\lim_{k\to\infty} \sum_{t_i \in \pi^k} H_{t_i}^{n,\delta,k}(A_{t_i+1}^{n,\delta} - A_{t_i}^{n,\delta}) = \int |dA_s^{n,\delta}|,$$

and we deduce $\int_0^{t\wedge T^{n,k}} |dA_s^{n,\delta}|$ is stochastically bounded for each K. Since the jumps of $A^{n,\delta}$ (and hence also of $|A^{n,\delta}|$) are bounded by 2δ, it follows that we have UCV.

Next suppose $(X^n)_{n\geq 1}$ satisfies UCV. Since $X^n \Rightarrow X$, there exists $\delta > 0$ such that $J_\delta(X^n)$ is stochastically bounded. By UCV we also have that

$$\left\{ \int_0^t |dA_s^{n,\delta}|, [M^{n,\delta}, M^{n,\delta}]_t; n \geq 1 \right\}$$

is stochastically bounded. This implies that $X^n - M^{n,\delta}$ satisfies the property:

$$\left\{ \int K_{s-}^n d(X^n - M^{n,\delta})_s; K^n \in \mathbf{D}; |K^n| \leq 1; n \geq 1 \right\}$$

is stochastically bounded. Now let $\varepsilon > 0$. There exists K such that $P^n([M^{n,\delta}, M^{n,\delta}]_t > K) < \varepsilon$ for all n. Define

$$T^n = \inf\{s : [M^{n,\delta}, M^{n,\delta}]_s > K\} \wedge t.$$

Then $P^n(T^n < t) < \varepsilon$. Next let $H^n \in \mathbf{D}$, $|H^n| \leq 1$. Then we have

$$\begin{aligned}
P^n(|(H_-^n \cdot M^{n,\delta})_t| > K) &\leq P^n\left(\sup_{s\leq t} |(H_-^n \cdot M^{n,\delta})_s| > K\right) \\
&\leq P^n(\sup_{s\leq T^n} |(H_-^n \cdot M^{n,\delta})_s| > K) + \varepsilon \\
&\leq \frac{1}{K^2} E\{\sup_{s\leq T^n} ((H_-^n \cdot M^{n,\delta})_s)^2\} + \varepsilon \\
&\leq \frac{1}{K^2} E\{[M^{n,\delta}, M^{n,\delta}]_{T^n}\} + \varepsilon \leq \frac{K + 4\varepsilon^2}{K^2} + \varepsilon.
\end{aligned}$$

This last quantity can be made arbitrarily small, and thus $M^{n,\delta}$ satisfies UT as well. □

We note that without the hypothesis that $X^n \Rightarrow X$, we have that if (X^n) satisfies UT, then it satisfies UCV, but if it satisfies UCV we need the extra hypotheses that $J_\delta(X^n)$ is stochastically bounded to prove it satisfies UT.

Next we give some general conditions that imply UT:

If $(X^n)_{n\geq 1}$ is a sequence of supermartingales such that

$$\inf_n(\inf_s X^n_s) \geq b \qquad (b \in \mathbf{R})$$

then (X^n) satisfies UT (cf [9]); (7.10)

If $(X^n)_{n\geq 1}$ is a sequence of local martingales and if for each $t < \infty$ one has

$$\sup_n E^n\{\sup_{s\leq t} |\Delta X^n_s|\} < \infty$$

then (X^n) satisfies UT (cf [9]). (7.11)

Clearly the condition UCV gives conditions which are easy to verify in practice. We give two examples:

– Let $(X^n)_{n\geq 1}$ be a sequence of semimartingales with decompositions

$$X^n = M^n + A^n \text{ such that } \sup_n\{E^n\{[M^n, M^n]_t\} + E^n\{\int_0^t |dA^n_s|\}\} < \infty, \tag{7.12}$$

each $t > 0$. Then $(X^n)_{n\geq 1}$ satisfies UCV.

– Let $(X^n)_{n\geq 1}$ be a sequence of semimartingales with decompositions

$$X^n = M^n + A^n \text{ such that } \sup_n\{\operatorname{Var}(M^n_t) + E^n\{\int_0^t |dA^n_s|\}\} < \infty, \tag{7.13}$$

each $t > 0$. Then $(X^n)_{n\geq 1}$ satisfies UCV. (Here $\operatorname{Var}(M^n_t)$ refers to the variance of the random variable M^n_t).

Note that (7.12) and (7.13) are trivially equivalent. Combining (7.11) and (7.12) we get: let $X^n \Rightarrow X$ and suppose X^n has decompositions

$$X^n = M^n + A^n$$

such that

$$\sup_n\left\{E^n\left\{\sup_{s\leq t}|\Delta M^n_s|\right\} + E^n\left\{\int_0^t |dA^n_s|\right\}\right\} < \infty. \tag{7.14}$$

Then $(X^n)_{n\geq 1}$ satisfies UT and UCV.

Theorem 7.7. *Let $(X^n)_{n\geq 1}$ be a sequence of vector valued semimartingales, X a vector valued process in $\mathbf{D}$, and assume $X^n \Rightarrow X$ and that $(X^n)_{n\geq 1}$ is a good sequence. Then $(X^n)_{n\geq 1}$ satisfies UT and UCV.*

Proof. We treat the scalar case. By Theorem 7.6 it suffices to show that UT holds. Suppose $(X^n)_{n\geq 1}$ is good but UT does not hold. Then there must exist $H^n \in \mathbf{S}$, $|H^n| \leq 1$, and constants c_n increasing to ∞ such that for some $\varepsilon > 0$,

$$\liminf_{n\to\infty} P^n\left\{\int H^n_{s-}dX^n_s \geq c_n\right\} \geq \varepsilon.$$

But this implies that

$$\liminf_{n\to\infty} P^n\left\{\int \frac{1}{c_n} H^n_{s-}\,dX^n_s \geq 1\right\} \geq \varepsilon \tag{7.15}$$

as well. Since $|H^n| \leq 1$ we have that $\frac{1}{c_n}H^n$ converges uniformly in distribution to the zero process. The goodness of $(X^n)_{n\geq 1}$ then implies that $\int \frac{1}{c_n} H^n_{s-}\,dX^n_s$ converges in distribution to 0, which contradicts (7.15). □

The next theorem is a key step to showing that each of UT and UCV imply goodness. For a sequence of vector processes (H^n, X^n) each defined on a space Θ^n, we let $\mathbf{H}^n = (\mathcal{H}^n_t)_{t\geq 0}$ denote the smallest filtration making (H^n, X^n) adapted and also satisfying the usual hypotheses. $\mathbf{H} = (\mathcal{H}_t)_{t\geq 0}$ is analogous for the limiting process (H, X).

Theorem 7.8. *Let (H^n, X^n), (H, X) be vector processes, each in* $\mathbf{D}$, *and suppose $(H^n, X^n) \Rightarrow (H, X)$. Assume UT or equivalently UCV holds. Then X is an* $\mathbf{H}$ *semimartingale.*

Proof. $X \in \mathbf{D}$ by hypothesis. If $H^m \in \mathbf{S}(\mathbf{H})$, $|H^n| \leq 1$, and $\lim_{m\to\infty} H^m = 0$ uniformly, we need to show $\lim_{m\to\infty} H^m \cdot X = 0$ in probability. Note that for $H^m \in \mathbf{S}$, $H^m \cdot X$ is well defined for any $X \in \mathbf{D}$. Since the limit is 0 — a constant — it suffices to show that $H^m \cdot X$ converges to 0 in distribution. Thus it suffices to show that

$$\{H^n \cdot X, H^m \in \mathbf{S}(\mathbf{H}), |H^m| \leq 1\} \tag{7.16}$$

is stochastically bounded.

Let

$$j(H,X) = \{s \geq 0 : P(\Delta H_s \neq 0 \text{ or } \Delta X_s \neq 0) > 0\}. \tag{7.17}$$

One can check fairly easily (cf, eg, [8, p. 313]) that $j(H, X)$ is at most countable. Therefore, $Q = \mathbf{R}_+ \setminus j(H, X)$ is dense. Since $(H^n, X^n) \Rightarrow (H, X)$, we have that the finite dimensional distributions, restricted to Q-valued tuples, of (H^n, X^n) converge to (H, X). (Typically this is denoted $(H^n, X^n) \overset{\mathcal{L}(Q)}{\Rightarrow} (H, X)$). This fact, together with the UT property of $(X^n)_{n\geq 1}$, is enough to conclude, using simple approximation arguments, that (7.16) is stochastically bounded. □

The next theorem (Theorem 7.10) is the key result in the theory of weak convergence of stochastic integrals. One can prove it using the UT approach (see [9]) or the UCV approach (see [10]). The UT approach is fairly intuitive given our definition of a semimartingale as a good integrator, but it is a little complicated to execute. The UCV approach is intuitively very simple, as is the proof. The main disadvantage is that we need a technical result concerning the Skorohod topology. Note that we will generalize this approach in Section 4,5,6 of Part II of these notes (at the end of this volume). To motivate the argument of the proof let us first make an observation.

Let $(x_n)_{n\geq 1}$, x, $(y_n)_{n\geq 1}$, y be functions in D where $(x_n, y_n) \to (x, y)$ in the Skorohod topology (that is $D_{\mathbb{R}^2}[0,\infty)$, *not* $D_{\mathbb{R}}[0,\infty) \times D_{\mathbb{R}}[0,\infty)$!). Assume further that y_n is piecewise constant and the number of discontinuities of y_n in a bounded time interval is uniformly bounded in n. Assuming all terms make sense, we then have

$$\left(x_n, y_n, \int x_n(s-)dy_n(s), \int y_n(s-)dx_n(s)\right) \to \left(x, y, \int x_{s-}dy_s, \int y_{s-}dx_s\right) \tag{7.18}$$

in the Skorohod topology $D_{\mathbb{R}^4}[0,\infty)$. In view of (7.18), it makes sense to try to approximate the processes involved by piecewise constant processes, but in such a way that they converge along with the approximating processes. Before giving Theorem 7.10 we establish a lemma that plays an essential role in its proof.

Suppose (E, ρ) is a metric space, and let $(\theta_k)_{k\geq 1}$ be a sequence of i.i.d. random variables, uniform on $[\frac{1}{2}, 1]$. Fix $z \in D$, $\varepsilon > 0$, and define inductively

$$T_0 = 0,$$
$$T_{k+1} = \inf\{t > T_k : \rho(z_t, z_{T_k}) \vee \rho(z_{t-}, z_{T_k}) \geq \varepsilon\theta_k\}$$

and let $y_k(z) = z_{T_k}$. (Note that $T_k = T_k(z)$; that is for each z we get a different sequence of times T_k.) We define

$$I^\varepsilon(z)_t = y_k(z) \text{ if } T_k \leq t < T_{k+1}. \tag{7.19}$$

Then $\rho(z_t, I^\varepsilon(z)_t) \leq \varepsilon$, for all t. The role of the $(\theta_k)_{k\geq 1}$ is to "spread" ε over an interval which then ensures the almost sure convergence of the $T_k^n = T_k(z^n)$ when z^n converges to z.

Lemma 7.9. *For I^ε defined as in (7.19), if $\lim_{n\to\infty} z_n = z$ in the Skorohod topology $D_E[0,\infty)$, then $(z_n, I^\varepsilon(z_n)) \to (z, I^\varepsilon(z))$ a.s. in the Skorohod topology $D_{E^2}[0,\infty)$.*

We refer the reader to [10, p. 1067] for a proof.

Theorem 7.10. *If (H^n, X^n) defined on Θ^n converges in distribution in the Skorohod topology to (H, X) and if $(X^n)_{n\geq 1}$ are semimartingales satisfying UCV (or equivalently UT), there exists a filtration* **H** *such that X is an* **H** *semimartingale and moreover*

$$(H^n, X^n, H_-^n \cdot X^n) \Rightarrow (H, X, H_- \cdot X). \tag{7.20}$$

That is, the sequence $(X^n)_{n\geq 1}$ is good.

Proof. That (UT) and (UCV) are equivalent under these hypotheses is Theorem 7.6. That X is an **H** semimartingale is Theorem 7.8. Thus it remains to establish (7.20).

Recall that for $x \in D$ and $\delta > 0$, we defined $J_\delta(x)$ in (7.2) as an operator that is used to shrink the large jumps to the size δ. Then $X^{n,\delta} = X^n - J_\delta(X^n)$ is a semimartingale with jumps bounded by δ. We define

$$Z^n = (H^n, X^n, J_\delta(X^n), X^{n,\delta}).$$

Let I^ε be as defined in (7.19). Then $I^\varepsilon(Z^n)$ is adapted to a filtration $\mathbf{K}^n = \mathbf{H}^n \vee \mathbf{U}$, where $\mathbf{U}$ is independent of $\mathbf{H}^n$. (By the independence, we note that X^n remains a semimartingale for the larger filtration $\mathbf{K}^n$.) Let $H^{n,\varepsilon}$ denote the first component of Z^n (which is M^{kd} valued in general, where M^{kd} represents $k \times d$ matrices). Then $|H^n - H^{n,\varepsilon}| \leq \varepsilon$ and moreover

$$(H^n, X^n, J_\delta(X^n), X^{n,\delta}, H^{n,\varepsilon}) \Rightarrow (H, X, J_\delta(X), X^\delta, H^\varepsilon).$$

Next define:

$$\begin{aligned}
U^n &= \int H^n_{s-} dX^n_s; \\
U^{n,\varepsilon} &= \int H^{n,\varepsilon}_{s-} dX^{n,\delta}_s + \int H^n_{s-} dJ_\delta(X^n)_s; \\
U &= \int H_{s-} dX_s; \\
U^\varepsilon &= \int H^\varepsilon_{s-} dX^\delta_s + \int H_{s-} dJ_\delta(X)_s.
\end{aligned}$$

Then it follows as in (7.18) that

$$(H^n, X^n, U^{n,\varepsilon}) \Rightarrow (H, X, U^\varepsilon).$$

Finally we let

$$R^{n,\varepsilon} = U^n - U^{n,\varepsilon}.$$

Then

$$\begin{aligned}
R^{n,\varepsilon} &= \int (H^n_{s-} - H^{n,\varepsilon}_{s-}) dX^{n,\delta}_s \\
&= \int (H^n_{s-} - H^{n,\varepsilon}_{s-}) dM^{n,\delta}_s + \int (H^n_{s-} - H^{n,\varepsilon}_{s-}) dA^{n,\varepsilon}_s
\end{aligned}$$

where $X^{n,\delta} = M^{n,\delta} + A^{n,\delta}$ is a decomposition of $X^{n,\delta}$ into the sum of a local martingale and a finite variation process. Using Doob's maximal quadratic inequality we have that for any stopping time T,

$$E\left\{\sup_{s \leq t\wedge T} |R^{n,\varepsilon}_s|\right\} \leq \varepsilon \left\{2E\left\{[M^{n,\delta}, M^{n,\delta}]^{\frac{1}{2}}_{t\wedge T}\right\} + 2E\left\{\int_0^{t\wedge T} |dA^{n,\delta}_s|\right\}\right\}.$$

An analogous estimate holds for $U - U^\varepsilon$. We now apply the UCV hypothesis to conclude that $(H^n, X^n, U^n) \Rightarrow (H, X, U)$. □

We wish to make several remarks. First note that if we combine Theorems 7.7 and 7.10, we have that if $(H^n, X^n) \Rightarrow (H, X)$, then $(X^n)_{n \geq 1}$ is *good if and only if UCV* (or equivalently UT) holds. Second, if convergence in distribution is replaced by convergence in probability in the hypothesis of Theorem 7.10 (in this case of course all processes are defined on the same space), then convergence in probability will also hold in the conclusion.

Third, we can use Theorem 7.10 to prove some nice properties of goodness (Theorems 7.11 through 7.13). The first theorem shows that goodness is inherited via stochastic integration. The proof is similar to the proof of Theorem 7.7.

Theorem 7.11. *Suppose* $(H^n, X^n) \Rightarrow (H, X)$, *and* $(X^n)_{n\geq 1}$ *is a good sequence of semimartingales. Then* $Y^n = H^n_- \cdot X^n$ *is also a good sequence of semimartingales.*

Proof. We treat the scalar case. By Theorem 7.10 it suffices to show $(Y^n)_{n\geq 1}$ satisfies UT. Suppose it does not. Then as in the proof of Theorem 7.7 there exists a sequence $K^n \in \mathrm{S}$, $|K^n| \leq 1$, and constants c_n such that for some $\varepsilon > 0$,

$$\liminf_{n\to\infty} P^n\left\{ \int K^n_{s-} dY^n_s \geq c_n \right\} \geq \varepsilon,$$

or equivalently

$$\liminf_{n\to\infty} P^n\left\{ \int \frac{1}{c_n} K^n_{s-} H^n_s dX^n_s \geq 1 \right\} \geq \varepsilon. \tag{7.21}$$

The hypothesis that $(X^n)_{n\geq 1}$ is good implies that the integrals in (7.21) converge to 0, which is a contradiction. □

Theorem 7.12. *Let* $(X^n, Y^n)_{n\geq 1}$ *be a sequence of semimartingales such that* $(X^n, Y^n) \Rightarrow (X, Y)$, *and both* $(X^n)_{n\geq 1}$ *and* $(Y^n)_{n\geq 1}$ *are good. Then*

$$(X^n, Y^n, [X^n, Y^n]) \Rightarrow (X, Y, [X, Y]) \tag{7.22}$$

and also $[X^n, Y^n]$ *and* $X^n Y^n$ *are good.*

Proof. Integration by parts yields

$$X^n Y^n = \int X^n_{s-} dY^n_s + \int Y^n_{s-} dX^n_s + [X^n, Y^n],$$

and (7.22) follows trivially. By Theorem 7.11 it suffices to show that $[X^n, Y^n]$ is good. But goodness implies UCV by Theorem 7.7, and the goodness of $[X^n, Y^n]$ follows easily. □

Theorem 7.13. *Let* $(X^n)_{n\geq 1}$ *be good, and suppose* $f : \mathbf{R}^d \times \mathbf{R}_+ \to \mathbf{R}$ *is* C^2 *on* $\mathbf{R}^d$ *and* C^1 *on* $\mathbf{R}_+$. *Let* $Y^n_t = f(X^n_t, t)$. *Then* $(Y^n)_{n\geq 1}$ *is also a good sequence.*

Proof. One need only apply the change of variables formula, and Theorems 7.12 and 7.11. □

We remark that the convergence in (7.22) is not as robust as it might seem. The next example, due to Jacod [6, p. 395], shows that one can have $X^n \Rightarrow X$, but $[X^n, X^n] \not\Rightarrow [X, X]$; thus a condition such as goodness is truly needed for the convergence of the quadratic variations.

Example 7.14. Let X^n be the non-random process $X^n_t = \sum_{i=1}^{[n^2 t]}(-1)^i \frac{1}{n}$. Then $|X^n| \leq \frac{1}{n}$, whence $X^n \Rightarrow 0$. On the other hand,

$$[X^n, X^n]_t = \sum_{i=1}^{[n^2 t]} \left(\frac{1}{n}\right)^2 = \frac{[n^2 t]}{n^2}$$

which converges to t. Since $[X, X] = 0$ trivially, we have $[X^n, X^n] \not\Rightarrow [X, X]$.

One of the primary uses of Theorem 7.10 is to the study of stochastic differential equations, which is the topic of §8..

8. Weak Convergence of Stochastic Differential Equations

In this section we consider stochastic differential equations in a form similar to those of § 5.. Let $(X^n)_{n\geq 1}$ be a sequence of semimartingales, and $(J^n)_{n\geq 1}$ be a sequence of processes in $\mathbf{D}$. Let

$$F, F^n : D_{\mathbf{R}^k}[0,\infty) \to D_{\mathbf{M}^{km}}[0,\infty)$$

have property (i) of Definition 5.1: that is, for $t > 0$ we have $F^n(X)_t = F^n(X^t)_t$ and $F(X)_t = F(X^t)_t$, which is a non-anticipation requirement. Note that we do not make the Lipschitz hypothesis ((5.1)(ii)). We will study equations of the type

$$X^n_t = J^n_t + \int_0^t F^n(X^n)_{s-} dZ^n_s \tag{8.1}$$

and give conditions that imply $X^n \Rightarrow X$, where X is a solution of the limiting equation

$$X_t = J_t + \int_0^t F(X)_{s-} dZ_s. \tag{8.2}$$

Note that without the Lipschitz assumption on $(F^n)_{n\geq 1}$ nor F we do not have uniqueness of solutions either for (8.1) or for (8.2). If we are willing to assume that *a priori* the solutions (8.1) are relatively compact, we have the following simple result:

Theorem 8.1. *Suppose that (J^n, X^n, Z^n) satisfies equation (8.1), that (J^n, X^n, Z^n) is relatively compact in the Skorohod topology for $D_{\mathbf{R}^{2k+m}}[0,\infty)$, and that $(J^n, Z^n) \Rightarrow (J, Z)$ and that $(Z^n)_{n\geq 1}$ is good. Assume further that F^n, F satisfy*

$$\begin{array}{l} \text{if } (x_n, y_n) \to (x, y) \text{ in the Skorohod topology, then} \\ (x_n, y_n, F^n(x_n)) \to (x, y, F(x)) \text{ in the Skorohod topology.} \end{array} \tag{8.3}$$

Then any limit point of the sequence $(X^n)_{n\geq 1}$ satisfies (8.2).

Proof. Suppose a subsequence of $(X^n)_{n\geq 1}$ converges in distribution. Then along a further subsequence, the triple (J^n, X^n, Z^n) will converge in distribution, to a process (J, X, Z). Theorem 7.10 then gives that X satisfies (8.2). □

The assumption (8.3) is that F^n and F are Skorohod continuous. Some examples of such are the following:

Example 8.2. Let $g : \mathbf{R}^k \times \mathbf{R}_+ \to \mathbf{M}^{km}$ and $h : \mathbf{R}_+ \to \mathbf{R}_+$ be continuous. The following functionals are non-anticipating and Skorohod continuous.

(i) $F(x)_t = g(x_t, t)$

(ii) $F(x)_t = \int_0^t h(t-s)g(x_s, s)ds$

If $k = m = 1$, then:

(iii) $F(x)_t = \sup_{s\leq t} h(t-s)g(x_s, s)$

(iv) $F(x)_t = \sup_{s\leq t} h(t-s)g(\Delta x_s, s)$.

Before stating our main result we need to make some definitions.

Definition 8.3. *(X, T) is a* local solution *of (8.2) if there exists a filtration* **F** *for which X, J, and Z are adapted, Z is a semimartingale, T is a stopping time, and such that*

$$X_{t\wedge T} = J_{t\wedge T} + \int_0^{t\wedge T} F(X)_{s-}dZ_s. \tag{8.4}$$

We say that we have strong local uniqueness *if any two solutions (X^1, T^1), (X^2, T^2) satisfy $X_t^1 = X_t^2$, $0 \leq t \leq T^1 \wedge T^2$.*

To define *weak* local uniqueness (that is, local uniqueness in the sense of distributions), we need the stopping times to be functions of the solutions.

Definition 8.4. *A tuple $(\widehat{J}, \widehat{Z}, \widehat{X}, \widehat{T})$ is a* weak local solution *if $(\widehat{J}, \widehat{Z})$ is a version of (J, Z), and (8.4) holds with $(\widehat{J}, \widehat{Z}, \widehat{X}, \widehat{T})$ replacing (J, Z, X, T). We say that* weak local uniqueness *holds for (8.2) if for any two weak local solutions (J^1, Z^1, X^1, T^1) and (J^2, Z^2, X^2, T^2) with $T^1 = h_1(X^1)$ and $T^2 = h_2(X^2)$ for measurable h_1, h_2, then $(X^1, (h_1 \wedge h_2)(X^1))$ and $(X^2, (h_1 \wedge h_2)(X^2))$ have the same distribution.*

We need to make some technical assumptions on the functional coefficients which are stronger than simple Skorohod continuity. Nevertheless Examples 8.2 can be shown to satisfy Condition 8.5 below.

Condition 8.5. Let Λ denote the collection of increasing maps λ of $\mathbf{R}_+$ to $\mathbf{R}_+$ with $\lambda_0 = 0$ and $\lambda_{t+h} - \lambda_t \leq h$, all t, $h \geq 0$. Assume that there exist mappings G^n, $G : D_{\mathbf{R}^k}[0,\infty) \times \Lambda \to D_{\mathbf{M}^{km}}[0,\infty)$ such that $F^n(x) \circ \lambda = G^n(x \circ \lambda, \lambda)$, and $F(x) \circ \lambda = G(x \circ \lambda, \lambda)$. Assume further

(a) For each compact subset $\mathcal{H} \subset D_{\mathbf{R}^k}[0,\infty) \times \Lambda$ and $t > 0$, (8.5)
$$\lim_{n\to\infty} \sup_{(x,\lambda)\in\mathcal{H}} \sup_{s\leq t} |G^n(x,\lambda)_s - G(x,\lambda)_s| = 0,$$

(b) For $(x_n, \lambda_n) \in D_{\mathbf{R}^k}[0,\infty) \times \Lambda$, if $\lim_{n\to\infty} \sup_{s\leq t} |x_n(s) - x(s)| = 0$ and $\lim_{n\to\infty} \sup_{s\leq t} |\lambda_n(s) - \lambda(s)| = 0$ for each $t > 0$,

then $\lim_{n\to\infty} \sup_{s\leq t} |G(x_n, \lambda_n)_s - G(x,\lambda)_s| = 0$.

Theorem 8.6. *Suppose $(J^n, Z^n) \Rightarrow (J, Z)$ where J^n, $J \in \mathbf{D}$; Z^n are semimartingales, and $(Z^n)_{n\geq 1}$ is good. Suppose F^n, F have representations in terms of G^n, G satisfying Condition 8.5. For $b > 0$, let*

$$\eta_n^b = \inf\{t > 0 : |F^n(X^n)_t| \vee |F^n(X^n)_{t-}| \geq b\}$$

and let $X^{n,b}$ denote the solution of

$$X_t^{n,b} = J_t^n + \int_0^t 1_{[0,\eta_n^b)}(s-) F^n(X^{n,b})_{s-} dZ_s^n$$

that agrees with X^n on $[0, \eta_n^b)$. Then $(J^n, X^{n,b}, Z^n)_{n\geq 1}$ is relatively compact and any limit point (J, X^b, Z) gives a local solution (X^b, T) of (8.2) with $T = \eta^c$, for any $c < b$.

Moreover if there exists a global solution X of (8.2) and weak local uniqueness holds, then $(J^n, X^n, Z^n) \Rightarrow (J, X, Z)$.

The proof of Theorem 8.6 involves some technical points, and we refer the reader to [10].

We next give two examples to show how Theorem 8.6 can be used.

Example 8.7 (Duffie-Protter). We can use Theorem 8.6 to help to derive and to justify models in continuous time stochastic finance theory as limiting cases of discrete models. As an example, let $(Y_i^n)_{i\geq 1}$ be the periodic rate of return on a security (such as a stock) with initial price S_0. After k periods the price of the security will be

$$S_k^n = S_0^n \prod_{i=1}^{k} (1 + Y_i^n).$$

Let $Z^n_t = \sum_{i=1}^{[nt]} Y^n_i$ and $S^n_t = S^n_{[nt]}$. Since $S^n_{k+1} - S^n_k = S^n_k Y^n_k$, we can write

$$S^n_t = S^n_0 + \int_0^t S^n_{s-} dZ^n_s.$$

If $(Z^n)_{n\geq 1}$ is good with $Z^n \Rightarrow Z$ then the limiting equation is

$$S_t = S_0 + \int_0^t S_{s-} dZ_s$$

which is the stochastic exponential and has a unique global solution, and thus if $S^n_0 \Rightarrow S_0$, by Theorem 8.6, $S^n \Rightarrow S$. Moreover by Theorem 7.11 we also thus know that $(X^n)_{n\geq 1}$ is good, hence if $(\theta^n_k)_{k\geq 1}$ represents a trading strategy and

$$G^n_t = \int_0^t \theta^n_{s-} dS^n_s$$

where $\theta^n_t = \theta^n_{[nt]}$ represents the resulting "gain" from the strategy θ^n, and if $(Z^n, \theta^n, S^n_0) \Rightarrow (Z, \theta, S_0)$ with $(Z^n)_{n\geq 1}$ still assumed to be good, we have $G^n \Rightarrow G$, with G given by

$$G_t = \int_0^t \theta_{s-} dS_s.$$

Many naturally occurring models have the property that $(Z^n)_{n\geq 1}$ is good.

Example 8.8. Emery [4] has discovered a class of martingales that have the Chaos Representation Property (CRP). A necessary condition to have this property, if $\langle M, M\rangle_t = t$, is that the local martingale M satisfy an equation of the type $[M, M]_t = t + \int_0^t \varphi_s dM_s$. A special case is:

$$d[X, X]_t = dt + f(X_{t-})dX_t, \tag{8.6}$$

where $f : \mathbf{R} \to \mathbf{R}$ is continuous. Therefore it is of interest to know when solutions of (8.6) exist. One can show existence for any such f by defining a sequence of discrete time martingales and then showing the sequence is relatively compact and that the limit satisfies (8.6). If one sets $\Delta Y^n_k = Y^n_{k+1} - Y^n_k$ and assumes $Y^n_0 = 0$, then the discrete time analogue of (8.6) becomes

$$(\Delta Y^n_k)^2 = \frac{1}{n} + f(Y^n_k)\Delta Y^n_k. \tag{8.7}$$

One then solves (8.7) for ΔY^n_k:

$$\Delta Y^n_k = \frac{f(Y^n_k) \pm \sqrt{f(Y^n_k)^2 + 4/n}}{2}.$$

Call the solutions Z^+ and Z^-. In order for Y^n to be a martingale we are forced to choose

$$P(\Delta Y_k^n = Z^+) = \frac{Z^-}{Z^- - Z^+}$$

and

$$P(\Delta Y_k^n = Z^-) = 1 - P(\Delta Y_k^n = Z^+).$$

We then define $X_t^n = Y_{[nt]}^n$ and show it is relatively compact and that the limit satisfies (8.6). See [10, p. 1044] for details. The above argument also applies to the more general equation

$$d[X,X]_t = dt + F(X)_{t-}dX_t, \tag{8.8}$$

where F satisfies Condition 8.5. Note that a martingale is called a ***normal martingale*** if $\langle M,M\rangle_t = t$ and it has CRP, thus solving equations (8.6) or (8.7) is a way to generate candidates for normal martingales. Emery [4] has shown that if M satisfies

$$d[M,M]_t = dt + \beta M_{t-}dM_t; \quad M_0 = x,$$

then M is a normal martingale for $-2 \le \beta \le 0$. Note that $\beta = 0$ corresponds to Brownian motion, $\beta = -1$ is Azéma's martingale, and $\beta = -2$ is the parabolic martingale ($|M_t| = \sqrt{t}$).

In many situations the approximating semimartingales are *not* good. We give an example to illustrate this phenomenon.

Example 8.9. Let $W = (W_t)_{t\ge0}$ be a standard Wiener process (that is, standard Brownian motion). Let us approximate W with an "approximate identity" as follows:

$$W_t^n = n\int_{t-\frac{1}{n}}^t W_s ds.$$

Then W^n is defined on the same space as W, W^n is adapted to the same filtration, and $\lim_{n\to\infty} W^n = W$ a.s., uniformly on compacts. However (W^n) is not good. Indeed, consider the equations

$$X_t^n = x + \int_0^t X_s^n dW_s^n. \tag{8.9}$$

Then $X_t^n = x\exp(W_t^n)$. But for the limiting equation

$$X_t = x + \int_0^t X_s dW_s$$

we have $X_t = x\exp(W_t - \frac{1}{2}t)$. Thus $W^n \Rightarrow W$, but $X^n \not\Rightarrow X$. This could not happen if $(W^n)_{n\ge1}$ were good by Theorem 8.6.

If in Example 8.9 we rewrite W^n as $W^n = W + (W^n - W) = Y^n + Z^n$, then we have $Y^n \Rightarrow W$ and $(Y^n)_{n\geq 1}$ is good (in this case $Y^n = W$ for all n, so the convergence and goodness are trivial), and Z^n is a sequence of semimartingales converging to 0. Equation (8.8) can be re-written

$$X_t^n = X + \int_0^t X_s^n dY_s^n + \int_0^t X_s^n dZ_s^n.$$

This idea allows us to handle naturally arising situations where goodness does not apply. It generalizes a well-known approach due originally to E. Wong and M. Zakai.

Theorem 8.10. *Let Y^n, Z^n be semimartingales on Θ^n, and let $f : \mathbf{R}^d \to \mathbf{M}^{dm}$ be C^2 and bounded with bounded derivatives of first and second order. Define matrices of processes in $\mathbf{D}$ by*

$$H_t^n = \int_0^t Z_{s-}^n dZ_s^n \quad ^6$$

and

$$K_t^n = [Y^n, Z^n]_t.$$

Assume $(Y^n)_{n\geq 1}$ $(H^n)_{n\geq 1}$ are good, and $Z^n \Rightarrow 0$. Moreover assume

$$\mathbf{A}^n = (X_0^n, Y^n, Z^n, H^n, K^n) \Rightarrow \mathbf{A} = (X, Y, 0, H, K).$$

Let X^n be the solution of:

$$X_t^n = X_0^n + \int_0^t f(X_{s-}^n)dY_s^n + \int_0^t f(X_{s-}^n)dZ_s^n. \tag{8.10}$$

Then $(\mathbf{A}^n, X^n)$ is relatively compact and any limit point $(\mathbf{A}, X)$ satisfies

$$X_t = X_0 + \int_0^t f(X_{s-})dY_s + \sum_{i,j,k} \int_0^t \partial_i f^j(X_{s-}) f^{ik}(X_{s-}) d(H_s^{kj} - K_s^{kj}), \tag{8.11}$$

where ∂_i denotes the partial derivative with respect to the ith variable, f^j denotes the jth column of f, etc.

Before we prove Theorem 8.10, we remark that the boundedness assumptions on f and its derivatives can be dropped, and X_0^n can be replaced with an exogenous process $(J_t^n)_{t\geq 0}$. See [10]. Also since $Z^n \Rightarrow 0$, it follows that H and K will *a fortiori* be continuous. Also, we have not assumed that $(K^n)_{n\geq 1}$ is a good sequence in Theorem 8.10; however the hypotheses will imply that $(K^n)_{n\geq 1}$ is also good (see (8.14) in the proof to follow).

Proof of Theorem 8.10. The proof will follow from the change of variables formula and integration by parts. First observe that

[6] Matrix entries: $(H_t^n)^{ij} = \int_0^t Z_{s-}^{n,i} dZ_s^{n,j}$; $(K_t^n)^{ij} = [Y^{n,i} Z^{n,j}]_t$

$$[Z^{n,i}, Z^{n,j}]_t = Z_t^{n,i} Z_t^{n,j} - Z_0^{n,i} Z_0^{n,j} - \int_0^t Z_{s-}^{n,i} dZ_s^{n,j} - \int_0^t Z_{s-}^{n,j} dZ_s^{n,i},$$

and therefore if $I_t^{n,i,j} = [Z^{n,i} Z^{n,j}]_t$, it follows that

$$I^{n,i,j} \Rightarrow -(H^{ij} + H^{ji}). \tag{8.12}$$

Since $I^{n,i,i}$ is non-decreasing and converges in distribution to a continuous process, it follows that $I^{n,i,i}$ is good. Moreover we can estimate the increments of $I^{n,i,j}$ by the increments of $I^{n,i,i}$ and $I^{n,j,j}$ to deduce that $(I^n)_{n\geq 1}$ itself is a good sequence.

Since f is assumed to be C^2, letting X^n be the solution of (8.9), by the change of variables formula we have

$$f^{ij}(X_t^n) = f^{ij}(X_0^n) + \sum_k \int_0^t \partial_k f^{ij}(X_{s-}^n) dX_s^k + R_t^{n,i,j},$$

where the increments of $R^{n,i,j}$ are dominated by a linear combination of the increments of $[Y^{n,k}, Y^{n,k}]$ and $[Z^{n,k}, Z^{n,k}]$, whence $(R^n)_{n\geq 1}$ is good.

Next we integrate by parts to obtain:

$$\begin{aligned}
\int_0^t f^{ij}(X_{s-}^n) dZ_s^j &= f^{ij}(X_t^n) Z_t^{n,j} - f^{ij}(X_0^n) Z_0^{n,j} \\
&\quad - \sum_k \int_0^t \partial_k f^{ij}(X_{s-}^n) Z_{s-}^{n,j} dX_s^{n,k} \\
&\quad - \int_0^t Z_{s-}^{n,j} dR_s^{n,i,j} \\
&\quad - \sum_k \int_0^t \partial_k f^{i,j}(X_{s-}^n) d[X^{n,k}, Z^{n,j}]_s \\
&\quad + [R^{n,i,j}, Z^{n,j}]_t \\
&= U_t^n - \sum_{k,\ell} \int_0^t \partial_k f^{i,j}(X_{s-}^n) f^{k,\ell}(X_{s-}^n) Z_{s-}^{n,j} dZ_s^{n,\ell} \\
&\quad - \sum_{k,\ell} \int_0^t \partial_k f^{i,j}(X_{s-}^n) f^{k,\ell}(X_{s-}^n) d([Y^{n,\ell}, Z^{n,j}]_s \\
&\quad + [Z^{n,\ell}, Z^{n,j}]_s),
\end{aligned}$$

where $U^n \Rightarrow 0$. Continuing:

$$= U_t^n - \sum_{k,\ell} \int_0^t \partial_k f^{ij}(X_{s-}^n) f^{k,\ell}(X_{s-}^n) d(H_s^{n,j,\ell} + K_s^{n,\ell,j} + I_s^{n,\ell,j}). \tag{8.13}$$

We have already seen that $(I^n)_{n\geq 1}$ is good, and we calculated its limit in (8.12). Note further that

$$\left|K^{n,j,k}_{t+h} - K^{n,j,k}_t\right| \leq \frac{1}{2}\left\{[Y^{n,j},Y^{n,j}]_{t+h} + I^{n,k,k}_{t+h} - [Y^{n,j},Y^{n,j}]_t - I^{n,k,k}_t\right\} \tag{8.14}$$

and it follows that $(K^n)_{n\geq 1}$ is good. Therefore it remains only to substitute (8.13) into (8.9) to complete the proof. □

9. Applications to Numerical Analysis of SDE's

Let us consider a simple stochastic differential equation driven by a semimartingale Y:

$$X_t = X_0 + \int_0^t f(X_{s-})dY_s \tag{9.1}$$

where f is a continuous function (not necessarily Lipschitz). One is often interested in estimating quantities of the form $E\{g(X_a)\}$ for a fixed time a. One could use a Monte Carlo method if the law of $g(X_a)$ or of X_a were known, but in general it is not. Therefore one uses the structure of the SDE (9.1) to estimate the law of X_a. The simplest method is the Euler method. (There are more complicated numerical schemes that converge faster, but we intend to combine our results with a Monte Carlo procedure, and since Monte Carlo convergence is slow, we do not consider them here.) A straightforward extension of the Euler method of ordinary differential equations leads to an Euler scheme of the type

$$\overline{X}^0_{t_{k+1}} = \overline{X}^0_{t_k} + f(\overline{X}^0_{t_k})(Y_{t_{k+1}} - Y_{t_k}) \tag{9.2}$$

where $\pi^n = \{0 = t_0 < t_1 < \ldots < t_{k_n} = a\}$ is a sequence of partitions of $[0,a]$ such that $\lim_{n\to\infty} \operatorname{mesh}(\pi^n) = 0$. We denote $\overline{X}^0$ to be the approximation (the dependence on n is implicit) to distinguish it from a solution X of (9.1).

It is convenient in this context to use a different scheme than the naive one (9.1). However note that for our scheme (9.4) below, the two schemes will agree on the partition points $(t_k)_{k\geq 1}$ of $\pi^n[0,a]$. We define

$$\eta(t) = t_k \text{ for } t_k \leq t < t_{k+1} \tag{9.3}$$

for a partition $\pi^n[0,a] = \{0 < t_0 < \ldots < t_{k_n} = a\}$ where again the dependence on n is implicit. Let $\overline{X}$ satisfy the equation:

$$\overline{X}_t = X_0 + \int_0^t f(\overline{X}_{\eta(s-)})dY_s, \tag{9.4}$$

so that the integrands in the stochastic integral are piecewise constant. Note that $\overline{X}_{t_k} = \overline{X}^0_{t_k}$ for partition points $(t_k)_{k\geq 1}$. We can put more general assumptions on η, as the next lemma shows.

Lemma 9.1. *Let (Y^n) be a sequence of semimartingales, $X^n \in \mathbf{D}$, and $\eta_n \in \mathbf{D}$, nondecreasing, $\eta_n(t) \le t$, and $\lim_{n\to\infty} \eta_n(t) = t$ for all $t \ge 0$. Assume also $(Y^n)_{n\ge1}$ is good and that $(X^n, Y^n) \Rightarrow (X, Y)$. Then*

$$\int X^n_{\eta_n(s-)} dY^n_s \Rightarrow \int X_{s-} dY_s.$$

Proof. Recall the notation J_δ introduced in §7. (equation (7.2)). Then for $\delta > 0$ we have

$$\int \left(J_\delta(X^n)_{\eta^n(s-)} - J_\delta(X^n)_{s-}\right) dY^n_s \Rightarrow 0.$$

Therefore there exists a sequence $(\delta_n)_{n\ge1}$ tending to 0 such that

$$\int \left(J_{\delta_n}(X^n)_{\eta^n(s-)} - J_{\delta_n}(X^n)_{s-}\right) dY^n_s \Rightarrow 0.$$

However the asymptotic continuity of X^{n,δ_n} implies that $(X^{n,\delta_n}_t - X^{n,\delta_n}_{\eta_n(t)})_{t\ge0} \Rightarrow 0$, whence

$$\int \left(X^{n,\delta_n}_{s-} - X^{n,\delta_n}_{\eta_n(s-)}\right) dY^n_s \Rightarrow 0,$$

and therefore

$$\int (X^n_{s-} - X^n_{\eta_n(s-)}) dY^n_s \Rightarrow 0$$

and the lemma is proved. □

Theorem 9.2. *Let (Y^n) be a good sequence of semimartingales such that $Y^n \Rightarrow Y$ Let η_n be as in Lemma 9.1. Let $f : \mathbf{R}^d \to \mathbf{M}^{km}$ be bounded and continuous, and let $\overline{X}^n$ satisfy*

$$\overline{X}^n_t = X_0 + \int_0^t f\left(\overline{X}^n_{\eta_n(s-)}\right) dY^n_s. \tag{9.5}$$

Then $(\overline{X}^n, Y^n)$ is relatively compact and any limit point satisfies

$$X_t = X_0 + \int_0^t f(X_{s-}) dY_s. \tag{9.6}$$

If all Y^n are defined on the same sample space as Y and if Y^n converges to Y in ucp and pathwise uniqueness holds for (9.6), then $\overline{X}^n$ converges to X in ucp as well.

Proof. The relative compactness is complicated and we refer the reader to Kurtz and Protter [10].

The fact that any limit point satisfies (9.6) then follows from Lemma 9.1.

Under the assumptions of the final assertion, we can treat $(\overline{X}^n, X)$ as a solution of a single system. The uniqueness then gives us that $(\overline{X}^n, X) \Rightarrow (X, X)$, whence $\overline{X}^n - X \Rightarrow 0$, and the conclusion follows. □

We note that often in applications we will have $Y^n = Y$ for all n; such a sequence is of course trivially good.

In certain circumstances we can give an analysis of the error in the Euler scheme; that is, we can determine the asymptotic distribution of the normalized error. Here weak convergence is essential, as we will see by considering the Brownian case in Example 9.4.

Theorem 9.3. *Let Y be a given* $\mathbf{F}$ *semimartingale and let f be a C^1 $\mathbf{M}^{dm}$ matrix valued function. Let $0 = T_0^n < T_1^n < \ldots$ be* $\mathbf{F}$ *stopping times and define*

$$\eta_n(t) = T_k^n \text{ if } T_k^n \leq t < T_{k+1}^n,$$

and let $\overline{X}^n$ satisfy (9.5). Let α_n be a sequence of positive constants tending to ∞, and set

$$U^n = \alpha_n(\overline{X}^n - X)$$

and define

$$Z_t^{n,i,j} = \alpha_n \int_0^t \left(Y_{s-}^i - Y_{\eta_n(s-)}^i\right) dY_s^j.$$

Assume that $(Z^n)_{n\geq 1}$ is good and that $(Y, Z^n) \Rightarrow (Y, Z)$. Then $U^n \Rightarrow U$, where U satisfies

$$U_t = \sum_i \int_0^t \nabla f_i(X_{s-})U_{s-}dY_s^i + \sum_{i,j} \int_0^t \sum_k \partial_k f^i(X_{s-})f^{j,k}(X_{s-})dZ_s^{i,j}. \quad (9.7)$$

Proof. The hypothesis imply that (9.6) has a unique solution, hence

$$(\overline{X}^n, X, Y, Z^n) \Rightarrow (X, X, Y, Z).$$

Let us treat only the scalar use ($d = m = 1$). Observe that

$$\overline{X}_{s-}^n - \overline{X}_{\eta_n(s-)}^n = f(\overline{X}_{\eta_n(s-)}^n)(Y_{s-} - Y_{\eta_n(s-)}).$$

Therefore

$$\begin{aligned}
U_t^n &= \int_0^t \alpha_n \left(f(\overline{X}_{s-}^n) - f(X_{s-})\right) dY_s - \int_0^t \alpha_n \left(f(\overline{X}_{s-}^n) - f(\overline{X}_{\eta_n(s-)}^n)\right) dY_s \\
&= \int_0^t \frac{f(\overline{X}_{s-}^n) - f(X_{s-})}{\overline{X}_{s-}^n - X_{s-}} U_{s-}^n dY_s \\
&\quad - \int_0^t f\left(\overline{X}_{\eta_n(s-)}^n\right) + f\left(\overline{X}_{\eta_n(s-)}^n\right) \left(Y_{s-} - Y_{\eta_n(s-)}\right) \\
&\quad - f\left(\overline{X}_{\eta_n(s-)}^n\right) \left(Y_{s-} - Y_{\eta_n(s-)}\right)^{-1} dZ_s^n.
\end{aligned}$$

Next let $T^{n,a} = \inf\{t > 0 : |U_t^n| > a\}$. Then $U_{t\wedge T^{n,a}}^n$ is relatively compact, and any limit point will satisfy (9.7) on $[0, T^a]$, where $T^a = \inf\{t > 0 : |U_t| > a\}$. But $\lim_{a\to\infty} T^a = \infty$ a.s., so $U^n \Rightarrow U$. □

Example 9.4. Let us take $Y_t = \begin{pmatrix} W_t \\ t \end{pmatrix}$ in Theorem 9.3, where W is an $n-1$ dimensional standard Wiener process (or Brownian motion). Let $\eta_n(t) = \frac{[nt]}{n}$. Then taking $\alpha_n = \sqrt{n}$, we have $(Y, Z^n) \Rightarrow (Y, Z)$, where Z is independent of Y. Moreover $Z^{im} = Z^{mi} = 0$, and since Z^{ij} are continuous local martingales with $[Z^{ij}, Z^{k\ell}]_t = \begin{cases} 0 & ij \neq k\ell \\ \frac{1}{2}t & ij = k\ell \end{cases}$ we conclude that Z is also a Brownian motion, independent of W. Note that since Z is independent of W, it need not "live" on the same space as W. Thus the limiting process U could appear only through weak convergence, in general.

References

1. Ahn, H., Protter, P., "A Remark on Stochastic Integration", *Sém. de Proba. XXVII*, LNM 1583, 312-315 (1994).
2. Dellacherie, C., *Capacités et Processus Stochastiques*, Springer, Berlin (1972).
3. Duffie, D., Protter, P., " From discrete to continuous time finance: weak convergence of the financial gain process", *J. Math. Finance* 2, 1-15 (1992).
4. Emery, M., "On the Azéma Martingales", *Sém. de Proba. XXIII*, LNM 1372, 66-87 (1989).
5. Ethier, S., Kurtz, T. G., *Markov Processes: Characterization and Convergence*, Wiley, New York (1986).
6. Jacod, J., "Theoremes Limites pour les Processus", *Ecole d'Eté de Proba. de St. Flour XIII*, LNM 1117, 299-409 (1985).
7. Jacod, J., Protter, P., "A Remark on the Weak Convergence of Processes in the Skorohod Topology", *J. Theoretical Probability*, 6, 463-472 (1993).
8. Jacod, J., Shiryaev, A. N., *Limit Theorems for Stochastic Processes*, Springer, Berlin (1987).
9. Jakubowski, A., Mémin, J., Pagès, G., "Convergence en Loi des Suites d'Intégrales Stochastiques sur l'Espace $\mathbf{D}^1$ de Skorohod", *Proba. Th. Rel. Fields* 81, 111-137 (1989).
10. Kurtz, T. G., Protter, P., "Weak Limit Theorems for Stochastic Integrals and Stochastic Differential Equations", *Ann. Proba.* 19, 1035-1070 (1991).
11. Kurtz, T. G., Protter, P., "Characterizing the Weak Convergence of Stochastic Integrals", in *Stochastic Analysis* (M. Barlow and N. Bingham, eds.), Cambridge U. P., 255-259 (1991).
12. Kurtz, T. G., Protter, P., "Wong-Zakai, Corrections, Random Evolutions, and Numerical Schemes for SDEs", in *Stochastic Analysis*, Academic Press, 331-346 (1991).
13. Mémin, J., Slominski, L., "Condition *UT* et Stabilité en Loi des Solutions d'Equations Différentielles Stochastiques", *Sém. de Proba. XXV*, LNM 1485, 162-177 (1991).
14. Pratelli, M., "La Classe des Semimartingales qui Permettent d'Intégrer les Processus Optionnels", *Sém. de Proba. XVII*, LNM 986, 311-320 (1983).
15. Protter, P., *Stochastic Integration and Differential Equations*, Springer, Berlin (1990).
16. Slominski, L., "Stability of Strong Solutions of Stochastic Differential Equations", *Stoch. Processes and Their Appl.* 31, 173-202 (1989).

Asymptotic Behaviour of some interacting particle systems; McKean-Vlasov and Boltzmann models

Sylvie Méléard

Université Paris 10, MODAL'X, UFR SEGMI, 200 avenue de la République, 92000 Nanterre, France, and Laboratoire de Probabilités (URA CNRS 224), Université Paris 6, 4 place Jussieu, 75252 Paris, France

1. Introduction

For about ten years, models of stochastic particle systems with mean field interaction have been studied in detail, about their asymptotic behaviour when the size of the system becomes very large. Mean field interaction means that the system acts over one fixed particle through the empirical measure of the system. If it is a smooth function of the empirical measure, we will call it weak interaction; for certain convolutions of the interaction with a smooth function, one speaks of moderate interaction (cf. Oelschläger [37] and Méléard-Roelly [30]), but we will not consider this situation in these notes.

Let us first quote the paper of A.S. Sznitman about Boltzmann's equation [44] where for the first time a rigorous scheme of study for such systems is described. Sznitman shows an asymptotic behaviour specific to mean field interactions, when the size of the system grows. This behaviour is called propagation of chaos and will be described in details in the following. Sznitman generalizes ideas first mathematically formulated by Kac [23] and then by McKean [28]. For ten years, many authors have given contributions about the propagation of chaos phenomenon by studying various, more or less general, systems. For example Tanaka [49], Léonard [25], Sznitman [44, 46, 47], Oelschläger [36, 37], Gärtner [12], Graham [15], Méléard-Roelly [30, 31], Graham-Méléard [17, 18], and all the references therein.

Let us first describe quickly what propagation of chaos means: if the initial particles are i.i.d. with distribution P_0, we say that there is propagation of chaos for the system of interacting particles if the law of k fixed particles tends to the distribution $P^{\otimes k}$ of k independent particles with same law P when the size of the system goes to infinity. One can prove (cf Sznitman, Tanaka) that for exchangeable systems, propagation of chaos is equivalent to the convergence in law of the empirical measure of the system towards P, when the size of the system tends to infinity, thus we have a law of large numbers. The distribution P is usually characterized as the unique solution of a nonlinear martingale problem defined on the path space. More precisely, the canonical process X satisfies the following martingale problem: for nice test functions φ,

$$\varphi(X_t) - \varphi(X_0) - \int_0^t L(P_s)\varphi(X_s)ds$$

is a P-martingale, where $P_s = P \circ X_s^{-1}$ gives the nonlinearity, P_0 is the given initial distribution and $L(.)$ is an operator defined through the behaviour of the particle system.

By taking expectations in every term of the martingale problem, one obtains an evolution equation in a weak form for $(P_t)_{t\geq 0}$:

$$\frac{d}{dt} < P_t, \varphi > = < P_t, L(P_t)\varphi > .$$

This property of propagation of chaos can thus be seen as describing the asymptotic behaviour of a stochastic interacting particle system, but also as giving a stochastic approximation of solutions of certain nonlinear PDE's, and justifies Monte-Carlo algorithms.

To illustrate this theory, we will choose two nonlinear PDE's related to the behaviour of particles of gas. One is the Vlasov equation, modelling gas with a large density of particles and the other one is the Boltzmann equation which models rarefied gas. In both situations, the nonlinear equation is obtained as the limit of deterministic equations when the number of particles tends to be very large. In the first situation (Vlasov equation), collisions are so close that at the limit, the collision terms are replaced by an interacting diffusing phenomenon with a drift term describing attraction or repulsion of particles. Conversely, in the second situation (Boltzmann equation), the interaction range decreases so to have a limiting macroscopic mean free-path. Thus, one obtains in the limiting equation a free transport and a collision term. In more complex models, the two equations can be mixed; for example if we take into account grazing collisions in the Boltzmann equation, one can model it by adding an interacting diffusion term to the free transport (cf. Cercigniani [6]).

To study this last situation and more general models, we present the propagation of chaos for general interacting diffusions with jumps in Section 4. The interaction appears in the coefficients of the diffusion process and also in the jump measure. The coefficients of the diffusion are Lipschitz continuous and then we do not work with general hypotheses on these coefficients as in Gărtner [12] but our model includes jumps (with a bounded jump measure). We show in this general context a propagation of chaos on the path space. To do that, we prove the convergence in law of the empirical measures considered as probability measures on the path space by the classical trilogy: tightness, identification of the limit values, uniqueness of the limit value. We deduce from it, because of the exchangeability of the laws of particles, the convergence in law of every fixed k-uplet of particles. We also deduce the convergence of the flow of time-marginals of the empirical measures and then obtain a functional law of large numbers.

Before considering the general situation of Section 4, we describe in Section 2 the equations we are interested in, we introduce the associated nonlinear martingale problems and some associated particle systems. The Boltzmann case presents very hard difficulties well known to numericians, because of the localization of the interaction. So we will consider mollified Boltzmann equations, where the interaction is delocalized.

In Section 3, we consider specifically these mollified Boltzmann equations. Since the interaction just appears in the jump term, we prove a very strong property of propagation of chaos (cf Graham-Méléard [18]). More precisely, we show that when the number n of interacting particles goes to infinity, the law of a fixed k-uplet of particles converges to the k-fold product of a probability measure P solution of the limit nonlinear martingale problem, with a precise $O(\frac{1}{n})$ rate of convergence in variation norm on a time interval $[0,T]$. To do that, we build the law of the interacting particles on a random graph and the law of the nonlinear limit process on a stochastic tree called the Boltzmann tree. The convergence is proved by a probabilistic pathwise argument called coupling. These results are proved for particle systems moving between collisions following a very general non interacting Markov process, including diffusion processes, reflected diffusion processes, jump processes and of course free transport as in the Boltzmann model.

We then propose simulations of approximations of a solution of a mollified Boltzmann equation by simulating the n-particle system following the interacting Boltzmann model and by considering the empirical measure of this system as an approximation of the solution.

We have seen above that in exchangeable cases, propagation of chaos is equivalent to the convergence in law of the empirical measures. A natural question is to know the speed of this convergence and that is the aim of Section 5. We study the asymptotic behaviour of the fluctuations of the empirical measures, defined by $\eta^n = n^{1/2}(\mu^n - P)$ as n goes to infinity, in the continuous case. For every n, η^n is a signed measure. The space of signed measures endowed with the weak convergence is a "bad" (non metrizable) topological space, and the main problem to overcome is to find a suitable space in which to immerse η^n to which both η^n and its limit belong. Following Fernandez-Méléard [10] which develop an Hilbertian approach, we obtain for the McKean-Vlasov model the convergence of the fluctuation processes in a well chosen weighted Sobolev space. This space is in a certain sense the smallest space (in a class of weighted Sobolev spaces) in which we can obtain the convergence of fluctuations.

The limit process is characterized as a Gaussian Ornstein-Uhlenbeck process which is the solution of a generalized Langevin equation. These results generalize results obtained by Hitsuda and Mitoma [19] which have proved the convergence of fluctuations (for the one-dimensional McKean-Vlasov model) in a distribution space by nuclear spaces techniques. In [10], the new result is the tightness of the fluctuations as processes belonging to a Sobolev space, the convergence of the finite dimensional marginals being due to Sznitman [48]. Here we give a new proof for the uniqueness using an evolution equation and which can be adapted in other situations. This is detailed in [11] for a model of interacting system of particles with space-time random birth.

The jump case (which includes Boltzmann's models) is really harder. The pathwise couplings used in the continuous case are no more tractable and the study of fluctuations necessitates refined techniques (cf. Méléard [29]). This point is not developed in these notes.

NOTATIONS

– K denotes a constant which can change from line to line.

– For every integer $j \geq 1$, $C_b^j(\mathbb{R}^d)$ is the space of bounded functions of class C^j on $\mathbb{R}^d$ with bounded derivatives of each order less than or equal to j, and $C_b(\mathbb{R}^d)$ is the space of continuous bounded functions on $\mathbb{R}^d$.

– $\mathcal{P}(E)$ is the space of probability measures on a space E.

– For $p \in \mathcal{P}(E)$ and φ a function on E, $< p, \varphi >$ denotes $\int \varphi(x)p(dx)$.

– $\mathcal{L}(X)$ is the distribution of a random variable X.

2. The physical equations, the associated nonlinear martingale problems and some related particle systems

2.1 The McKean-Vlasov model

2.1.1 The McKean-Vlasov Equation and the associated nonlinear martingale problem. The nonlinear partial differential equation (called McKean-Vlasov Equation), is given in dimension d by

$$\frac{\partial P_t}{\partial t} = \frac{1}{2}\sum_{i,j=1}^{d} \frac{\partial^2}{\partial x_i \partial x_j}(a_{ij}[x, P_t]P_t) - \sum_{i=1}^{d} \frac{\partial}{\partial x_i}(b_i[x, P_t]P_t), \quad P_0 \text{ is given,} \tag{2.1}$$

where P_t is a probability measure on $\mathbb{R}^d$, and

$$\begin{aligned} b[x,p] &= \int_{\mathbb{R}^d} b(x,y)p(dy), \quad b(x,y) \text{ being a vector of } \mathbb{R}^d \\ a[x,p] &= \sigma[x,p]^t\sigma[x,p] \\ \sigma[x,p] &= \int_{\mathbb{R}^d} \sigma(x,y)p(dy), \quad \sigma(x,y) \text{ being a matrix of size (d, k),} \end{aligned}$$

for p a probability measure on $\mathbb{R}^d$.

This equation is understood in a weak sense: for nice test functions φ, we have:

$$\frac{\partial}{\partial_t} < P_t, \varphi > = < P_t, \frac{1}{2} \sum_{i,j=1}^{d} a_{ij}[x, P_t] \frac{\partial^2 \varphi}{\partial x_i \partial x_j}(x) + \sum_{i=1}^{d} b_i[x, P_t] \frac{\partial \varphi}{\partial x_i}(x) > . \quad (2.2)$$

This equation has been studied from a probabilistic point of view by several authors. In particular McKean [28], Sznitman [47], Tanaka [49], Léonard [25] and Gärtner [12] have proved that probabilistic methods allow to study solutions of (2.1).

The main idea consists in associating with (2.1) the second order differential generator $\mathcal{L}(p)\varphi(x)$ defined for φ in $C_b^2(\mathbb{R}^d)$ by

$$\mathcal{L}(p)\varphi(x) = \frac{1}{2} \sum_{i,j=1}^{d} a_{ij}[x, p] \frac{\partial^2 \varphi}{\partial x_i \partial x_j}(x) + \sum_{i=1}^{d} b_i[x, p] \frac{\partial \varphi}{\partial x_i}(x). \qquad (2.3)$$

The interest of probabilistic methods in studying weak solutions of some non linear evolution equations is to introduce underlying processes whose time marginals of the distribution are solutions of this equation. These processes are defined as solutions of a martingale problem.

Definition 2.1. *Let $\{X_t, t \in [0,T]\}$ be the canonical process on $C([0,T], \mathbb{R}^d)$. The probability measure P on $C([0,T], \mathbb{R}^d)$ is a solution of $(\mathcal{P}_V)$ issued from P_0 if for every $\varphi \in C_b^2(\mathbb{R}^d)$,*

$$\varphi(X_t) - \varphi(X_0) - \int_0^t \mathcal{L}(P_s)\varphi(X_s)ds \qquad (2.4)$$

is a P-martingale where $P_s = P \circ X_s^{-1}$ and $P_0 = P \circ X_0^{-1}$.

Important Remark: If we take expectations in (2.4), the flow $(P_t)_{t\geq 0}$ is a solution of the evolution equation (2.1). The martingale problem gives much more information than the evolution equation. It enables to consider multidimensional time-marginals as $P[X_s \in A, X_t \in B]$ or functionals depending on the whole process, as for example hitting times. So we consider the whole Markov process corresponding to the underlying physical model. This idea will be the key point in the following

Theorem 2.2. *If the coefficients σ and b are Lipschitz continuous on $\mathbb{R}^{2d}$, and if the law P_0 has a second order moment, there is existence and uniqueness of the solution of the previous nonlinear martingale problem.*

Proof. In fact, we prove here a stronger result since we prove that there is existence and uniqueness, pathwise (given $X_0 \in L^2(\Omega)$ and B) and in law, of the solution of the following nonlinear stochastic differential equation:

$$X_t = X_0 + \int_0^t \sigma[X_s, P_s]dB_s + \int_0^t b[X_s, P_s]ds \qquad (2.5)$$

where the nonlinearity appears through P_s, which is the marginal at time s of the law P of X, where B is a k-dimensional Brownian motion and X_0 is distributed according to P_0 and is independent of B.

This proof is completely detailed in Sznitman [47, Theorem 1.1, p.172] in the case $\sigma(\cdot) = 1$ and b bounded, and is based on a fixed point theorem. In our more general situation, one introduces the space $\mathcal{P}_2(C^T)$ of probability measures on $C^T = C([0,T], \mathbb{R}^d)$ with finite second order moment, on C^T ($P \in \mathcal{P}_2(C^T)$ iff $E_P(\sup_{t\leq T}|X_t|^2) < \infty$) endowed with the the Kantorovitch-Rubinstein or Vaserstein metric D_T (cf. Pollard [40] or Rachev [42]), which is defined for m_1, m_2 by:

$$D_T(m_1,m_2) = \inf\left\{\left(\left(\int_{C^T\times C^T} d_T(x,y)\right)^2 m(dx,dy)\right)^{1/2}, \right.$$
$$m \in \mathcal{P}_2(C([0,T],\mathbb{R}^d) \times C([0,T],\mathbb{R}^d))$$
$$\left.\text{having marginals } m_1 \text{ and } m_2\right\}.$$

Here, d_T denotes the uniform metric on $\mathcal{P}_2C([0,T], \mathbb{R}^d)$. The metric D_T is a complete metric on $\mathcal{P}(C^T)$ and $D_T(m_n, m) \to 0$ implies that (m_n) converges weakly to m (see Zolotarev [54] for example). One considers the mapping $\psi : \mathcal{P}_2(C^T) \to \mathcal{P}_2(C^T)$ which associates with $m \in \mathcal{P}_2(C^T)$ the law of X^m defined by:

$$X_t^m = X_0 + \int_0^t \sigma[X_s^m, m_s]dB_s + \int_0^t b[X_s^m, m_s]ds, \quad t \leq T.$$

It is easy to prove that if $X_0 \in L^2(\Omega)$ then $\mathcal{L}(X^m) \in \mathcal{P}_2(C^T)$. Besides l et us observe that if (X_t) is a solution of (2.5) then the law of (X_t) is a fixed point of the function ψ and conversely.

By pathwise considerations one proves that for $t \leq T$,

$$D_t^2(\psi(m^1), \psi(m^2)) \leq K \int_0^t D_u^2(m^1, m^2)du,$$

for $m^1, m^2 \in \mathcal{P}_2(C^T)$. Then one deduces from the fixed point theorem the existence and uniqueness of the solution P of the martingale problem (2.4) defined on $[0,T]$. Pathwise uniqueness follows immediately from the Lipschitz continuity of the coefficients. □

We introduce now some stochastic interacting particle systems which converge in a sense we precise in Theorem 2.3 to the solution of this martingale problem.

2.1.2 The associated stochastic interacting particle system. The nonlinearity in the previous martingale problem appears through the distribution of the process. If we introduce an approximating system of interacting particles, it is natural to replace this law by the empirical measure of the interacting particles. We consider the following triangular system: the system of n particles $(X^{1n}, X^{2n}, ..., X^{nn})$ is composed of diffusions, solutions of the stochastic differential equations: $\forall i \in \{1, ..., n\}$,

$$X_t^{in} = X_0^i + \int_0^t \sigma[X_s^{in}, \mu_s^n] dB_s^i + \int_0^t b[X_s^{in}, \mu_s^n] ds \tag{2.6}$$

where (B^i) are independent $\mathbb{R}^k$-valued Brownian motions, (X_0^i) are initial i.i.d. variables with law P_0 independent of (B^i) and μ^n is the empirical measure of the system, which is the probability measure on the path space defined by:

$$\mu^n = \frac{1}{n} \sum_{i=1}^n \delta_{X^{in}}, \quad (\delta \text{ denoting the Dirac measure}).$$

Under regularity hypotheses on the coefficients σ and b, this model is mathematically simple. It has been extensively studied as a "laboratory model". In this specific case, the nonlinear process X and the interacting particles X^{in} are continuous and are solutions of stochastic differential equations. One can then introduce a **coupling** between the system (X^{in}) and a system (X^i) of independent processes with same law P than X and defined on the same probability space than (X^{in}): $\forall i \in \{1, ..., n\}$,

$$X_t^i = X_0^i + \int_0^t \sigma[X_s^i, P_s] dB_s^i + \int_0^t b[X_s^i, P_s] ds.$$

One easily obtains a pathwise estimate comparing X^{in} and X^i (see also Lemma 5.2 in these notes):

Theorem 2.3. $\forall i \in \{1, ..., n\}$, *one has*

$$\begin{aligned} \sup_n E(\sup_{t \le T} |X_t^{in}|^2 &< +\infty, \\ \sup_n E(\sup_{t \le T} |X_t^i|^2 &< +\infty, \\ \sup_n nE(\sup_{t \le T} |X_t^{in} - X_t^i|^2) &< +\infty. \end{aligned}$$

This property obviously implies that the law of every fixed subsystem of k particles issued from (X^{in}) converges when n tends to infinity to the law $P^{\otimes k}$. This property is called propagation of chaos. We will prove in Section 4 such an asymptotic result for more general models defined in Subsection 2.3, including jumps and generalizing both the McKean-Vlasov model and the following Boltzmann model. Because of the jumps, the coupling does not give satisfying pathwise estimates and we will use other convergence techniques.

2.2 The Boltzmann equation and some associated particle systems

2.2.1 The physical equation. The Boltzmann equation is an evolution equation for the density $f(t,x,v)$ of particles at time t in position $x \in \mathbb{R}^3$ with speed $v \in \mathbb{R}^3$. $f(t,x,v)$ is positive and normalized so that $\int f(t,x,v)dxdv = 1$. The Boltzmann equation is given by:

$$\begin{aligned}&\partial_t f(t,x,v) + v \cdot \nabla_x f(t,x,v)\\ &= \int_{S^2} dn \int_{\mathbb{R}^3} dw\, q(v,w,n)(f(t,x,v^*)f(t,x,w^*) - f(t,x,v)f(t,x,w)) \quad (2.7)\end{aligned}$$

where v^* and w^* represent the post-collisional velocities of two particles of speeds v and w having collided in a position x in which their centers are on a line of direction given by the unit vector n. There is a statistical knowledge of the geometry of the impact given by n. The cross-section $q(v,w,n)$ quantifies the likelihood of interactions at point x of particles of speed v and w with impact parameter n and depends only on $\mid v-w \mid$ and $(v-w).n$. There can be various boundary conditions on x. All these models and kinds of collision are discussed in Cercigniani [6], Neunzert and al. [35], Babovsky and Illner [3] and Wagner [51, 52], Graham and Méléard [18].

Conservation of kinetic energy and momentum for binary collisions imply that v^* and w^* are obtained by exchanging the projections of the speeds on n:

$$v^* = v + ((w-v)\cdot n)n, \quad w^* = w + ((v-w)\cdot n)n.$$

As the McKean-Vlasov Equation, this equation has a probabilistic interpretation. We integrate test functions ϕ with respect to (2.7), which gives

$$\begin{aligned}&\int \phi(x,v)(\partial_t f(t,x,v) + v\cdot\nabla_x f(t,x,v))dxdv\\ &\quad = \int \phi(x,v)\int_{S^2} dn \int_{\mathbb{R}^3} dwq(v,w,n)\\ &\qquad\qquad (f(t,x,v^*)f(t,x,w^*) - f(t,x,v)f(t,x,w))dxdv.\end{aligned}$$

Then, we transpose the operators to have them operating on ϕ. Integrating by parts and setting $P_t(dx,dv) = f(t,x,v)dxdv$, (2.7) is rewritten as

$$\begin{aligned}&\partial_t\langle P_t,\phi\rangle - \langle P_t, v\cdot\nabla_x\phi(x,v)\rangle\\ &= \langle P_t(dx,dv), \int(\phi(x,v^*) - \phi(x,v))q(v,w,n)dnf(t,x,w)dw\rangle. \quad (2.8)\end{aligned}$$

This can be seen as the evolution equation for the flow of marginals of a nonlinear stochastic process in which the particle evolves according to the free flow and its velocity changes from v to v^* at point x and time t according to $q(v,w,n)f(t,x,w)dwdn$.

This situation is unpleasant, mainly because of the local interaction; the density $f(t,x,w)dw$ is quite misbehaved in terms of the law $P_t(dx,dv)$ from both a continuity and a boundedness point of view. This lack of continuity renders any probabilistic existence and uniqueness result difficult to establish. From an analytical point of view, it is also very difficult, and existence and uniqueness for the Boltzmann equation have thus been very elusive, and are usually valid under very restrictive assumptions. The best results of existence seem to be those of DiPerna-Lions [7].

2.2.2 The mollified model. To bypass these problems we consider mollified Boltzmann equations in which the interaction is no longer local in space. The whole measure $P_t(dx, dw) = f(t, x, w)dxdw$ and not only the local measure $f(t, x, w)dw$ intervenes in the jump term. We may for instance use a regularizing kernel $I(x, y)$ and approximate $f(t, x, w)$ by $\int I(x, y)f(t, y, w)dy$ in (2.8). These approximations are close to the statistical physics model and are suited for a probabilistic study.

An example is given by grid methods. Space is partitioned in cells of non-zero Lebesgue measure and particles in each cell interact as if they had the same location. Let $\{\Delta\}$ be the set of cells, and for a point x let Δ_x be the cell in which it lies. We approximate $f(t, x, w)$ by $\frac{\int_{\Delta_x} f(t,y,w)dy}{|\Delta_x|}$, which can be written symmetrically in x and y, using the kernel $I(x, y) = \sum_\Delta \frac{1}{|\Delta|} 1_\Delta(x) 1_\Delta(y)$.

Another approximation is as follows: consider an interaction radius δ and denoting by V_δ the volume of the ball of radius δ, replace $f(t, x, w)$ by $\int 1_{|x-y|<\delta} f(t, y, w)dy/V_\delta$, with the kernel $I(x, y) = \frac{1}{V_\delta} 1_{|x-y|<\delta}$.

The cell size or δ can vary in space in order to allow to refine the grid near the points where the density varies most.

This leads to the mollified equation with delocalized cross-section

$$q(x, v, y, w, n) = I(x, y)q(v, w, n)$$

and (2.8) is replaced by

$$\begin{aligned} &\partial_t\langle P_t, \phi\rangle - \langle P_t, v \cdot \nabla_x \phi(x, v)\rangle \\ &\quad = \langle P_t(dx, dv)P_t(dy, dw), \int (\phi(x, v^*) - \phi(x, v))q(x, v, y, w, n)dn\rangle \end{aligned} \tag{2.9}$$

This has a simple mean-field interpretation in which a particle evolves in an ocean of infinitely many independent particles distributed according to the measure P_t, chooses at random another particle according to the above law, performs a collision with it and then forgets everything about it. This does not say what happens to the collision partner. Depending on what happens to this collision partner, different algorithms have been developed by Bird, Nanbu and others. A physical binary symmetric collision derivation for example leads to write the same collision term as

$$\begin{aligned} &\langle P_t(dx, dv)P_t(dy, dw), \\ &\quad \int \frac{1}{2}(\phi(x, v^*) - \phi(x, v) + \phi(y, w^*) - \phi(y, w))q(x, v, y, w, n)dn\rangle \end{aligned} \tag{2.10}$$

The unboundedness of the cross-section can also lead to serious difficulties. This may be taken care of by physical considerations (angular cut-off, speeds bounded by the speed of light). Here we shall assume that

$$\sup_{x,y,v,w} \int q(x,v,y,w,n)dn < \infty.$$

2.2.3 The underlying nonlinear martingale problem. As in the McKean-Vlasov model, we introduce some underlying processes whose flows of time-marginals are solutions of this mollified Boltzmann equation.

In fact we generalize the model of the previous section by considering an $\mathbb{R}^d$-valued process whose generator is the sum of a very general operator $\mathcal{L}$ describing the free-flow (without interaction), and a "jump" operator K describing the interactions.

The second operator K is defined by

$$K(p)\phi(z) = \int (\phi(z+h) - \phi(z))\Gamma(z,p,dh), \quad z \in \mathbb{R}^d, p \in \mathcal{P}(\mathbb{R}^d),$$

where Γ is a bounded positive kernel. The jumps occur at a rate which is the total mass $|\Gamma|$ and have an amplitude chosen according to the jump law $\frac{\Gamma}{|\Gamma|}$. Γ is uniquely determined if it does not have mass at 0, in which case it is called the Lévy kernel. Adding mass at zero to Γ does not change K: this is a central idea both in the interaction graphs we will use and in simulations. (In the Boltzmann model, $d = 6$, $z = (x, v)$ and Γ is the image measure of $\int_{y \in \mathbb{R}^3} q(x,v,y,w,n)p(dy,dw)dn$ by the mapping $(w,n) \to ((w-v).n)n$).

Assumptions on free-flow: $\mathcal{L}$ generates a Markov semi-group and operates on a sufficiently large domain $\text{Dom}(\mathcal{L})$ of the space of bounded functions on $\mathbb{R}^d$.

Assumptions on collisions:

- We assume for simplicity that Γ is linear in p and that there is no self interaction, which means that $\Gamma(z,p,dh) = \int \Gamma(z,a,dh)p(da)$ where for every $z \in \mathbb{R}^d$, $\Gamma(z,z,dh) = 0$.
- The only regularity assumption we take on the measure Γ is that Γ is bounded uniformly in z, a:

$$\sup_{z,a} \Gamma(z,a,\mathbb{R}^d) \leq \Lambda. \tag{2.11}$$

These assumptions are satisfied in the Boltzmann model (cf [18]).

Definition 2.4. *A solution to the nonlinear martingale problem $(\mathcal{P}_B)$ is a probability measure P on the Skorohod space $\mathbb{D}([0,T], \mathbb{R}^d)$, endowed with the canonical process X, such that for any function ϕ in $\text{Dom}(\mathcal{L})$,*

$$\phi(X_t) - \phi(X_0) - \int_0^t \mathcal{L}\phi(X_s) + K(P_s)\phi(X_s)ds = M_t^\phi \tag{2.12}$$

is a P-martingale, where the nonlinearity lies in the term $P \circ X_s^{-1}$.

Remark 2.5. 1) - By taking expectations in the nonlinear martingale problem we obtain the (mollified) Boltzmann equation which is the evolution equation of the flow of time-marginals

$$\partial_t < P_t, \phi > = < P_t, \mathcal{L}\phi + K(P_t)\phi > . \tag{2.13}$$

Here again, the martingale problem gives much more information than the evolution equation.
2) - The flow $t \to P_t$ is continuous as a mapping from $[0,T]$ into $\mathcal{P}(\mathbb{R}^d)$ since the underlying process is quasi-left continuous. (The Lévy measure is absolutely continuous with respect to Lebesgue measure).
3) - The uniqueness of the nonlinear martingale problem $(\mathcal{P}_B)$ is not proved for general Markov operators $\mathcal{L}$. Under more restrictive assumptions on $\mathcal{L}$, Graham [14] has proved some uniqueness result (cf. Theorem 2.8 below).

An important point which helps to understand the choice of the particle systems we introduce in Subsection 2.2.4 is the following. In (2.13), the nonlinearity appearing in the jump term is a usual mean field nonlinearity, as for example in the Vlasov model. That gives an obvious choice of approximating particle systems and leads to the Nanbu system. (Nanbu in [34] constructed in 1983 an algorithm for the Boltzmann equation by using this approach). But this choice is very far from the physical interpretation of the equation. It may be more convenient to construct approximating particle systems corresponding to the underlying statistical physics model by using a generalization of (2.10), as follows.

Let $\hat{\Gamma}(z,a,dh,dk)$ be a symmetrical-paired jump measure giving the joint amplitude of jumps of two particles entering in collision. We assume

$$\hat{\Gamma}(z,a,dh,\mathbb{R}^d) = \Gamma(z,a,dh); \quad \hat{\Gamma}(z,z,dh,dk) = 0.$$

(The two marginals are equal to Γ and there is no self-interaction).

We can rewrite the jump term of (2.13) as

$$< P_t(dz)P_t(da), \int \frac{1}{2}(\phi(z+h) - \phi(z) + \phi(a+k) - \phi(a))\hat{\Gamma}(z,a,dh,dk) > . \tag{2.14}$$

This new interpretation will lead us to the Bird approximating systems in reference to the historical Bird algorithm for the Boltzmann equation (cf. [51]).

2.2.4 Boltzmann particle systems. We construct a family of mean-field interacting particle systems which approximates the solution to the nonlinear martingale problem $(\mathcal{P}_B)$, corresponding to different choices of the jump measure $\hat{\Gamma}$ in (2.14). We consider here two particle systems which we call the Nanbu model and the Bird model. The first one is the simplest one which approximates the mollified Boltzmann equation and the second one is the closest one to the physical interpretation. In both cases, the n-particle system is a $(\mathbb{R}^d)^n$-valued Markov process whose generator is defined for functions ϕ belonging to Dom($\mathcal{L}_i$) for every i by:

Nanbu case (mean field case):

$$L^n\phi(z) = \sum_{i=1}^{n} \mathcal{L}_i\phi(z) + \frac{1}{n}\sum_{i,j=1}^{n}\int(\phi(z^n + e_i.h) - \phi(z^n))\Gamma(z_i, z_j, dh). \quad (2.15)$$

Bird case (binary interaction case):

$$L^n\phi(z) = \sum_{i=1}^{n} \mathcal{L}_i\phi(z) + \frac{1}{n}\sum_{i,j=1}^{n}\int\frac{1}{2}(\phi(z^n + e_i.h + e_j.k) - \phi(z^n))\hat{\Gamma}(z_i, z_j, dh, dk). \quad (2.16)$$

In both cases, $\mathcal{L}_i$ denotes the extension of $\mathcal{L}$ on $(\mathbb{R}^d)^n$ which operates only on the variable z_i, and the mapping e_i is defined from $\mathbb{R}^d$ into $(\mathbb{R}^d)^n$ by $e_i.h = (0, \ldots, 0, h, 0, \ldots, 0)$ where h is at the i-th place.

In the Markov process generated by the first generator, there are no simultaneous jumps of particles, while in the second model, two interacting particles jump simultaneously. The difference between these two behaviours appears in the Doob-Meyer processes associated with the martingale problems, which we write now.

Consider the Skorohod space $\mathbb{D}([0,T], (\mathbb{R}^d)^n)$ and denote by

$$X^n = (X^{1n}, .., X^{in}, .., X^{nn})$$

the canonical process. In both cases, for $\phi \in$ Dom($\mathcal{L}$),

$$\begin{aligned} &\phi(X_t^{in}) - \phi(X_0^{in}) - \int_0^t \mathcal{L}\phi(X_s^{in})ds \\ &-\frac{1}{n}\sum_{j=1}^{n}\int_0^t\int_{\mathbb{R}^d}(\phi(X_s^{in} + h) - \phi(X_s^{in}))\Gamma(X_s^{in}, X_s^{jn}, dh)ds \\ &= M_t^{\phi,in} \end{aligned} \quad (2.17)$$

is a martingale, with Doob-Meyer process given by:

$$\begin{aligned} &< M^{\phi,in}, M^{\phi,in} >_t \\ &= A_i^i(\phi) + \frac{1}{n}\sum_{j=1}^{n}\int_0^t\int(\phi(X_s^{in} + h) - \phi(X_s^{in}))^2\Gamma(X_s^{in}, X_s^{jn}, dh)ds \end{aligned} \quad (2.18)$$

where $A_i^i(\phi)$ is the Doob-Meyer process corresponding to $\mathcal{L}_i$. For $i \neq j$, the brackets are different in the Nanbu case and in the Bird case:

Nanbu case: for $i \neq j$, $< M^{\phi,in}, M^{\phi,jn} >= 0$.

Bird case:

$$\begin{aligned} &< M^{\phi,in}, M^{\phi,jn} >_t \\ &= \frac{1}{n}\int_0^t\int(\phi(X_s^{in}+h) - \phi(X_s^{in}))(\phi(X_s^{jn}+k) - \phi(X_s^{jn}))\hat{\Gamma}(X_s^{in}, X_s^{jn}, dh, dk)ds. \end{aligned} \quad (2.19)$$

2.3 The Boltzmann equation with Vlasov terms and generalized interacting systems of diffusions with jumps

In this third subsection, we consider the difficult case where both diffusion and jump terms are nonlinear. This arises for example in certain physical models mixing strong (Boltzmann) interaction with weak (Vlasov) interaction. See for instance Cercigniani [6]. The most intuitive idea is certainly to imagine grazing collisions. Particles can collide by rebounding, giving then a diffusing behaviour.

2.3.1 The nonlinear martingale problem. As before we consider probability measures on the path space, characterized as solutions of a nonlinear martingale problem and whose time-marginals will be solutions of a certain evolution equation. The free flow in the Boltzmann equation is replaced by the second order differentiable operator $\mathcal{L}(p)$ defined in (2.3), p being a probability measure on $\mathbb{R}^d$, and the collision operator is defined with respect to a jump measure $\Gamma(x,p,dh)$.

Definition 2.6. *A solution of the nonlinear martingale problem $(\mathcal{P}_{VB})$ is a probability measure P on $\mathbb{D}([0,T],\mathbb{R}^d)$, endowed with the canonical process X, such that for any ϕ in $C_b^2(\mathbb{R}^d)$,*

$$M_t^\phi = \phi(X_t)-\phi(X_0) - \int_0^t \left(\mathcal{L}(P_s)\phi(X_s) + \int (\phi(X_s+h)-\phi(X_s))\Gamma(X_s,P_s,dh)\right)ds \quad (2.20)$$

is a P-martingale, where $P_s = P\circ X_s^{-1}$.

Remark 2.7. 1) Taking the expectation in the two members of (2.20), we obtain

$$\begin{aligned}\langle P_t,\phi\rangle &= \langle P_0,\phi\rangle + \int_0^t \langle P_s, \mathcal{L}(P_s)\phi(.)\rangle ds \\ &\quad + \int_0^t \langle P_s, \int (\phi(\cdot+h)-\phi(\cdot))\Gamma(\cdot,P_s,dh)\rangle ds\end{aligned}$$

which is the weak form of a Boltzmann equation with Vlasov terms.
2) The Lévy measure of the process X is absolutely continuous with respect to the Lebesgue measure (in time), the process X is then quasi-left continuous and the flow $(P_t)_{t\geq 0}$ is continuous.

Assumptions on the coefficients
(H0): P_0 is a probability measure on $\mathbb{R}^d$ with finite second moments.
(H1): σ and b are Lipschitz continuous in (x,y).
(H2): $\sup_{(x,p)\in\mathbb{R}^d\times\mathcal{P}(\mathbb{R}^d)} |\Gamma(x,p)| \leq \Lambda, \quad \Lambda\in\mathbb{R}_+^*$.
(H3): let us denote by e the mean vector of Γ and by c the second moment matrix: $e(x,p)=\int h\Gamma(x,p,dh)$; $c_{i,j}(x,p)=\int h_ih_j\Gamma(x,p,dh)$. Then:

$$| e(x,p) |^2 + \mathrm{tr}(c(x,p)) \leq K(1+ | x |^2).$$

(H4): the measure $\Gamma(x,p,.)$ is Lipschitz continuous in (x,p) if we choose the complete Kantorovitch-Rubinstein (or Vaserstein) metric ρ on the space $\mathcal{P}_1(\mathbb{R}^d)$ of probability measures on $\mathbb{R}^d$ with finite first moment (cf. Rachev [42]):

$$\rho(\mu,\nu) = \inf\left\{ \int_{\mathbb{R}^d \times \mathbb{R}^d} | x-y | \ r(dx,dy) : r \in \mathcal{P}_1(\mathbb{R}^d \times \mathbb{R}^d) \right.$$
$$\left. \text{such that } r(dx \times \mathbb{R}^d) = \mu(dx), r(\mathbb{R}^d \times dy) = \nu(dy) \right\}.$$

This metric is extended by

$$\rho(m,n) = \sup\{< m,f > - < n,f >: \ |f(x) - f(y)| \leq | x-y |, f(0) = 0\}$$

to bounded jump measures m and n having a first moment.

Observe that our models do not include the case of discontinuous interaction kernels (as for example in problems related to the Burgers equation: see Sznitman [46], Bossy [5]).

Now we can state Theorem 2.2 in Graham [14]:

Theorem 2.8. *Assume (H0),..., (H4). Then there is a unique solution to the nonlinear martingale problem* $(\mathcal{P}_{VB})$ *starting at* P_0.

Graham's proof generalizes the proof of Theorem 2.2. The techniques are also related to a fixed point theorem, but the jumps introduce difficulties in the choice of couplings which is very delicate and nontrivial because of the complexity of the Skorohod topology.

Remark 2.9. The existence of a solution of the nonlinear martingale problem implies the existence of a solution of the nonlinear PDE. Conversely, if one assumes the uniqueness of the solution of the nonlinear PDE, one obtains uniqueness in the nonlinear martingale problem since the nonlinearity appears through the flow $(P_t)_{t\geq 0}$.

2.3.2 The interaction diffusion particle system with jumps. We consider for every given integer n a system of n diffusion processes with jumps $(X^{1n},...,X^{nn})$ in $(\mathbb{R}^d)^n$, whose time evolution is determined by a Markov generator L^n acting on $(\mathbb{R}^d)^n$, defined for every function ϕ in $C_b^2((\mathbb{R}^d)^n)$ and for $x^n = (x_1,...,x_n) \in (\mathbb{R}^d)^n$ by

$$L^n\phi(x^n) = \sum_{i=1}^n \mathcal{L}_i(\bar{x}^n)\phi(x^n) + \sum_{i=1}^n K_i(\bar{x}^n)\phi(x^n) \qquad (2.21)$$

where

$$\bar{x}^n = \frac{1}{n}\sum_{i=1}^n \delta_{x^i}$$

is the empirical measure of $(x_1, ..., x_n)$.

$\mathcal{L}_i(p)$ denotes the extension of the second order differential operator $\mathcal{L}(p)$ defined in (2.3), operating only on the variable x_i. The operators K_i are the jump operators defined by:

$$K_i(\bar{x}^n)\phi(x^n) = \int_{\mathbb{R}^d\setminus\{0\}} (\phi(x^n + e_i.h) - \phi(x^n))\Gamma(x_i, \bar{x}^n, dh). \qquad (2.22)$$

We are here in a situation generalizing the Nanbu model. The law of this process is characterized as the solution of the following martingale problem on the canonical space $D([0,T], (\mathbb{R}^d)^n)$:

If $(X^{in})_{i=1...n}$ denotes the canonical process and μ^n the corresponding empirical measure, for $\phi \in C_b^2(\mathbb{R}^d)$, $\forall i \in \{1, ..., n\}$,

$$\begin{aligned}\phi(X_t^{in}) \quad &- \quad \phi(X_0^{in}) - \int_0^t \mathcal{L}(\mu_s^n)\phi(X_s^{in}) \\ &+ \quad \int (\phi(X_s^{in} + h) - \phi(X_s^{in}))\Gamma(X_s^{in}, \mu_s^n, dh)ds = M_t^{\phi,in} \qquad (2.23)\end{aligned}$$

is a martingale. The Doob-Meyer bracket

$$< M^{\phi,in} >_t = A_t^i(\phi) + \int_0^t \int (\phi(X_s^{in} + h) - \phi(X_s^{in}))^2 \Gamma(X_s^{in}, \mu_s^n, dh)ds \quad (2.24)$$

(in which $A_t^i(\phi) = \int_0^t \mathrm{trace}(\nabla\phi(X_s^{in}))^*.a(X_s^{in}, \mu_s^n)\nabla\phi(X_s^{in})ds)$ can easily be deduced from (2.23). There are no simultaneous jumps and $< M^{\phi,in}, M^{\phi,jn} >= 0$ for $i \neq j$.

Existence and uniqueness of solutions: If we assume that the initial random variables (X_0^{in}) are i.i.d. according to a probability measure P_0 having second order moments, the boundedness of Γ and the Lipschitz continuity of the coefficients allow to obtain existence and uniqueness of the solution. The growth and moment assumptions prevent explosion and ensure L^2 stochastic integrands. We refer to Sznitman ([44], Appendix) and to Lepeltier-Marchal [27] for the uniqueness and the martingale properties.

3. Propagation of chaos and convergence rate in variation norm for the Boltzmann model

3.1 Propagation of chaos in variation norm

Propagation of chaos is a probabilistic limit result in which the law of a fixed number k of particles in an interacting system converges to the k-fold product of a limiting probability measure as the number of particles goes to infinity.

We consider in this section the Boltzmann model studied in Subsection 2.2. Because the interaction is restricted to the collision operator we can adapt the interaction graph techniques in Graham-Méléard [17], in which more general models of interacting particle systems are studied. The convergence results are on the path space and imply results on the flow of time marginals. We denote by $|\cdot|_T$ the variation norm on $\mathbb{D}([0,T],\mathbb{R}^d)$.

We take i.i.d. random variables of law P_0 as the initial values for the particle system in Subsection 2.2.4, but more general initial chaoticity conditions may work as in Theorem 1.4 in [16].

Theorem 3.1. *Assume $(X_0^{in})_{1\le i\le n}$ i.i.d. with law P_0. Consider the particle system $(X^{in})_{1\le i\le n}$ defined in Section 2.2.4 and suppose that $|\hat{\Gamma}(z,a,.)|\le \Lambda$ for a constant Λ. Then*

1) We have propagation of chaos: for any k-uplet $(i_1,..,i_k)$

$$|\mathcal{L}(X^{i_1n},\ldots,X^{i_kn})-\mathcal{L}(X^{i_1n})\otimes\ldots\otimes\mathcal{L}(X^{i_kn})|_T \le 2k(k-1)\frac{\Lambda T+\Lambda^2T^2}{n-1} \tag{3.1}$$

and there exists a law P defined uniquely (on a Boltzmann tree) such that

$$|\mathcal{L}(X^{in})-P|_T\le 6\,\frac{e^{\Lambda T}-1}{n-1} \tag{3.2}$$

and which solves the nonlinear martingale problem $(\mathcal{P}_B)$.

2) The empirical measures μ^n converge in law and in probability to P as probability measures on $\mathbb{D}([0,T],\mathbb{R}^d)$ with rate $O(1/\sqrt{n})$, for the weak convergence and for the Skorohod metric on $\mathbb{D}([0,T],\mathbb{R}^d)$.

Proof. **1) Sample path representation using interaction graphs.** The first step consists in giving for any $k\le n$ and distinct $i_1,\ldots,i_k$ in $\{1,\ldots,n\}$ a pathwise representation of the process $(X^{i_1n},\ldots,X^{i_kn})$ on the time interval $[0,T]$ using interaction graphs.

At time T, the evolution of the state of a particle has been directly affected by other particles with which it has collided, and recursively these particles have been affected by still other particles which thus influence indirectly our first particle, and so on... We now describe this construction on $[0,T]$ in reverse time so that we can build the path of our particle with the least amount of superfluous knowledge: this defines its past history, containing all particles that have influenced it through the chains of interactions described above. If we prove that the histories of two particles become disjoint in the limit, it will be possible by a coupling argument to show that they become independent.

These chains of interactions are described by interaction graphs, which are random subsets of $[0,T]\times\{1,\ldots,n\}$. Considering (2.16) let us first introduce independent Poisson processes $(N_{ij})_{1\le i<j\le n}$ of rate $\Lambda/(n-1)$ (Λ is a bound of $|\hat{\Gamma}|$). For $i<j$, N_{ij} are random clocks giving the epochs at which X^{in} and X^{jn} are authorized to jump simultaneously and possibly interact, and

$N_{ji} = N_{ij}$. We imagine then time as being vertical and directed upwards, and the indices of particles as being on a horizontal level. We work our way backward in time starting from time T to build a graph rooted on the given subset $\{i_1, ..., i_k\}$. Every time we encounter a jump of a Poisson process N_{ij} for an i (or j) already in the graph we include the index j (or i) in the graph at that time. This branching is deterministic given the Poisson processes and generates two branches: see Fig. 3.1.

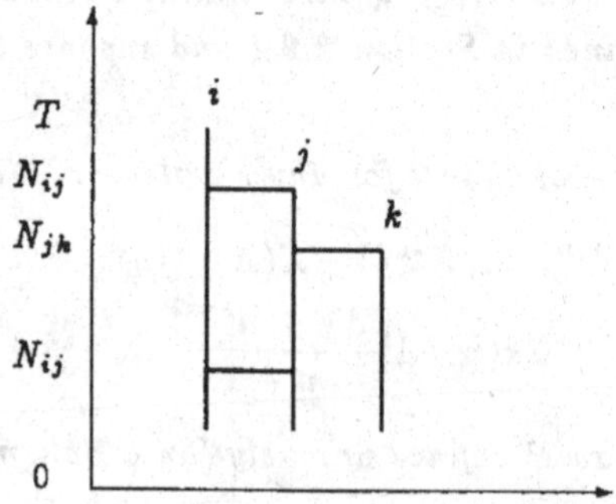

Fig. 3.1. Interaction graph rooted at i.

Once an index is selected, the whole vertical line from the point of its first selection down to zero belongs to the graph, and we proceed recursively from the selected indices to define the graph. We do not have a tree, since a particle may influence another one several times. We can thus build down from time T an interaction graph $G^n_{i_1,...,i_k}$ rooted on $(i_1, ..., i_k)$.

Once we reach time 0, we construct a pathwise representation of the process $(X^{i_1 n}, .., X^{i_k n})$ in direct time on $[0, T]$. The interaction graph represents all the information necessary to construct the process, and we only need to consider at time t the indices appearing then in the graph. We use independent variables of law P_0 at time 0. Then we follow the indices in the graph. Between the jumps of the N_{ij}'s we use a pathwise representation of the process generated by $\mathcal{L}$. At the jump times of N_{ij} we compute $\hat{\Gamma}$ at the position reached by (X^{in}, X^{jn}); with probability $1- \mid \hat{\Gamma} \mid /\Lambda$ we do nothing and with probability $\mid \hat{\Gamma} \mid /\Lambda$ we choose the joint amplitude of jumps according to the law $\frac{\hat{\Gamma}}{|\hat{\Gamma}|}$. All this must be done independently. This gives the correct evolution for the Markov process since $\frac{\Lambda}{n-1}\left(\frac{\hat{\Gamma}}{\Lambda} + \left(1 - \frac{\hat{\Gamma}}{\Lambda}\right)\delta_0\right)$ and $\frac{\hat{\Gamma}}{n-1}$ coincide up to the mass at zero.

2) The coupling.
We construct a set of independent particles, couple an interacting system to it, and show that the two sets do not differ much.

We could take any finite collection of indices, but for simplicity we restrict ourselves to the construction for two indices i and j. We take two independent copies of the Poisson processes and initial values as in the previous part of the proof, which we distinguish using superscripts i and j. We first construct as above two independent interaction graphs $G_i^{i,n}$ and $G_j^{j,n}$ and two independent processes $X_i^{i,n}$ and $X_j^{j,n}$.

Let us now describe the coupling. We want to build an interaction graph G_{ij}^n in such a way that the subgraph G_i^n stemming from i is as close as possible to $G_i^{i,n}$, and G_j^n to $G_j^{j,n}$, then we choose the initial values and jump variables among the two independent copies so that the process (X^{in}, X^{jn}) (obtained on G_{ij}^n) is as close as possible to $(X_i^{i,n}, X_j^{j,n})$, with $\mathcal{L}(X^{in}) = \mathcal{L}(X_i^{i,n})$ and $\mathcal{L}(X^{jn}) = \mathcal{L}(X_j^{j,n})$.

We must choose for any pair (k,l) of indices between the successive jumps of N_{kl}^i and N_{kl}^j in order to get N_{kl}. We go back in time from T. We start from indices i and j. For a pair of indices (i,k) with $k \neq j$ we only consider N_{ik}^i and for a pair (j,l) with $l \neq i$ we consider N_{jl}^j. Thus the subgraphs will grow like $G_i^{i,n}$ and $G_j^{j,n}$. There is a problem for the pair (i,j): we must then use a priority rule in order to choose between N_{ij}^i and N_{ij}^j. Many priority rules work. For symmetry reasons, we resolve every conflict by a fair toss-up. If i is chosen, we use Ni_{ij} and G_i^n is equal to $G_i^{i,n}$.

Every time we encounter a jump of a $N_{i,j}^i$ or $N_{i,j}^j$, the priority rule will force us either to use it or to discard it for both G_i^n and G_j^n. This introduces a difference between either G_i^n and $G_i^{i,n}$ or G_j^n and $G_j^{j,n}$. We call the occurrence of such a conflict a direct interaction between i and j. Since we go backward in time, we may simply consider successively the jumps of the processes needed for the past history of the particles. The subgraphs G_i^n and G_j^n are made to grow as $G_i^{i,n}$ and $G_j^{j,n}$, except that these conflicts and priority rules are recursively extended to all pairs of indices whose elements enter respectively the two interaction subgraphs.

Once the interaction graph G_{ij}^n is built, we must choose between the two independent copies of the initial values needed to build the particle system. If $G_i^n = G_i^{i,n}$ and $G_j^n = G_j^{j,n}$ the subgraphs are disjoint: on the first one we use the variables with superscript i and on the second with j. On the contrary event $A_{ij}^n = \{G_i^n \neq G_i^{i,n}\} \cup \{G_j^n \neq G_j^{j,n}\} = \{G_i^{i,n} \cap G_j^{j,n} \neq \emptyset\}$ we use the same priority rule as for the graphs.

The definition of the variation norm implies clearly that

$$| \mathcal{L}(X^{in}, X^{jn}) - \mathcal{L}(X^{in}) \otimes \mathcal{L}(X^{jn}) |_T \leq 2P(A_{ij}^n).$$

Similarly for k indices we have:

$$| \mathcal{L}(X^{i_1 n}, ..., X^{i_k n}) - \mathcal{L}(X^{i_1 n}) \otimes ... \otimes \mathcal{L}(X^{i_k n}) |_T \leq 2P\Big(\cup_{1 \leq p < q \leq k} A_{i_p i_q}^n \Big).$$

We now have to compute $P(A_{ij}^n)$.

3) The interaction chains and estimates on convergence.
A^n_{ij} happens only if we encounter a jump of a pair of conflicting Poisson processes before reaching 0. This event happens if there exists an **interaction chain** between i and j, which we now describe.

We go from T to 0. There are integers m and p and indices $i_1, \dots, i_m$ and $j_1, \dots, j_p$ such that in $G_i^{i,n}$, i branched on i_1,then after i_1 on i_2, and so on to i_m, and also in $G_j^{j,n}$, j branched on j_1, then j_1 on j_2, and so on to j_p in $G_j^{j,n}$. After the branchings on i_m and j_p these two indices underwent a direct interaction, meaning that either i_m branched on j_p in $G_i^{i,n}$ or j_p branched on i_m in $G_j^{j,n}$. This last step can be taken from both sides and thus happens at twice the branching rate. There may be many interaction chains and we choose one with the least indices involved, in which $i, j, i_1, \dots, i_m, j_1, \dots, j_p$ are pairwise distinct. we define the interaction chain length as $m + p + 1$: see Fig. 3.1.

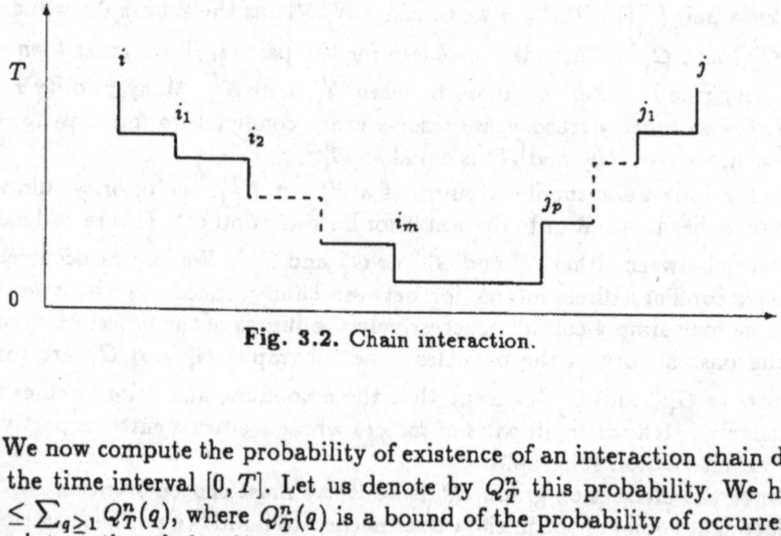

Fig. 3.2. Chain interaction.

We now compute the probability of existence of an interaction chain during the time interval $[0, T]$. Let us denote by Q_T^n this probability. We have $Q_T^n \leq \sum_{q \geq 1} Q_T^n(q)$, where $Q_T^n(q)$ is a bound of the probability of occurrence of an interaction chain of length q during $[0, T]$. We presently compute $Q_T^n(q)$ by induction on q.

We start with $q = 1$. For a direct interaction to happen between 0 and T, there is a route containing both i and j such that either N_{ij}^i or N_{ij}^j jumps in $[0, T]$. The probability of this event is less than

$$Q_T^n(1) \leq 1 - e^{-2\frac{\Lambda}{n-1}T} \leq 2\frac{\Lambda}{n-1}T.$$

We now assume we have a bound $Q_T^n(q-1)$ of the probability of occurence of an interaction chain of length $q - 1$, for $q \geq 2$. Then for an interaction chain of length q to happen, there must be first (backward from T) the birth of a new branch at either i or j after a waiting time $t \in [0, T]$, and this new branch must then be joined in time $T-t$ by an interaction chain of length $q-1$ to the one of i or j which did not branch first. The maximal rate of this new branch is 2Λ and thus:

$$Q_T^n(q) \leq \int_0^T Q_{T-t}^n(q-1)2\Lambda e^{-2\Lambda t}dt.$$

If we denote the exponential density of parameter 2Λ by $e_{2\Lambda}$, for $q \geq 2$,

$$Q_T^n(q) \leq Q_\cdot^n(q-1) * e_{2\Lambda}(T) \leq Q_\cdot^n(1) * e_{2\Lambda}^{*(q-1)}(T).$$

Since $e_{2\Lambda}^{*k}(t) = (2\Lambda)^k \frac{t^{k-1}}{(k-1)!} e^{-2\Lambda t}$ we obtain

$$\begin{aligned} Q_T^n &\leq Q_T^n(1) + 2\Lambda \int_0^T Q_{T-t}^n(1) \sum_{q\geq 2} \frac{(2\Lambda t)^{q-2}}{(q-2)!} e^{-2\Lambda t}dt \\ &\leq Q_T^n(1) + 2\Lambda \int_0^T (1 - e^{-2\frac{\Lambda}{n-1}(T-t)})dt \\ &\leq 2\frac{\Lambda T + \Lambda^2 T^2}{n-1}. \end{aligned}$$

We have seen that the event that the k interacting particles and the k independent particles differ is included in the event of at least an interaction chain between two of the k indices. Considering the $k(k-1)/2$ pairs of indices we obtain the first bound (3.1) in Theorem 3.1.

4) The limit Boltzmann tree.

A similar coupling argument between the interaction graph issued from one index for a given n and a limit Boltzmann tree, where the links are taken amongst an infinite supply of independent similar links, shows that the laws of the processes X^{in} converge to the law P of a process constructed similarly on the tree.

For given n and i we build an interaction tree T_i^n inspired from G_i^n. Each branch of the tree will have a distinct label, which is a finite sequence of $\{1, ..., n\}$ giving the successive filiation from the root. We need a sequence of independent Poisson processes of rate $\frac{\Lambda}{n-1}$ and independent initial values thus indexed. The construction is inspired from step 2 of the proof.

We start from a root labeled i and work backward in time from T. At the first encounter with a jump of a Poisson process N_{ij}^i the branch dies and two branches are born labeled ii and ij. Recursively, from a branch $ii_1...i_k$, at the first encounter with a jump of a Poisson process $N_{i_kj}^{ii_1...i_k}$, the branches born are labeled $ii_1...i_ki_k$ and $ii_1...i_kj$. Once zero is reached, we use the initial values indexed by the last indices of the labeling, and construct processes $\bar{X}^{ii_1...i_k,n}$ on each branch as in section 2. We then build a process $\tilde{X}^{in}$ on $[0,T]$ by collecting all the processes $\bar{X}^{i,n}, \bar{X}^{ii,n}, ..., \bar{X}^{i...i,n}, ...$ indexed by finite sequences consisting only of i. This Boltzmann process is thus without self-interactions.

To obtain the bound (3.2) we have then to couple an interaction graph G_i^n to the tree T_i^n and thus obtain a coupling of X^{in} and $\tilde{X}^{in}$.

Obtaining the graph from the tree is easy: we only need to retain the last index in the labeling. Naturally there are going to be conflicts to be resolved by priority rules. Conflicting Poisson processes have same subindices jl and are labeled by indices of branches alive at the same time and ending by j and l, and are assigned as priority as in step 2 of the proof. A conflict happens if two branches in the tree become linked by an interaction chain as in step 3, and we call this event an **interaction loop**.

We shall now bound the probability L_T^n of occurrence of an interaction loop during $[0,T]$. $L_T^n \leq \sum_{k\geq 1} L_T^n(k)$, where $L_T^n(k)$ is a bound of the probability to obtain a loop interaction with a chain joining two branches created in $k \geq 1$ steps from i, using the estimates of step 3.

$$
\begin{aligned}
L_T^n(1) &= \int_0^T Q_{T-t}^n \Lambda e^{-\Lambda t} dt, \\
L_T^n(k) &= 2\int_0^T L_{T-t}^n(k-1)\Lambda e^{-\Lambda t} dt \\
&= 2L_\cdot^n(k-1) * e_\Lambda(T) = 2^{k-1} L_\cdot^n(1) * e_\Lambda^{*(k-1)}(T) \\
&= 2^{k-1} Q_\cdot^n * e_\Lambda^{*k}(T).
\end{aligned}
$$

Thus by summing over $k \geq 1$,

$$
\begin{aligned}
L_T^n &\leq \Lambda \int_0^T Q_{T-t}^n \sum_{k\geq 1} \frac{(2\Lambda t)^{k-1}}{(k-1)!} e^{-\Lambda t} dt \\
&\leq \Lambda \int_0^T Q_{T-t}^n e^{\Lambda t} dt
\end{aligned}
$$

Using the bound found in step 3 for Q_{T-t}^n and discarding some negative terms, we finally find the bound (3.2).

5) The empirical measure and the nonlinear martingale problem. For a bounded function ϕ defined on $\mathbb{D}([0,T], \mathbb{R}^d)$, we have

$$
\begin{aligned}
E(\langle \phi, \mu^n - P\rangle^2) &= E\left(\left(\frac{1}{n}\sum_{i=1}^n \phi(X^{in}\right) - \langle \phi, P\rangle)^2\right) \\
&= \frac{1}{n} E((\phi(X^{1n}) - \langle \phi, P\rangle)^2) \\
&\quad + \frac{n-1}{n} E((\phi(X^{1n}) - \langle \phi, P\rangle)(\phi(X^{2n}) - \langle \phi, P\rangle)) \\
&= E(\phi(X^{1n})\phi(X^{2n})) - 2\langle \phi, P\rangle E(\phi(X^{1n})) + \langle \phi, P\rangle^2 \\
&\quad + O(1/n)
\end{aligned}
$$

which is $O(1/n)$ uniformly in T and $\|\phi\|_\infty$. This implies the convergence in law and in probability (since the limit is deterministic) of μ^n to P, (see details in the proof of Proposition 4.2), and gives an $O(1/\sqrt{n})$ rate for appropriate metrics.

6) The convergence of the law of one process X^{in} being in variation norm, there is no need of regularity to prove that the Boltzmann process solves the nonlinear martingale problem $(\mathcal{P}_B)$. We consider the martingale problem (2.17) satisfied by one particle in the interacting system together with (2.18) and (2.19) and the convergence of the empirical measures. We use a probabilistic characterization of martingales suited for taking the limit and show that the limit law P solves the limit nonlinear martingale problem. More details on these convergence techniques are given in Section 4.

The proof of Theorem 3.1 is now finished. □

Remark 3.2. If the marginal of rank k of the law of the n-particle system at time t has a density $f^n_{k,t}$ on $(\mathbb{R}^d)^k$ and if the limit law at time t has a density f_t, then

$$|\mathcal{L}(X^{1n}_t,\dots,X^{kn}_t)-(\mathcal{L}(X^{1n}_t))^{\otimes k})| \;=\; \|f^n_{k,t}-(f^n_{1,t})^{\otimes k}\|_1\,,$$
$$|\,\mathcal{L}(X^{in}_t)-P_t\,| \;=\; \|f^n_{1,t}-f_t\|_1$$

and

$$\sup_{0\le t\le T}\,|\,\mathcal{L}(X^{1n}_t,\dots,X^{kn}_t)-(\mathcal{L}(X^{1n}_t))^{\otimes k})\,|$$
$$\le|\,\mathcal{L}(X^{1n},\dots,X^{kn})-(\mathcal{L}(X^{1n}))^{\otimes k})\,|_T\,,$$
$$\sup_{0\le t\le T}\,|\,\mathcal{L}(X^{in}_t)-P_t\,|\le|\,\mathcal{L}(X^{in})-P\,|_T\,.$$

We can use the estimates in Theorem 3.1 to get results comparable to those in Pulvirenti et al. [41].

3.2 Algorithms for the Boltzmann equation

We have shown how the empirical measure of the particle system approximates the law of the Boltzmann equation. We now show how to simulate the particle systems in Section 2.2.4, which depend on the choice of the joint jump measure $\hat{\Gamma}$. This choice cannot come from the Boltzmann equation since it only gives the one-particle jump measure Γ which is the marginal of $\hat{\Gamma}$ up to mass at zero. External considerations must intervene, and we saw that Nanbu uses the least information available while Bird uses physical information.

The ideas of the graph construction in Section 3.1 can be used. The jump operator can be simulated exactly, by using an acceptance-rejection or fictitious-collisions method in a very simple way. This was already used in the proof in Theorem 3.1. The core idea is that adding mass at zero to the jump measure does not change the jump operator, and this involves no approximation whatsoever. Since the total mass of $\hat{\Gamma}$ is bounded by Λ, one can always add $(1- \mid \hat{\Gamma} \mid /\Lambda)\delta_0$ to $\hat{\Gamma}$ and obtain a positive measure of constant total mass Λ.

There are n particles and the total rate for the $n(n-1)/2$ pairs of possible interactions is $n\Lambda/2$ as seen in (2.16). A Poisson process of rate $n\Lambda/2$ gives the sequence of collision times. At each of these we choose uniformly the pair of particles which interact, update the states of these particles under $\mathcal{L}$, compute $\mid \hat{\Gamma} \mid$ at these states, discard the jump with probability $1- \mid \hat{\Gamma} \mid /\Lambda$ and with probability $\mid \hat{\Gamma} \mid /\Lambda$ choose the joint jump amplitude according to $\hat{\Gamma}/ \mid \hat{\Gamma} \mid$. All this is done independently. This method only necessitates at each step the evaluation of the cross-section of the interacting pair and not those of the $n(n-1)/2$ pairs.

The inter-arrival times of a Poisson process of rate θ are independent exponential variables of rate θ, which each have density $\theta e^{-\theta t} dt$ on $\mathbb{R}_+$. The memory-less property of exponential variables is essential. If U is a uniform random variable on $[0,1]$ then $-\mathrm{Log}(U)/\theta$ is exponential of rate θ. This exact simulation of the inter-collision times is a random time-discretization consistent with the Markovian evolution.

We do not compute a collision integral operator at fixed times using a point-set method, but must simulate the time evolution of a Markov process which has this operator as a generator. The random time-steps and jump choices automatically give the collision operator, which is not approximated in any way. Using quasi-Monte-Carlo methods, like low discrepancy sequences instead of random numbers, is very risky as stated by Neunzert et al. [35]: the essential properties of Markov processes, exponential times, and Poisson processes involve independence in a central way.

There is no splitting involved, and if the simulation of the evolution by $\mathcal{L}$ is exact (as in case of free-flow in Section 2.2.1) the whole process yields an exact simulation of the particle system. $\mathcal{L}$ can be approximated by many means including deterministic numerical methods. If $\mathcal{L}$ is a diffusion operator with simple boundary conditions we may use a continuous-time random walk with exponential or deterministic holding times. If the flow of the differential equation $\dot{z} = b(z)$ is well understood we should use it, and simulate the contribution of the second-order operator by a mean-zero random walk. Without diffusion terms a Runge-Kutta method seems a good choice. These approximations of the evolution of $\mathcal{L}$ can be superposed without any problem to the simulation of the jumps.

There clearly must be a global time counter for the system which is incremented by the exponentials. For a more precise description of the simulation, we can decide to update the particles in at least two fashions. The first way needs n auxiliary time counters, one for each particle, and we only update at each collision the two partners from the time of their last update. When the global time counter goes beyond T, no collision is performed and all particles are updated from their time of last update to time T. The second way has only the global counter and all n particles are updated at the times of collisions. The first way needs more memory but less computations.

Remark. If we want fixed deterministic time-steps, we approximate the linear Markov process of the particle system by a Markov chain. From an analytical point of view it represents a discretization of a linear evolution problem. These approximations of linear problems are well known and there is a huge probabilistic and analytical literature on these subjects including rates of convergence. This is the level of description in which fit the classical Nanbu and Bird algorithms. Simulating the exponential inter-collision times as we do could be more costly than using fixed time-steps equal to their mean as in Pulvirenti et al. [41], but less than the computation of the number of collisions at each fixed time-step as in Nanbu [34]. It is certainly more precise and avoids the issue of splitting.

4. Propagation of chaos for exchangeable mean field systems of diffusions with jumps

We are now interested in general models of interacting diffusions with jumps, as described in Section 2.3. The interaction graph techniques do not work and we shall use martingale problems techniques.

4.1 The propagation of chaos

In this context we use the method developed in Sznitman [47] in order to show an asymptotic result about the behaviour of the system $(X^{1n}, ..., X^{nn})$, for a large number of particles. This asymptotic property is called propagation of chaos. The exchangeability of $(X^{1n}, ..., X^{nn})$ will be fundamental in all what follows.

Let us recall here the definition of chaoticity.

Definition 4.1. *Let E be a Polish space, Q be a probability measure on E and Q^n be probability measures on E^n. The sequence $(Q^n)_n$ is Q-chaotic if for any fixed integer k and any continuous bounded functions $f_1, ..., f_k$ on E,*

$$\lim_{n\to\infty} < Q^n, f_1 \otimes ... \otimes f_k \otimes 1^{n-k} >= \prod_{i=1}^{k} < Q, f_i > .$$

That means that when n tends to infinity, any fixed finite number of coordinates become independent with the same distribution Q.

This will be applied only when each Q^n is a symmetrical measure. This additional assumption is a key point in this approach since therefore Q-chaoticity is equivalent to a law of large numbers for the empirical measures of the coordinate variables:

Proposition 4.2. *If $(Q^n)_n$ is a sequence of symmetrical probability measures on E^n, it is Q-chaotic if and only if the empirical measures $\mu^n = \frac{1}{n}\sum_{i=1}^n \delta_{X_i}$ (where (X_i) are the canonical coordinates on E^n) converge in law (and in probability) as $\mathcal{P}(E)$-valued variables under Q^n, to the deterministic probability measure Q.*

Proof. This proof can be found in Sznitman [47] p.177.

i) Suppose first that (Q^n) is Q-chaotic and take f in $C_b(E)$.

$$E_n(< \mu^n - Q, f >^2) = \frac{1}{n^2}\sum_{i,j=1}^n E_n(f(X_i)f(X_j)) - \frac{2}{n} < Q, f > \sum_{i=1}^n E_n(f(X_i)) + < Q, f >^2 .$$

Using the symmetry, we find:

$$\frac{1}{n}E_n(f(X_1)^2) + \frac{n-1}{n}E_n(f(X_1)f(X_2)) - 2 < Q, f > E_n(f(X_1)) + < Q, f >^2$$

which tends to zero by definition of the Q-chaoticity, when n tends to infinity.

This implies that (μ^n) converges in probability to the deterministic probability measure Q. Indeed, since E is a Polish space, there exists a countable set V of continuous bounded functions on E which is convergence determining. See for example Ethier-Kurtz [8, p.111-112]. Here $< \mu^n, f >$ tends in probability to $< Q, f >$ for every function of V. Then for all subsequence n_k, there exists a further subsequence n'_k such that $< \mu^{n'_k}, f >$ tends almost surely to $< Q, f >$, for every function f in V. But then $(\mu^{n'_k})$ converges (weakly) almost surely to Q. That implies the convergence in probability of (μ^n) to Q.

ii) Conversely, suppose that the sequence (μ^n) converges in law to the deterministic measure Q (under the laws Q^n) and consider $f_i \in C_b(E)$ bounded by K. We have

$$|< Q^n, f_1 \otimes ... \otimes f_k \otimes 1^{n-k} > - \prod_{i=1}^k < Q, f_i >|$$

$$\leq |< Q^n, f_1 \otimes ... \otimes f_k \otimes 1^{n-k} > - < Q^n, \prod_{i=1}^k < f_i, \mu^n >>| + |< Q^n, \prod_{i=1}^k < f_i, \mu^n >> - \prod_{i=1}^k < Q, f_i >| .$$

The second term in the right side goes to zero because of the hypothesis of convergence in law of (μ^n) towards Q. To study the first term, we need symmetry. We write it as:

$$|< Q^n, \frac{1}{n!} \sum_{\sigma \in S_n} f_1(X_{\sigma(1)})...f_k(X_{\sigma(k)}) - \prod_{i=1}^{k} < \mu^n, f_i, >>|$$

$$\leq \quad K^k \left[1 - \frac{n(n-1)...(n-k+1)}{n^k} + \frac{n^k - n(n-1)...(n-k+1)}{n^k} \right]$$

(for this we have distinguished the terms which correspond to injections from $[1, k]$ into $[1, N]$ from the others), and that tends to zero when n tends to infinity. □

The asymptotic behaviour of the system is described in the following theorem:

Theorem 4.3. *Assume that the initial particles X_0^{in} are independent and distributed according to P_0. Then, under the hypotheses (H0),..., (H4) (cf. Section 2.3.1), the sequence of probability measures P^n on $\mathbb{D}([0,T],(\mathbb{R}^d)^n)$ solving the n-particle martingale problem (2.23) is P-chaotic, where P is the unique solution (on $\mathbb{D}([0,T],\mathbb{R}^d)$) of the nonlinear martingale problem $(\mathcal{P}_{VB})$ starting at P_0 and described in (2.20).*

Remark: In fact, in this theorem it would suffice to assume that the initial distribution ν^n of $(X_0^{in})_{1\leq i\leq n}$ is exchangeable for each n, and that the sequence $(\nu^n)_n$ is P_0-chaotic. We have chosen an i.i.d. assumption for simplicity.

Owing to Proposition 4.2, the proof of Theorem 4.3 will be a consequence of the following result:

Theorem 4.4. *Let $\mu^n = \frac{1}{n}\sum_{i=1}^n \delta_{X^{in}}$ be the empirical measure of the system, considered as a probability measure over the path space $\mathbb{D}([0,T],\mathbb{R}^d)$. Let π^n be the law of μ^n. Then, under (H0),..., (H4), the sequence (π^n) converges for the weak topology to δ_P, where P is the unique solution of the non linear martingale problem $(\mathcal{P}_{VB})$ described in (2.20), and thus μ^n converges in law and in probability to the deterministic limit P.*

The proof will be the aim of the following subsection.

4.2 Proof of Theorem 4.4

To prove this theorem, we follow the classical trilogy of arguments:
1) Tightness of (π^n) in $\mathcal{P}(\mathcal{P}(\mathbb{D}([0,T],\mathbb{R}^d)))$.
2) Identification of the limiting values of (π^n) as solutions of (2.20).
3) Uniqueness of the solution of (2.20).
The third point is given by Theorem 2.8.

Lemma 4.5. *The tightness of the laws of $\mu^n \in \mathcal{P}(\mathcal{P}(\mathbb{D}([0,T],\mathbb{R}^d)))$ is equivalent to the tightness of the distributions of the semimartingales $X^{1n} \in \mathcal{P}(\mathbb{D}([0,T],\mathbb{R}^d))$.*

This lemma is a consequence of the exchangeability of $(X^{in})_{1\leq i\leq n}$ and of the following proposition.

Proposition 4.6. *Let E be a Polish space and (m^n) be a sequence of probability measures on $\mathcal{P}(E)$. Then the sequence (m^n) is tight if and only if the sequence of the intensity measures $I(m^n) \in \mathcal{P}(E)$, defined by*

$$\forall f \in B_b(E), < I(m^n), f >= \int_{\mathcal{P}(E)} < \mu, f > dm^n(\mu)$$

is tight.

Proof of Proposition 4.6. This proof is contained for example in [47, p.179]: The map $m \longrightarrow I(m)$ is clearly continuous for the weak topology. It suffices then to prove that if $(I(m^n))$ is tight, then (m^n) is tight.

For every $\varepsilon > 0$, denote by K_ε a compact subset of E, with $I(m^n)(K_\varepsilon^c) \leq \varepsilon$, for every n. For ε, $\eta > 0$, and any n, Bienaymé-Tchebytcheff inequality gives $m^n(\{\mu, \mu(K_{\varepsilon\eta}^c) \geq \eta\} \leq \frac{1}{\eta} I(m^n)(K_{\varepsilon\eta}^c) \leq \varepsilon$. It follows that

$$m^n(\bigcup_{k\geq 1} \{\mu, \mu(K^c_{\varepsilon\frac{2^{-k}}{k}}) \geq \frac{1}{k}\}) \leq \sum_{k\geq 1}^{\infty} \varepsilon 2^{-k} = \varepsilon.$$

So m^n puts a mass greater or equal to $1 - \varepsilon$ on the compact subset

$$\bigcap_{k\geq 1} \{\mu, \mu(K^c_{\varepsilon\frac{2^{-k}}{k}}) \leq \frac{1}{k}\}$$

of $\mathcal{P}(E)$. This proves that the sequence (m^n) is tight. □

Proof of Theorem 4.4. By Lemma 4.5, we first have to prove the tightness of the distributions of (X^{1n}) which will imply the tightness of the distributions π^n of (μ^n). The proof exactly follows Section 3.2 in Joffe-Métivier [22] which studies the tightness of sequences of semimartingales with a locally square integrable martingale part. The process X^{1n} is such a semimartingale since X^{1n} itself is locally square integrable. The main difference with the general situation studied in [22] is that X^{1n} is defined considering the whole system $(X^{in})_{1\leq i\leq n}$ as in Section 2.3 and the empirical measure intervenes, but all the growth assumptions are uniform w.r.t. the probability measures, and the proofs can be adapted without difficulty. We just give the main ideas.

The process $X^{1n}_.$ has the semimartingale decomposition $X^1_0 + M^{1n}_. + A^{1n}_.$, and the processes $< M^{1n} >_.$ and $A^{1n}_.$ can easily be deduced from (2.23) and (2.24).

One first proves that

$$E(\sup_{0\leq t\leq T} |X^{1n}_t|^2) < \infty \;\; ; \;\; E(\sup_{0\leq t\leq T} |M^{1n}_t|^2) < \infty$$

(see [22, Lemma 3.2.2]). It is somewhat technical because of the jumps. Hypotheses (H0)-(H4) imply the linear growth of the coefficients and one concludes by Gronwall's lemma.

The second step is to use the criterions of Aldous and Rebolledo. Let us recall these two criterions.

Aldous criterium ([2], [22]): Let (X^n) be a sequence of cadlag processes defined respectively on their own probability spaces $(\Omega^n, (F_t^n), P^n)$ taking values in $\mathbb{R}^d$. Assume:
1) For each t in a dense subset of $[0,T]$, the laws of (X_t^n) are tight in $\mathcal{P}(\mathbb{R}^d)$.
2) $\forall \varepsilon > 0, \eta > 0$ there exist $\delta > 0$ and n_0 such that for any sequence of stopping times τ_n,

$$\sup_{n \geq n_0} \sup_{\theta \leq \delta} P^n(| X^n_{\tau_n + \theta} - X^n_{\tau_n} | > \eta) \leq \varepsilon.$$

Then the sequence of laws of (X^n) is tight in $\mathcal{P}(\mathbb{D}([0,T], \mathbb{R}^d)$.

Rebolledo's criterium ([43], [22]): If (X^n) is a sequence of finite-dimensional semimartingales of the form $X^n = M^n + A^n$ with A^n a finite variation process and M^n a locally square-integrable martingale and if the sequences (A^n) and $(< M^n >)$ satisfy the assumptions of Aldous criterium, the laws of the sequence (X^n) form a tight family.

By following Joffe-Métivier and applying these two results, we obtain that the sequence $(\mathcal{L}(\mu^n))_{n \geq 1}$ is relatively compact.

We now characterize its accumulation points by a martingale problem technique.

Let $\Pi^\infty \in \mathcal{P}(\mathcal{P}(\mathbb{D}([0,T], \mathbb{R}^d)))$ be an accumulation point of (π^n). It is the limit of a subsequence, which is still denoted by π^n. For $\phi \in C_b^2(\mathbb{R}^d)$, $0 \leq s_1, \ldots, s_q \leq s \leq t$, $g_1, \ldots, g_q \in C_b(\mathbb{R}^d)$, $Q \in \mathcal{P}(\mathbb{D}([0,T], \mathbb{R}^d))$, set

$$F_{sts_1\ldots s_q}(Q) = \left\langle Q, \left(\phi(X_t) - \phi(X_s) - \int_s^t \mathcal{L}(Q_u)\phi(X_u) + K(Q_u)\phi(X_u)\, du\right) \right.$$
$$\left. g_1(X_{s_1}) \ldots g_q(X_{s_q}) \right\rangle.$$

$F_{sts_1\ldots s_q}$ is not continuous since the projections $X \to X_t$ are not continuous for the Skorohod metric. However for any $Q \in \mathcal{P}(\mathbb{D}([0,T], \mathbb{R}^d))$, $X \to X_t$ is Q-a.s. continuous for all t outside an at most countable set D_Q, and thus F is continuous at point Q if $s, t, s_1, ..., s_q$ are not in D_Q: we use here the continuity and boundedness of $\phi, g_1, .., g_q$, and also the fact that $(q, x) \to \mathcal{L}(q)\phi(x) + K(q)\phi(x)$ is continuous on $\mathcal{P}(\mathbb{D}([0,T], \mathbb{R}^d)) \times \mathbb{R}^d$.

Now, one can show that the set D of all t for which $\Pi^\infty(Q : t \in D_Q) > 0$ is again at most countable. Thus if $s, t, s_1, ..., s_q$ are in D^c, $F_{sts_1\ldots s_q}$ is Π^∞-a.s. continuous.

$$< \pi^n, F^2_{sts_1\ldots s_q} > = E(F_{sts_1\ldots s_q}(\mu^n)^2)$$

$$= E\left(\left(\frac{1}{n}\sum_{i=1}^{n}(M_t^{\phi,in} - M_s^{\phi,in})g_1(X_{s_1}^{in})\dots g_q(X_{s_q}^{in})\right)^2\right)$$

$$= \frac{1}{n}E\left(\left((M_t^{\phi,1n} - M_s^{\phi,1n})g_1(X_{s_1}^{1n})\dots g_q(X_{s_q}^{1n})\right)^2\right)$$

$$+\frac{n-1}{n}E\Big((M_t^{\phi,1n} - M_s^{\phi,1n})(M_t^{\phi,2n} - M_s^{\phi,2n})$$

$$g_1(X_{s_1}^{1n})\dots g_q(X_{s_q}^{1n})g_1(X_{s_1}^{2n})\dots g_q(X_{s_q}^{2n})\Big).$$

The first term goes to zero because of the uniform integrability given by the L^2 bounds and the second term is null because $\langle M^{\phi,1n}, M^{\phi,2n}\rangle = 0$. Hence the sequence $F_{sts_1\dots s_q}(\mu^n)$ is uniformly integrable and by the a.s. continuity of $F_{sts_1\dots s_q}$ we get

$$< \Pi^\infty, |F_{sts_1\dots s_q}| >= \lim E^n(|F_{sts_1\dots s_q}(\mu^n)|) = 0.$$

Thus Π^∞-a.s., $F_{sts_1\dots s_q}(Q)$ is zero for every $s, t, s_1, \dots, s_q$ outside the at most countable set D. The process X being càdlàg, that is sufficient to show that Π^∞-a.s., Q satisfies the nonlinear martingale problem $(\mathcal{P}_{VB})$. This problem has a unique solution P by Theorem 2.8, and Π^∞ is the Dirac mass at P. Therefore $(\mathcal{L}(\mu^n))_{n\geq 1}$ converges to the Dirac mass at P.

4.3 Empirical measures as probability measures on the path space or as processes with probability measure values

Theorem 4.4 gives a result of convergence on the path space. That means that we consider the empirical measures as probability measures on the path space, and we obtain asymptotic results for the distribution of all the paths of the particles, the limit law being described as solution of a martingale problem on $\mathbb{D}([0,T],\mathbb{R}^d)$.

This convergence is stronger than the convergence for the flow of one-dimensional time-marginals, as seen in Theorem 4.7 below. The property of convergence of $(\mu_t^n)_{t\geq 0}$ is usually called **functional law of large numbers.** It gives of course less information since we could not deduce from it the asymptotic behaviour for functionals on the path space of the particles.

Theorem 4.7. *Under the assumptions (H0)...(H4), the convergence of (μ^n) to P proved in Theorem 4.4 implies the convergence in law and in probability of the probability measure-valued processes $(\mu_t^n)_{t\geq 0}$ towards the flow $(P_t)_{t\geq 0}$, in the space $\mathbb{D}([0,T],\mathcal{P}(\mathbb{R}^d))$ endowed with the uniform topology.*

The converse of this theorem is false, as it is proved by a counter-example given in [47, p.182]. In this example, two mean field interacting particle systems are exhibited, which are chaotic with respect to two different probability measures P and Q, but their empirical measure processes are equal.

Proof of Theorem 4.7. The flow $(P_t)_{t\geq 0}$ is deterministic and continuous. Then the convergence to $(P_t)_{t\geq 0}$ is the same for the Skorohod and the uniform topologies. The key point is that the laws $(P_t)_{t\geq 0}$ are the time-marginal distributions of a quasi-left continuous process whose Lévy's measure is absolutely continuous with respect to Lebesgue measure. This fact appears in the following lemma, where we adapt an intermediary result appearing in the proof of Lemma 2.8 in Léonard [26].

Lemma 4.8. *Let us consider a sequence of random probability measures (μ^n) on $\mathbb{D}([0,T],\mathbb{R}^d)$. Suppose that the following assumptions hold:*
(i) (μ^n) converge in law to Q in $\mathcal{P}(\mathbb{D}([0,T],\mathbb{R}^d))$
(ii) The measure Q is deterministic
(iii) If we denote by $(X_t)_{t\geq 0}$ the canonical process on $\mathbb{D}([0,T],\mathbb{R}^d)$,

$$\lim_{\alpha\to 0}\sup_{0\leq t\leq T} E^Q\big(\sup_{t-\alpha<s<t+\alpha} |\Delta X_s| \wedge 1\big) = 0.$$

Then we have convergence in probability of the flow $(\mu_t^n)_{t\geq 0}$ to $(Q_t)_{t\geq 0}$ in $\mathbb{D}([0,T],\mathcal{P}(\mathbb{R}^d))$ endowed with the uniform topology.

Proof. Let ρ be the Vaserstein distance on $\mathbb{R}^d$. We have to prove that $\sup_{0\leq t\leq T}\rho(\mu_t^n, Q_t)$ tends in probability to zero when n tends to infinity. The convergence in probability is equivalent to the fact that from every subsequence one can extract a subsubsequence which converges almost surely.

Since Q is deterministic, the sequence (μ^n) converges in probability to Q. Then for every sequence, there exists a subsequence n_k such that the sequence (μ^{n_k}) converges almost surely to Q. Let us now fix an ω such that $\mu^{n_k}(\omega)$ converges to Q. By Skorohod's representation, there exist a probability space $(W^\omega, \mathcal{F}^\omega, \Pi^\omega)$ and $\mathbb{D}([0,T],\mathbb{R}^d)$-valued random variables $Z_\omega^{n_k}$ and Z_ω such that $\Pi^\omega\circ(Z_\omega^{n_k})^{-1} = \mu^{n_k}(\omega)$, $\Pi^\omega\circ Z_\omega^{-1} = Q$ and $(Z_\omega^{n_k})$ converge to Z_ω Π^ω-almost surely in $\mathbb{D}([0,T],\mathbb{R}^d)$. Making use of Vaserstein distance, it is then immediate to see that it suffices to show that

$$\lim_{k\to+\infty}\sup_{0\leq t\leq T} E_{\Pi^\omega}(|Z_t^{n_k} - Z_t| \wedge 1) = 0.$$

Let $x\in\mathbb{D}([0,T],\mathbb{R}^d)$ and $S\subset[0,T]$. As in Billingsley [4], we set:

$$w_x(S) = \sup\{|x_s - x_t|, s,t\in S\},\quad w_x'(\delta) = \inf_{\{t_i\}}\max_{1\leq i\leq r} w_x([t_{i-1},t_i))$$

where the infimum extends over the finite sets $\{t_i\}$ of points such that

$$0 = t_0 < t_1 < \ldots < t_r = T\ ;\ t_i - t_{i-1} > \delta,\quad \forall i\in\{1,\ldots,r\}.$$

One can easily remark as in [26] (2.13) that:

$$w_x([a,b]) \leq 2w_x'(b-a) + \sup_{a\leq s\leq b}|\Delta x_s| \tag{4.1}$$

for every $0 \leq a \leq b \leq T$.

On the other hand, let $\alpha > 0$, $0 \leq t < T$, and $x, y \in \mathbb{D}([0,T], \mathbb{R}^d)$ be such that $d_{sko}(x,y) \leq \alpha$, where d_{sko} is the Skorohod metric, then for any $\varepsilon > 0$, one can find a time change λ such that

$$\sup_{0\leq s\leq T} \mid \lambda(s) - s \mid \leq \alpha + \varepsilon \; ; \; \sup_{0\leq s\leq T} \mid y(s) - x_{\lambda(s)} \mid \leq \alpha + \varepsilon$$

and deduce that

$$d_{sko}(x,y) \leq \alpha \Rightarrow \mid x_t - y_t \mid \leq \alpha + w_x([t-2\alpha, t+2\alpha]).$$

Then by (4.1), this leads us for all $\alpha > 0$ to

$$\begin{aligned} E_{\Pi^\omega}(|Z_t^{n_k} - Z_t| \wedge 1) &\leq \Pi^\omega(d_{sko}(Z^{n_k}, Z) > \alpha) \\ &+ E_{\Pi^\omega}(\mid Z_t^{n_k} - Z_t \mid \wedge 1, d_{sko}(Z^{n_k}, Z) \leq \alpha) \\ &\leq \Pi^\omega(d_{sko}(Z^{n_k}, Z) > \alpha) + \alpha + E_{\Pi^\omega}(w_Z[t-2\alpha, t+2\alpha] \wedge 1) \\ &\leq \Pi^\omega(d_{sko}(Z^{n_k}, Z) > \alpha) + \alpha \\ &\quad + E_{\Pi^\omega}(\sup_{s\in[t-2\alpha,t+2\alpha]} \mid \Delta Z_s \mid \wedge 1 + 2w'_Z(4\alpha) \wedge 1). \end{aligned}$$

Since (Z^{n_k}) tends to Z in $D([0,T], \mathbb{R}^d)$ Π_ω a.s. and since for every $x \in \mathbb{D}([0,T], \mathbb{R}^d)$, $w'_x(\delta)$ tends to zero with δ, it suffices to prove that

$$\lim_{\alpha\to 0} \sup_{0\leq t\leq T} E_{\Pi^\omega}(\sup_{s\in[t-2\alpha,t+2\alpha]} \mid \Delta Z_s \mid \wedge 1) = 0.$$

But

$$\sup_{0\leq t\leq T} E_{\Pi^\omega}(\sup_{s\in[t-2\alpha,t+2\alpha]} \mid \Delta Z_s \mid \wedge 1) \;=\; E_Q(\sup_{s\in[t-2\alpha,t+2\alpha]} \mid \Delta X_s \mid \wedge 1)$$

where X is the canonical process on $\mathbb{D}([0,T], \mathbb{R}^d)$. Property (iii) allows to conclude. □

End of the proof of Theorem 4.7. We have to prove the third assumption of the lemma in our context:

$$\lim_{\alpha\to 0} \sup_{0\leq t\leq T} E_P(\sup_{s\in[t-\alpha,t+\alpha]} \mid \Delta X_s \mid \wedge 1) = 0.$$

But

$$\begin{aligned} E_P(\sup_{s\in[t-\alpha,t+\alpha]} \mid \Delta X_s \mid \wedge 1) &\leq P(\exists \text{a jump between } t-\alpha \text{ and } t+\alpha) \\ &\leq 1 - \exp(-2\Lambda\alpha) \end{aligned}$$

which tends to zero when α tends to 0, uniformly in $t \in [0,T]$. (Λ is defined in (H2) (Section 2.3.1)). □

5. Convergence of the fluctuations for the McKean-Vlasov model

We have just above seen that the flow $(\mu_t^n)_{t\geq 0}$ converges to the flow $(P_t)_{t\geq 0}$. We are now interested in the behaviour of the fluctuations around this limit and in the speed of convergence. We study here the McKean-Vlasov model described in Subsection 2.1. The difficulty to generalize the following techniques to models with jumps lies in Lemma 5.2 which gives precise pathwise estimates between the interacting particles and coupled independent particles. In Méléard [29] a fluctuation result is obtained in the Boltzmann case using evolution equations and Sobolev imbeddings. We will not develop it here.

In the following we prove that in the McKean-Vlasov case the fluctuations belong uniformly (in n and t) to the weighted Sobolev space $W_0^{-(1+D),2D}$ and converge in $C([0,T], W_0^{-(2+2D),D})$ to an Ornstein-Uhlenbeck process obtained as the solution of a generalized Langevin equation in $W_0^{-(4+2D),D}$, where D is equal to $1+[d/2]$ ($[d/2]$ being the integer part of $d/2$). All what we need about these weighted Sobolev spaces is described in the Appendix (Section 6). Some tricky inequalities concerning the norm of some mappings are also stated.

5.1 Some pathwise estimations

We consider the interacting particle system defined in (2.6) and prove some pathwise results under the following hypotheses.
(K0): $E(|X_0^1|^{8D}) < +\infty$, where $D = [d/2] + 1$.
(K1): The functions $\sigma(x, y)$ and $b(x, y)$ have bounded partial derivatives of every order less than $D + 1$.

We first recall that condition (K0) propagates uniformly in n and t.

Lemma 5.1. *Assume (H1) (cf. Subsection 2.3.1) and (K0). Then,*

$$\sup_n E(\sup_{s\leq T} |X_s^{in}|^{8D}) < +\infty, \quad 1 \leq i \leq n; \quad E(\sup_{s\leq T} |X_s|^{8D}) \; < \; +\infty,$$

where X is defined in (2.5).

The proof uses classical pathwise arguments and Gronwall's inequality.□

Let us now define the flow of fluctuations we are interested in. For every integer n, let η^n be the fluctuation process defined by

$$\eta_t^n = n^{1/2}(\mu_t^n - P_t),$$

and let us introduce the coupling. Let X^i, $i \geq 1$ be independent copies of X (the limit process given in (2.5)); more precisely (X^i) satisfies

$$X_t^i = X_0^i + \int_0^t \sigma[X_s^i, P_s]dB_s^i + \int_0^t b[X_s^i, P_s]ds \tag{5.1}$$

where P_t is the law of X_t, $(B^i)_{i\geq 1}$ are the k-dimensional Brownian motions and $(X_0^i)_{i\geq 1}$ the i.i.d. $I\!R^d$-valued random variables with law P_0 and independent of $(B^i)_{i\geq 1}$ given in (2.6).

The following estimate is the technical key point for the proof of Proposition 5.3.

Lemma 5.2. *Assume (H1) and (K0). Then,*

$$E(\sup_{t\leq T} |X_t^{in} - X_t^i|^4) \leq \frac{K}{n^2}.$$

Proof. Let us give a brief proof. It approximately follows the same steps as in Lemma 1 in Hitsuda and Mitoma [19] (see also Sznitman [48]). Owing to Lipschitz continuity of σ and b, Schwarz's and Gronwall's inequalities, we obtain easily, by introducing the empirical measure $\mu_.$ of the system $(X_.^i)$, that

$$\begin{aligned}
&E(\sup_{s\leq t} |X_s^{in} - X_s^i|^4) \\
&\leq K\Big(E(\sum_{j=1}^d \sum_{h=1}^k \int_0^t E((\sigma_{jh}[X_s^{in}, \mu_s^n] - \sigma_{jh}[X_s^i, P_s])^4)ds \\
&\quad + \sum_{j=1}^d \int_0^t E((b_j[X_s^{in}, \mu_s^n] - b_j[X_s^i, P_s])^4)ds\Big) \\
&\leq K\Big(E(\int_0^t E(\sup_{u\leq s} |X_u^{in} - X_u^i|^4)ds \\
&\quad + \sum_{j=1}^d \sum_{h=1}^k \int_0^t E((\sigma_{jh}[X_s^i, \mu_s] - \sigma_{jh}[X_s^i, P_s])^4)ds \\
&\quad + \sum_{j=1}^d \int_0^t E((b_j[X_s^i, \mu_s] - b_j[X_s^i, P_s])^4)ds\Big) \\
&\leq K\Big(E(\int_0^t E(\sup_{u\leq s} |X_u^{in} - X_u^i|^4)ds \\
&\quad + \sum_{j=1}^d \sum_{h=1}^k \int_0^t E((\frac{1}{n}\sum_{J=1}^n \sigma_{jh}(X_s^i, X_s^J) - \sigma_{jh}[X_s^i, P_s])^4)ds \\
&\quad + \sum_{j=1}^d \int_0^t E((\frac{1}{n}\sum_{J=1}^n b_j(X_s^i, X_s^J) - b_j[X_s^i, P_s])^4)ds\Big).
\end{aligned}$$

A convexity inequality is not sufficient to obtain a good estimate of the second and the third terms of the last expression. One remarks that since the variables $(X^i)_{1\leq i\leq n}$ are i.i.d. with law P, then

$$E(\sigma_{jh}(X^i_s, X^J_s)|X^r_s, \text{for all } r \neq J) = \sigma_{jh}[X^i_s, P_s], \quad \text{for } i \neq J,$$

and then, for $J \neq k_1, k_2, k_3$ and $i \neq J$,

$$\begin{aligned} E((\sigma_{jh}(X^i_s, X^J_s) - \sigma_{jh}[X^i_s, P_s]) \cdot (\sigma_{jh}(X^i_s, X^{k_1}_s)) - \sigma_{jh}[X^i_s, P_s]) \\ \cdot(\sigma_{jh}(X^i_s, X^{k_2}_s) - \sigma_{jh}[X^i_s, P_s]) \cdot (\sigma_{jh}(X^i_s, X^{k_3}_s) - \sigma_{jh}[X^i_s, P_s])] = 0. \end{aligned}$$

Thus, we just have to consider in the second term the n terms of the form $(\sigma_{jh}(X^i_s, X^J_s) - \sigma_{jh}[X^i_s, P_s])^4$ and the $\frac{n(n-1)}{2}$ terms of the form

$$(\sigma_{jh}(X^i_s, X^J_s) - \sigma_{jh}[X^i_s, P_s])^2 \cdot (\sigma_{jh}(X^i_s, X^k_s) - \sigma_{jh}[X^i_s, P_s])^2 \quad k \neq J.$$

But Lemma 5.1 implies that the n first terms give an estimate in $\frac{1}{n^3}$ and the $\frac{n(n-1)}{2}$ other terms an estimate in $\frac{1}{n^2}$. We do the same reasoning for the third term. Gronwall's lemma finally allows to obtain $\frac{1}{n^2}$ as speed of convergence. □

We presently show that η^n_t belongs uniformly in n and t to the weighted Sobolev space $W_0^{-(1+D),2D}$ (defined in Section 6). This will be one of the keys to obtain the tightness result. We use the two previous lemmas and tricky inequalities about mappings appearing in the semimartingale representation of η^n. These inequalities are developed in Appendix Lemma 6.1 and their use was inspired by a paper of Ferland-Fernique-Giroux [9].

Define $S^n_t(\varphi)$, $T^n_t(\varphi)$ by

$$\begin{aligned} S^n_t(\varphi) &= n^{1/2}\left(\frac{1}{n}\sum_{i=1}^{n}(\varphi(X^{in}_t) - \varphi(X^i_t))\right); \\ T^n_t(\varphi) &= n^{1/2}\left(\frac{1}{n}\sum_{i=i}^{n}(\varphi(X^i_t) - \langle P_t, \varphi\rangle)\right). \end{aligned}$$

Clearly $\langle \eta^n_t, \varphi\rangle = S^n_t(\varphi) + T^n_t(\varphi)$.

Proposition 5.3. *Under* (**K0**), *the family* $(\eta^n_t)_{n\in\mathbb{N}, t\in[0,T]}$ *is bounded in* $W_0^{-(1+D),2D}$:

$$\sup_n \sup_{t\leq T} E(\|\eta^n_t\|^2_{-(1+D),2D}) < +\infty.$$

Proof. Let $\{\varphi_p\}_{p\geq 1}$ be a complete orthonormal system in $W_0^{1+D,2D}$, then

$$\sum_{p\geq 1}\langle \eta^n_t, \varphi_p\rangle^2 \leq 2(\sum_{p\geq 1} S^n_t(\varphi_p)^2 + \sum_{p\geq 1} T^n_t(\varphi_p)^2). \tag{5.2}$$

Since the mapping $x \mapsto x^2$ is convex we have

$$S_t^n(\varphi_p)^2 = n(\frac{1}{n}\sum_{i=1}^{n}(\varphi_p(X_t^{in}) - \varphi_p(X_t^i)))^2 \leq n\frac{1}{n}\sum_{i=1}^{n} D^2_{X_t^{in}X_t^i}(\varphi_p),$$

where the mapping $D_{X_t^{in}X_t^i}$ is defined in Appendix, Lemma 6.1. Since the variables (X_t^{in}, X_t^i), $i = 1, ..., n$ are exchangeable we have

$$\begin{aligned} E(\sup_{t\leq T}\sum_{p\geq 1} S_t^n(\varphi_p)^2) &\leq E(\sup_{t\leq T}\sum_{i=1}^{n}\|D_{X_t^{in}X_t^i}\|^2_{-(1+D),2D}) \\ &\leq nK\left(E(\sup_{t\leq T}|X_t^{1n} - X_t^1|^2(1+|X_t^{1n}|^{4D} + |X_t^1|^{4D}))\right) \\ &\qquad \text{(from (6.4))} \\ &\leq nK\left((E(\sup_{t\leq T}|X_t^{1n} - X_t^1|^4))^{1/2}\right. \\ &\qquad \left.(E(\sup_{t\leq T}(1+|X_t^{1n}|^{8D} + |X_t^1|^{8D})))^{1/2}\right). \end{aligned}$$

Thus, from Lemmas 5.1 and 5.2 we have

$$\sup_n E(\sup_{t\leq T}\sum_{p\geq 1} S_t^n(\varphi_p)^2) < +\infty. \tag{5.3}$$

On the other hand

$$\begin{aligned} E(\sum_{p\geq 1} T_t^n(\varphi_p)^2) &= \frac{1}{n}\sum_{p\geq 1}\sum_{i=1}^{n}\left(E(\varphi_p^2(X_t^i)) - (E(\varphi_p(X_t^i)))^2\right) \\ &\leq \frac{1}{n}\sum_{i=1}^{n}\sum_{p\geq 1} E(\varphi_p^2(X_t^i)) = \frac{1}{n}\sum_{i=1}^{n} E(\sum_{p\geq 1} D^2_{X_t^i}(\varphi_p)) \\ &\leq KE(1+|X_t^1|^{4D}) < +\infty. \end{aligned} \tag{5.4}$$

from Lemma 5.1 and (6.5), the mapping $D_{X_t^i}$ being defined in Lemma 6.1. We have used that the variables X_t^i, $i = 1, ..., n$ are i.i.d. with distribution P_t. Thus from (5.2) (5.3) (5.4) and Parseval's identity:

$$\sup_n \sup_{t\leq T} E(\|\eta_t^n\|^2_{-(1+D),2D}) = \sup_n \sup_{t\leq T} E(\sum_{p\geq 1}\langle\eta_t^n, \varphi_p\rangle^2) < +\infty.$$

□

Remark 5.4. Observe that since $\|\cdot\|_{-(2+2D),D} \leq K\|\cdot\|_{-(1+D),2D}$ (cf. Appendix), we have $\sup_n \sup_{t\leq T} E(\|\eta_t^n\|^2_{-(2+2D),D}) < +\infty$ and in particular

$$\sup_n E(\|\eta_0^n\|^2_{-(2+2D),D}) < +\infty \quad \text{under (K0)}.$$

Remark 5.5. For every fixed n, we have

$$E(\sup_{t\leq T} \| \eta_t^n \|^2_{-(1+D),2D}) < +\infty. \tag{5.5}$$

Proof. It suffices to prove that $E(\sup_{t\leq T}\sum_{p\geq 0} T_t^n(\varphi_p)^2) < \infty$. A rough bound on the above expectation and Lemmas 5.1 and 6.1 imply

$$E(\sup_{t\leq T}\sum_{p\geq 0} T_t^n(\varphi_p)^2) \leq Kn.$$

□

We will now study the semimartingale representation of η^n and properties of the operator $L(\mu_s^n)$ which appears in the drift term. Applying Itô's formula to (2.6), we obtain that η^n satisfies the following martingale property. For every $\varphi \in C_b^2(\mathbb{R}^d)$ and $t > 0$,

$$M_t^n(\varphi) = \langle \eta_t^n, \varphi\rangle - \langle \eta_0^n, \varphi\rangle - \int_0^t \langle \eta_s^n, L(\mu_s^n)\varphi\rangle ds, \tag{5.6}$$

is a continuous martingale with quadratic variation process

$$\langle M^n(\varphi)\rangle_t = \int_0^t \langle \mu_s^n, \sum_{h=1}^k \Big(\sum_{j=1}^d \frac{\partial\varphi}{\partial x_j}\sigma[\cdot,\mu_s^n]\Big)^2\rangle ds \tag{5.7}$$

where

$$L(\mu_s^n)\varphi(x) = \mathcal{L}(\mu_s^n)\varphi(x) + G(\mu_s^n)\varphi(x), \tag{5.8}$$

$$G(\mu_s^n)\varphi(x) = \langle P_s, \sum_{j=1}^d \frac{\partial\varphi}{\partial x_j} b_j(\cdot,x)\rangle + \langle P_s, \frac{1}{2}\sum_{j,l=1}^d \frac{\partial^2\varphi}{\partial x_j\partial x_l} A_{jl}(\cdot,x,\mu_s^n,P_s)\rangle$$

and

$$A_{jl}(y,x,\mu_s^n,P_s) = \sum_{h=1}^k \Big(\sigma_{jh}(y,x)\sigma_{hl}[y,\mu_s^n] + \sigma_{hl}(y,x)\sigma_{jh}[y,P_s]\Big).$$

Lemma 5.6. *Under (K0) and (K1), for every n the random operator $L(\mu_s^n)$ defined by (5.8) is a linear continuous mapping from $W_0^{2+2D,D}$ into $W_0^{1+D,2D}$ and for all $\varphi \in W_0^{2+2D,D}$, uniformly in n and in ω,*

$$\|L(\mu_s^n)\varphi\|2_{(1+D),2D} \;\leq\; K\|\varphi\|_{(2+2D),D} \tag{5.9}$$

Proof. This lemma is technical. We just study here the term $\langle P_s, \sum_{j=1}^d \frac{\partial\varphi}{\partial x_j} b_j(\cdot,x)\rangle$. For the others, the arguments are of the same nature. Using the fact that one can differentiate under the integral sign and that the derivatives of b are bounded, we have:

$$\begin{aligned}
\|\langle P_s, \sum_{j=1}^d \frac{\partial\varphi}{\partial x_j} b_j(\cdot,x)\rangle\|_{1+D,2D}^2 &= \sum_{k=0}^{1+D} \int_{\mathbb{R}^d} \frac{\left(D^k(\langle P_s, \sum_{j=1}^d \frac{\partial\varphi}{\partial x_j} b_j(\cdot,x)\rangle)\right)^2}{1+|x|^{4D}} dx \\
&\le K \int_{\mathbb{R}^d} \frac{\langle P_s, \sum_{j=1}^d \frac{\partial\varphi}{\partial x_j}\rangle^2}{1+|x|^{4D}} dx \\
&\le \int_{\mathbb{R}^d} \frac{K dx}{1+|x|^{4D}} \int_{\mathbb{R}^d} \left(\sum_{j=1}^d \frac{\partial\varphi}{\partial x_j}\right)^2 (y) P_s(dy) \\
&\le K\|\varphi\|_{2+2D,D}^2 \int_{\mathbb{R}^d} (1+|y|^{2D}) P_s(dy).
\end{aligned}$$

We can easily conclude since P_s has finite moments of order 2D (from Lemma 5.1) and $4D > d$. The above inequalities show how the choice of the weighted Sobolev spaces and Hypothesis (K1) appear. □

5.2 Tightness of the Fluctuation Process

We will assume in all what follows that conditions (K0), (K1) are fulfilled.

In this section, we prove some pathwise estimates for η^n and M^n (the martingale part of η^n), we deduce from them that the paths of η^n and M^n are almost surely strongly continuous in $W_0^{-(2+2D),D}$ and $W_0^{-(1+D),2D}$ respectively, and finally obtain the tightness of the laws of the fluctuation processes in the space $C([0,T], W_0^{-(2+2D),D})$.

Proposition 5.7. *The process* (M_t^n) *is a* $W_0^{-(1+D),2D}$*- valued martingale and it satisfies*

$$\sup_n E(\sup_{t\le T} \|M_t^n\|_{-(1+D),2D}^2) < +\infty.$$

Proof. Let $\{\varphi_p\}_{p\ge1}$ be a complete orthonormal system in $W_0^{1+D,2D}$ of functions of class C^∞ with compact support. We show the stronger property

$$\sup_n \sum_{p\ge1} E(\sup_{t\le T}(M_t^n(\varphi_p))^2) < +\infty. \tag{5.10}$$

Since σ is bounded we obtain:

$$\sum_{p\ge1} E((M_T^n(\varphi_p))^2) = \sum_{p\ge1} E\left(\int_0^T \langle \mu_s^n, \sum_{h=1}^k \left(\sum_{j=1}^d \frac{\partial\varphi_p}{\partial x_j}\sigma_{jh}[\cdot,\mu_s^n]\right)^2\rangle ds\right)$$

$$\leq K\sum_{p\geq 1}\int_0^T E\Big(\langle\mu_s^n,\Big(\sum_{j=1}^d \frac{\partial\varphi_p}{\partial x_j}\Big)^2\rangle\Big)ds$$
$$= K\sum p\geq 1\int_0^T E\Big(\frac{1}{n}\sum_{i=1}^n\Big(\sum_{j=1}^d \frac{\partial\varphi_p}{\partial x_j}(X_s^{in})\Big)^2\Big)ds$$
$$= K\int_0^T E\Big(\sum_{p\geq 1}\Big(\sum_{j=1}^d \frac{\partial\varphi_p}{\partial x_j}(X_s^{1n})\Big)^2\Big)ds$$
$$= KE\Big(\int_0^T \|H_{X_s^{1n}}\|^2_{-(1+D),2D}ds\Big),$$

where the mapping $H_{X_s^{1n}}$ is defined in Appendix, Lemma 6.1. Doob's inequality and (6.6) imply then that:

$$\sup_n\sum_{p\geq 1}E(\sup_{t\leq T}(M_t^n(\varphi_p))^2) \leq K\sup_n E(\sup_{s\leq T}(1+|X_s^{1n}|^{4D}))$$
$$< +\infty \quad \text{(by Lemma 5.1).}$$

□

Remark 5.8. Observe that since $W_0^{-(1+D),2D} \hookrightarrow W_0^{-(2+2D),D}$, M_t^n is also a martingale in $W_0^{-(2+2D),D}$, which satisfies

$$\sup_n E(\sup_{t\leq T}\|M_t^n\|^2_{-(2+2D),D}) < +\infty.$$

Proposition 5.9. *Let $\{\psi_p\}_{p\geq 1}$ be a complete orthonormal system in $W_0^{2+2D,D}$. Then,*

$$\sup_n E(\sum_{p\geq 1}\sup_{t\leq T} < \eta_t^n,\psi_p >^2) < +\infty, \tag{5.11}$$

and in particular

$$\sup_n E(\sup_{t\leq T}\|\eta_t^n\|^2_{-(2+2D),D}) < +\infty. \tag{5.12}$$

Proof. Let $\{\psi_p\}_{p\geq 1}$ be a complete orthonormal system in $W_0^{2+2D,D}$ of functions of class C^∞ on $\mathbb{R}^d$ with compact support. Then by (5.6) and Schwarz's inequality,

$$\langle\eta_t^n,\psi_p\rangle^2 \leq 4\Big(\langle\eta_0^n,\psi_p\rangle^2 + (\int_0^t\langle\eta_s^n, L(\mu_s^n)\psi_p\rangle ds)^2 + M_t^n(\psi_p)^2\Big)$$
$$\leq 4\Big(\langle\eta_0^n,\psi_p\rangle^2 + T\int_0^T\langle\eta_s^n, L(\mu_s^n)\psi_p\rangle^2 ds + M_t^n(\psi_p)^2\Big).$$

Thus, by Doob's inequality,

$$E(\sum_{p\geq 1}\sup_{t\leq T}\langle\eta_t^n,\psi_p\rangle^2)\leq 4\Big(E\|\eta_0^n\|^2_{-(2+2D),D} + TE\int_0^T\sum_{p\geq 1}\langle\eta_s^n,L(\mu_s^n)\psi_p\rangle^2 ds + 2\sum_{p\geq 1}E(M_T^n(\psi_p)^2)\Big). \quad (5.13)$$

We estimate the second right-hand term. Let us consider the linear mapping H defined from $W_0^{2+2D,D}$ into $\mathbb{R}$ by $H(\psi)=\langle\eta_s^n,L(\mu_s^n)\psi\rangle$. ¿From Lemma 5.6, we deduce

$$|\langle\eta_s^n,L(\mu_s^n)\psi\rangle|\leq K\|\eta_s^n\|_{-(1+D),2D}\|\psi\|_{2+2D,D}.$$

Thus by Parseval's identity

$$\|H\|^2_{-(2+2D),D}=\sum_{p\geq 1}\langle\eta_s^n,L(\mu_s^n)\psi_p\rangle^2\leq K^2\|\eta_s^n\|^2_{-(1+D),2D}.$$

Then

$$\begin{aligned}E(\sum_{p\geq 1}\int_0^T\langle\eta_s^n,L(\mu_s^n)\psi_p\rangle^2 ds) &\leq K^2\int_0^T E(\|\eta_s^n\|^2_{-(1+D),2D})ds\\ &\leq K^2T\sup_n\sup_{s\leq T}E(\|\eta_s^n\|^2_{-(1+D),2D})\\ &< +\infty \quad \text{by Proposition 5.3.}\end{aligned}$$

Then from (5.13), Remark 5.4 and Remark 5.8 we have (5.11). □

Proposition 5.10. *For every integer n, the processes M^n and η^n are almost surely strongly continuous respectively in $W_0^{-(1+D),2D}$ and in $W_0^{-(2+2D),D}$.*

Proof. We follow the same argument as in [9, Proposition 3.5]. Let $\{\psi_p\}_{p\geq 1}$ be a complete orthonormal system in $W_0^{2+2D,D}$, then by (5.11) we can find for every n a set Ω_0^n such that $P(\Omega_0^n)=1$ and for all $\omega\in\Omega_0^n$,

$$\sum_{p\geq 1}\sup_{t\leq T}\langle\eta_t^n(\omega),\psi_p\rangle^2<+\infty.$$

We choose M_0 such that $\sum_{p>M_0}\sup_{t\leq T}\langle\eta_t^n(\omega),\psi_p\rangle^2<\frac{\varepsilon^2}{6}$. Let $\{t_m\}_{m\geq 1}$ be a sequence in $[0,T]$ such that $t_m\to t$. Then

$$\|\eta_{t_m}^n(\omega)-\eta_t^n(\omega)\|_{-(2+2D),D}=\sum_{p\geq 1}\langle\eta_{t_m}^n(\omega)-\eta_t^n(\omega),\psi_p\rangle^2$$

$$\leq \sum_{p=1}^{M_0} \langle \eta^n_{t_m}(\omega) - \eta^n_t(\omega), \psi_p\rangle^2 + 2\sum_{p>M_0} \{\langle \eta^n_{t_m}(\omega), \psi_p\rangle^2 + \langle \eta^n_t(\omega), \psi_p^2\rangle\}$$

$$< \sum_{p=1}^{M_0} \frac{\varepsilon^2}{3M_0} + \frac{2\varepsilon^2}{6} + \frac{2\varepsilon^2}{6} = \varepsilon^2.$$

The majoration of the first term is due to the fact that for every function ψ_p, the process $< \eta^n_t, \psi_p >$ is continuous. Thus the mapping $t \mapsto \eta^n_t(\omega)$ is continuous in $W_0^{-(2+2D),D}$. The proof of continuity of M^n is similar.

To prove the tightness of the sequence of the laws of (η^n) in $\mathbb{D}([0,T], W_0^{-(2+2D),D})$, we use the Hilbert semimartingale characterization of (η^n). The properties of Hilbert-valued semimartingales we use can be found in the book of Métivier [32].

Applying Itô's formula to $\varphi(X^{in}_t)\psi(X^{jn}_t)$ we obtain that η^n is solution of the following stochastic differential equation in $W_0^{-(2+2D),D}$:

$$\eta^n_t - \eta^n_0 - \int_0^t L(\mu^n_s)^* \eta^n_s ds = M^n_t \tag{5.14}$$

where $L(\mu^n_s)^*$ is the adjoint operator of $L(\mu^n_s)$, and M^n_t is a square integrable martingale belonging to $W_0^{-(1+D),2D}$ with Doob-Meyer process $\ll M^n \gg_t$ taking values in $\mathcal{L}(W_0^{1+D,2D}, W_0^{-(1+D),2D})$ and defined by:

For every $\varphi, \psi \in W_0^{1+D,2D}$

$$\ll M^n \gg_t \cdot\varphi(\psi) = \langle M^n(\varphi), M^n(\psi)\rangle_t$$

$$= \sum_{h=1}^{k} \int_0^t \langle \mu^n_s, \left(\sum_{j=1}^{d} \frac{\partial\varphi}{\partial x_j}\sigma_{jh}[\cdot,\mu^n_s]\right)\left(\sum_{j=1}^{d} \frac{\partial\psi}{\partial x_j}\sigma_{jh}[\cdot,\mu^n_s]\right)\rangle ds. \tag{5.15}$$

Proposition 5.11. *The integral term $\int_0^t L(\mu^n_s)^*\eta^n_s ds$ in (5.14) is for every n well defined as a Bochner integral in $W_0^{-(2+2D),D}$, for all $t \in [0,T]$.*

Proof. Following Yosida [53, p.132] and since $W_0^{-(2+2D),D}$ is separable, it suffices to verify that:
1) for every $\varphi \in W_0^{2+2D,D}$, the mapping $s \to < L(\mu^n_s)^*\eta^n_s, \varphi >$ is measurable,
2) $\int_0^T \|L(\mu^n_s)^*\eta^n_s\|_{-(2+2D),D} ds < +\infty$.

The first assertion is immediate. For the second one it suffices to consider a function φ in $W_0^{2+2D,D}$ and to notice that

$$\begin{aligned} |< L(\mu^n_s)^*\eta^n_s, \varphi >| &= |< \eta^n_s, L(\mu^n_s)\varphi >| \\ &\leq \| \eta^n_s \|_{-(1+D),2D} \| \, \|L(\mu^n_s)\varphi\|_{1+D,2D} \\ &\leq K \| \eta^n_s \|_{-(1+D),2D} \| \, \|\varphi\|_{2+2D,D} \quad \text{by Lemma 5.6.} \end{aligned}$$

Then $\|L(\mu_s^n)^*\eta_s^n\|_{-(2+2D),D} \leq K \sup_{s\leq T} \| \eta_s^n \|_{-(1+D),2D}$ which is finite almost surely owing to Remark 5.5. The integral $\int_0^T \|L(\mu_s^n)^*\eta_s^n\|_{-(2+2D),D}ds$ is thus finite almost surely.

We recall a criterion of tightness for Hilbert-valued processes (cf. [22, p. 35]):

A sequence of adapted continuous processes $(Y^n)_{n\geq 1}$ on the filtered spaces (Ω^n, F_t^n, P^n), taking values in a Hilbert space H is tight in $C([0,T], H)$ if both following conditions hold:

I: There exists a Hilbert space H_0 such that $H_0 \hookrightarrow_{H.S.} H$ and such that for each $t \leq T$,

$$\sup_n E^n(\|Y_t^n\|_{H_0}^2) < +\infty.$$

II: (Aldous condition) For every $\varepsilon_1, \varepsilon_2 > 0$ there exist $\delta > 0$ and an integer n_0 such that for every (F_t^n)-stopping time $\tau_n \leq T$,

$$\sup_{n\geq n_0} \sup_{\theta\leq\delta} P^n(\|Y_{\tau_n}^n - Y_{\tau_n+\theta}^n\|_H \geq \varepsilon_1) \leq \varepsilon_2.$$

We are now in a position to prove the following results.

Theorem 5.12. *The sequence of the laws of $(M^n)_{n\geq 1}$ and $(\eta^n)_{n\geq 1}$ are tight in the space $C([0,T], W_0^{-(2+2D),D})$.*

Proof. ¿From Propositions 5.3 and 5.7, condition **I** is satisfied with $H_0 = W_0^{-(1+D),2D}$ and $H = W_0^{-(2+2D),D}$ since the embedding $W_0^{-(1+D),2D} \hookrightarrow W_0^{-(2+2D),D}$ is of Hilbert-Schmidt type (cf. Appendix).

Condition **II** will hold for (M^n) if it holds for the processes A^n, where A_t^n is the trace in $W_0^{-(2+2D),D}$ of $\ll M^n \gg_t$ (Rebolledo's Theorem, cf. [22, p.40]). If further **II** holds also for the processes $\int_0^{\cdot} L(\mu_s^n)^*\eta_s^n ds$, then it holds for (η^n) as well. The estimates previously obtained readily imply these properties. □

Remark 5.13. Proposition 5.9 implies that every limit point η of the sequence η^n satisfies

$$E(\sup_{t\leq T} \|\eta\|_{-(2+2D),D}) < +\infty.$$

5.3 Uniqueness and characterization of the fluctuation limit process

We prove in this section the uniqueness of the fluctuation limit process. In [10] we use a result of Sznitman who proves the convergence of the finite-dimensional marginals of the fluctuations allowing to conclude for the McKean-Vlasov model. We present here a new approach, inspired by a work of Mitoma [33] about such equations on nuclear spaces, in the linear case. We first obtain the limit fluctuation processes as solutions of a stochastic differential equation. A difficulty comes from the fact that as n tends to infinity, it is not possible to close the equation (5.14) at the limit, since the drift term makes sense if η^n is considered as a process taking values in $W_0^{-(1+D),2D}$ and η^n is not tight in this space. We will then consider processes η^n as taking values in $W_0^{-(4+2D),D}$, and the limit equation will be obtained in this space under more restrictive conditions on σ and b than in Section 5.2. The second (and stronger) difficulty is that the linear operator $L(P_s)$ appearing in the drift term is not bounded and it is then impossible to apply as usual Gronwall's Lemma to prove uniqueness of the solution of this equation.

Working with weighted Sobolev spaces allows us to obtain simpler and better results with weaker regularity assumptions on the coefficients than in the work of Mitoma. We express the limit values of η^n as solutions of an evolution equation in the dual space $C^{-(4+2D)}$ of $C_b^{4+2D}(\mathbb{R}^d)$. To do that, the key idea is to consider the operator $L(P_s)$ as the second order differential operator $\mathcal{L}(P_s)$ perturbed by $G(P_s)$. We use estimates given by Kunita about the flow of solutions of the stochastic differential equation associated with $\mathcal{L}(P_s)$ to show that the semigroup appearing in the evolution equation is continuous in $C^{-(6+2D)}$. We prove the uniqueness of the limit fluctuation process in the later space and since it contains the weighted Sobolev space $W_0^{-(4+2D),D}$ and then $W_0^{-(2+2D),D}$, we conclude to the convergence of the sequence η^n and to the characterization of the limit.

The proof is given in several steps. In the first step we prove the convergence of the sequence of martingales (M^n) to a Gaussian process and characterize its covariance process. Using this convergence and properties of the operators $\mathcal{L}(P_t), G(P_t)$ defined in (5.8) and considered here as operators on the space $C_b^j(\mathbb{R}^d)$ (cf. Lemma 5.17), we obtain a semimartingale representation for the accumulation points of η^n in $C^{-(4+2D)}$. We introduce the evolution operator U(t,s) generated by $\mathcal{L}(P_t)$ and define the associated stochastic flow of diffeomorphisms on $\mathbb{R}^d$. We derive Kolmorogov's forward and backward equations and deduce some estimates on $U(t,s)$. In particular, the linear operator $U(t,s)$ is continuous on $C^{-(4+2D)}$. This allows us to prove the uniqueness of the limit evolution equation by a Gronwall argument.

Theorem 5.14. *Under the hypotheses (K0) and (K1), the sequence $(M^n)_{n\geq 1}$ converges in law in $C([0,T], W_0^{-(2+2D),D})$ to a continuous Gaussian process W with covariance given by:*

$$\forall \varphi_1, \varphi_2 \in W_0^{2+2D,D}, \ \forall s,t \in [0,T],$$

$$E(W_t(\varphi_1)\cdot W_s(\varphi_2)) = \int_0^{s\wedge t} \sum_{h=1}^{k} \langle P_s, \Big(\sum_{j=1}^{d} \frac{\partial \varphi_1}{\partial x_j}\sigma_{jh}[\cdot,P_s]\Big)\Big(\sum_{j=1}^{d} \frac{\partial \varphi_2}{\partial x_j}\sigma_{jh}[\cdot,P_s]\Big)\rangle ds \quad (5.16)$$

Proof. It is easy to prove that the covariance process of M^n (cf. (5.15)) converges in probability to the function defined by (5.16) for $s=t$, by using the convergence of μ^n to P (as a consequence of the propagation of chaos). The limit points of the sequence (M^n) are thus characterized as square integrable continuous martingales with deterministic covariance process and are then uniquely characterized as the Gaussian process with covariance given in (5.16). □

Let us now give the theorem which characterizes the limit points of the sequence of fluctuation processes.

Theorem 5.15. *Assume (K0), (K1). Assume moreover that* σ and b *are in* $C_b^{2+2D}(\mathbb{R}^{2d})$. *Then if* η *is a limit point of the sequence* (η^n), *it is an Ornstein-Uhlenbeck process solution of the following stochastic differential equation in* $W_0^{-(4+2D),D}$

$$\eta_t - \eta_0 - \int_0^t L(P_s)^* \eta_s ds = W_t, \tag{5.17}$$

where W *is the Gaussian process defined in Theorem 5.14,* $L(P_s)^*$ *is the adjoint of* $L(P_s)$, *and* η_0 *is well defined.*

Remark 5.16. Remark 5.13 and the same considerations as in Proposition 5.11 imply that the integral $\int_0^t L(P_s)^*\eta_s ds$ appearing in (5.17) is well defined as a Bochner integral in $W_0^{-(4+2D),D}$.

Proof of Theorem 5.15. We consider the equation (5.14). We have explained in the introduction of this section why we are obliged to immerse the fluctuation processes η^n in $W_0^{-(4+2D),D}$, in which they are tight (there is a continuous embedding of $W_0^{-(2+2D),D}$ into $W_0^{-(4+2D),D}$).

If we assume that σ and b are in $C_b^{2+2D}(\mathbb{R}^{2d})$, we can easily prove that the operators $L(\mu_s^n)$ are continuous from $W_0^{(4+2D),D}$ into $W_0^{(2+2D),D}$, and more precisely

$$\begin{aligned}
||L(\mu_s^n)\varphi||_{2+2D,D} &\leq K||\varphi||_{4+2D,D} && (5.18)\\
||L(P_s)\varphi||_{2+2D,D} &\leq K||\varphi||_{4+2D,D} && (5.19)
\end{aligned}$$

K being uniform on n and on the randomness as in Lemma 5.6. We deduce from it that η^n is a solution of the equation (5.14) considered in $W_0^{-(4+2D),D}$. Thus, we are now in a position to prove that every limit point η is a solution of (5.17).

For this, it clearly suffices to prove that for every $\varphi \in W_0^{(4+2D),D}$, $\langle\eta_t^n,\varphi\rangle - \langle\eta_0^n,\varphi\rangle - \int_0^t\langle\eta_s^n, L(\mu_s^n)\varphi\rangle ds - M_t^n(\varphi)$ tends to $\langle\eta_t,\varphi\rangle - \langle\eta_0,\varphi\rangle - \int_0^t\langle\eta_s, L(P_s)\varphi\rangle ds - W_t(\varphi)$.

Let us remark that η_0 is well defined as limit in $W^{-(4+2D),D}$ of $\sqrt{n}(\mu_0^n - P_0)$. Indeed, we have proved the tightness of η_0^n (by projection at time 0), and we can characterize the values $<\eta_0,\varphi>$ for every function φ of $W_0^{(4+2D),D}$ by applying the central limit theorem to the sequence of real-valued random variables $< \sqrt{n}(\mu_0^n - P_0), \varphi >$.

Observe that the function $\alpha \to \langle \alpha_t, \varphi\rangle - \langle \alpha_0, \varphi\rangle - \int_0^t \langle \alpha_s, L(P_s)\varphi\rangle ds$ is continuous on $C([0,T], W_0^{-(4+2D),D})$. Then for every $\varphi \in W_0^{(4+2D),D}$, $\langle \eta_t^n, \varphi\rangle - \langle \eta_0^n, \varphi\rangle - \int_0^t \langle \eta_s^n, L(P_s)\varphi\rangle ds$ converges to $\langle \eta_t, \varphi\rangle - \langle \eta_0, \varphi\rangle - \int_0^t \langle \eta_s, L(P_s)\varphi\rangle ds$ when n tends to infinity. It remains thus to prove that $\int_0^t \langle \eta_s^n, L(\mu_s^n)(\varphi) - L(P_s)(\varphi)\rangle ds$ tends in law to zero when n tends to infinity. It suffices to prove that it tends to zero in $L^1(\Omega)$.

But $E(|\int_0^t \langle \eta_s^n, L(\mu_s^n)\varphi - L(P_s)\varphi\rangle ds|)$ is bounded by $E\left(\int_0^t |\langle \eta_s^n, (L(\mu_s^n) - L(P_s))\varphi\rangle| ds\right)$. ¿From Schwarz's inequality , we obtain:

$$\begin{aligned}
E\left(\int_0^t |\langle \eta_s^n, (L(\mu_s^n) - L(P_s))\varphi\rangle| ds\right)
&\leq E\left(\int_0^t \|\eta_s^n\|_{-(2+2D),D} \|(L(\mu_s^n) - L(P_s))\varphi\|_{2+2D,D} ds\right) \\
&\leq \int_0^t (E\|\eta_s^n\|^2_{-(2+2D),D})^{1/2} (E(\|(L(\mu_s^n) - L(P_s))\varphi\|^2_{2+2D,D}))^{1/2} ds \\
&\leq K \int_0^t (E(\|(L(\mu_s^n) - L(P_s))\varphi\|^2_{2+2D,D}))^{1/2} ds,
\end{aligned}$$

where in the last step we use Proposition 5.9.
We introduce for every $\varphi \in W_0^{4+2D,D}$ the function $R_s^n(\varphi)$ defined by

$$\begin{aligned}
R_s^n(\varphi)(x) &= (L(\mu_s^n) - L(P_s))\varphi(x) \\
&= \frac{1}{2}\sum_{j,l=1}^{d} \frac{\partial^2 \varphi}{\partial x_j \partial x_l}(x)(a_{jl}[x, \mu_s^n] - a_{jl}[x, P_s]) \\
&\quad + \sum_{j=1}^{d} \frac{\partial \varphi}{\partial x_j}(x)(b_j[x, \mu_s^n] - b_j[x, P_s]) \\
&\quad + \frac{1}{2}\langle P_s, \sum_{j,l=1}^{d} \frac{\partial^2 \varphi}{\partial x_j \partial x_l}(A_{jl}(\cdot, x, \mu_s^n, P_s) - A_{jl}(\cdot, x, P_s, P_s))\rangle
\end{aligned}$$

Then, by definition of the norm,

$$\|R_s^n(\varphi)\|^2_{2+2D,D} = \sum_{\overline{p}=0}^{2+2D} \int_{\mathbb{R}^d} \frac{|D^p(R_s^n(\varphi)(x))|^2}{1+|x|^{2D}} dx.$$

We deal only with one term; for the others the proof is similar. (We choose the last term which is more unusual). Using the fact that one can differentiate under the integral sign and that the derivatives of σ are bounded, we have:

$$\|\langle P_s, \sum_{j,l=1}^{d} \frac{\partial^2 \varphi}{\partial x_j \partial x_l}(A_{jl}(\cdot, x, \mu_s^n, P_s) - A_{jl}(\cdot, x, P_s, P_s))\rangle\|_{2+2D,D}^2$$

$$= \sum_{\bar{p}=0}^{2+2D} \int_{\mathbb{R}^d} \frac{|D^p(\langle P_s, \sum_{j,l=1}^{d} \frac{\partial^2\varphi}{\partial x_j \partial x_l}(\sum_{h=1}^{k} \sigma_{jh}(\cdot, x)(\sigma_{hl}[\cdot, \mu_s^n] - \sigma_{hl}[\cdot, P_s]))\rangle)|^2}{1+|x|^{2D}} dx$$

$$\leq K \int_{\mathbb{R}^d} \frac{\langle P_s, \sum_{j,l=1}^{d} \frac{\partial^2\varphi}{\partial x_j \partial x_l}(\sum_{h=1}^{k}(\sigma_{hl}[\cdot, \mu_s^n] - \sigma_{hl}[\cdot, P_s]))\rangle^2}{1+|x|^{2D}} dx$$

$$\leq K \int_{\mathbb{R}^d} \frac{1}{1+|x|^{2D}} dx \int_{\mathbb{R}^d} \left(\sum_{j,l=1}^{d} \frac{\partial^2 \varphi}{\partial x_j \partial x_l}(y) \sum_{h=1}^{k} (\sigma_{hl}[y, \mu_s^n] - \sigma_{hl}[y, P_s]) \right)^2 P_s(dy)$$

$$\leq K\|\varphi\|_{2+2D,D}^2 \int_{\mathbb{R}^d} (1+|y|^{2D}) \left(\sum_{l=1}^{d} \sum_{h=1}^{k} (\sigma_{hl}[y, \mu_s^n] - \sigma_{hl}[y, P_s]) \right)^2 P_s(dy).$$

Now we prove that

$$E\left(\sup_{s\leq T} \left(\sum_{l=1}^{d} \sum_{h=1}^{k} (\sigma_{hl}(y, \mu_s^n] - \sigma_{hl}[y, P_s]) \right)^2 \right) \leq \frac{K}{n}.$$

Let $(X^i), n \geq i \geq 1$ be the n independent copies of (X) defined in (5.1). We can write

$$\begin{aligned}(\sigma_{hl}[y, \mu_s^n] - \sigma_{hl}[y, P_s])^2 &\leq 2(\frac{1}{n}\sum_{i=1}^{n}(\sigma_{hl}(y, X_s^{in}) - \sigma_{hl}(y, X_s^i))^2 \\ &+ (\frac{1}{n}\sum_{i=1}^{n}(\sigma_{hl}(y, X_s^i) - \sigma_{hl}[y, P_s]))^2)\end{aligned}$$

then

$$E(\sup_{s\leq T}\left(\sum_{h=1}^{k}\sum_{l=1}^{d}(\sigma_{hl}[y, \mu_s^n] - \sigma_{hl}[y, P_s]) \right)^2)$$

$$\leq \frac{K}{n}(\sum_{i=1}^{n} E(\sup_{s\leq T}(X_s^{in} - X_s^i)^2) + \frac{K}{n}\sum_{h=1}^{k}\sum_{l=1}^{d} \mathrm{Var}(\sup_{s\leq T}\sigma_{hl}(y, X_s^1))$$

by convexity inequality and Lipschitz continuity of σ

$$\leq \frac{K_1}{n} + \frac{K_2}{n} \quad \text{owing to Lemma 5.2 and to the boundedness of } \sigma$$

Since P_s has moments of order $2D$ and by the same type of arguments for the other terms of $R_s^n(\varphi)(x)$, we obtain finally that

$$E(\sup_{s\leq T} \|R_s^n(\varphi)\|_{2+2D,D}^2)^{\frac{1}{2}} \leq \frac{K\|\varphi\|_{2+2D,D}}{\sqrt{n}}$$

and we deduce from it that $E\left(\int_0^t |\langle \eta^n_s, (L(\mu^n_s) - L(P_s))\varphi\rangle| ds \right)$ tends to zero when n tends to infinity. We have then proved that every limit value of η^n is solution of (5.17).

The next step of this work is to prove the uniqueness of solutions of (5.17). If η_1 and η_2 are two limit points of the sequence η^n in $C([0,T], W_0^{-(4+2D),D})$, the difference $\tilde{\eta} = \eta_1 - \eta_2$ is solution of

$$\tilde{\eta}_t = \int_0^t L(P_s)^* \tilde{\eta}_s ds.$$

The operator $L(P_s)$ is linear but not bounded in $W_0^{-(4+2D),D}$ and Gronwall's arguments do not work to prove that $\tilde{\eta}_t = 0, \forall t \in [0,T]$. Here the trick is to prove that $\tilde{\eta}$ is solution of an evolution equation and to prove that this equation has a unique solution in the bigger space $C^{-(6+2D)}$. The operator $L(P_s)$ is not associated with a semigroup but we consider it as a perturbation of the second order differential operator $\mathcal{L}(P_s)$ by the perturbation $G(P_s)$. Let us now study in detail the evolution system $U(t,s)$ associated with $\mathcal{L}(P_s)$.

Let B_t be a standard Brownian motion. Assume that $\sigma, b \in C_b^{j+1}(\mathbb{R}^{2d})$ for j a positive integer, then (see for example Kunita [24, p.227]) the flow $(X_{st}(x))$ defines a C_b^j-diffeomorphism, where $(X_{st}(x))$ is the unique solution of the Itô stochastic differential started from $x \in \mathbb{R}^d$:

$$X_{st}(x) \quad = \quad x + \int_s^t \sigma[X_{sr}(x), P_r] dB_r + \int_s^t b[X_{sr}(x), P_r] dr, \quad t \geq s.$$

We have moreover Itô's forward and backward formulae for the stochastic flows (cf. Kunita [24, p.256]). If $\sigma, b \in C_b^j(\mathbb{R}^{2d})$ for $j \geq 3$, for every test function $\varphi \in C_b^2(\mathbb{R}^d)$, $\forall x \in \mathbb{R}^d$,

$$\varphi(X_{st}(x)) - \varphi(x) = \int_s^t \nabla\varphi(X_{sr}(x))^* \sigma[X_{sr}(x), P_r] dB_r + \int_s^t \mathcal{L}(P_r)\varphi(X_{sr}(x)) dr,$$

$$\varphi(X_{st}(x)) - \varphi(x) = \int_s^t \nabla_x(\varphi(X_{rt}(x)))^* \sigma[x, P_r] dB_r + \int_s^t \mathcal{L}(P_r)\varphi(X_{rt}(x)) dr.$$

For any $\varphi \in C_b^2(\mathbb{R}^d)$, let $U(t,s)\varphi$ be defined by

$$(U(t,s)\varphi)(x) = E(\varphi(X_{st}(x))).$$

Then, taking expectations of both sides of Itô's forward and backward formulae and using Fubini's theorem and the fact that

$$\sup_{0 \leq s \leq t \leq S} E(| D^i X_{st}(x) |^2) < +\infty, \ \forall i \leq j \ , \ \forall x \in \mathbb{R}^d$$

(cf. Gihman-Skorohod [13, p. 61]), we obtain the backward and forward equations for every φ in $C_b^j(\mathbb{R}^d), j \geq 3, \forall x \in \mathbb{R}^d$:

$$\frac{\partial}{\partial t}(U(t,s)\varphi)(x) = (U(t,s)\mathcal{L}(P_t)\varphi)(x) \tag{5.20}$$
$$\frac{\partial}{\partial s}(U(t,s)\varphi)(x) = -\mathcal{L}(P_s)(U(t,s)\varphi)(x). \tag{5.21}$$

We will show that the two above equations (5.20) and (5.21) are satisfied in the Banach space $C_b^{4+2D}(\mathbb{R}^d)$, by giving a sense to the integrals $\int_s^t U(r,s)(\mathcal{L}(P_r)\varphi)dr$ and $\int_s^t \mathcal{L}(P_r)U(t,r)\varphi dr$ in C_b^{4+2D}.

We need estimates given in the Banach spaces $C_b^j(\mathbb{R}^d)$. The justification of the introduction of these spaces is Lemma 5.18 below, in which we prove that the semigroup appearing in the evolution equation satisfied by the limit fluctuation processes is a bounded operator on $C_b^j(\mathbb{R}^d)$. Let us remark that since $\|.\|_{C^{-(2+2D)}} \leq K\|.\|_{-(2+2D),D}$ (cf. Appendix), the sequence (η^n) is tight in $C([0,T], C^{-(2+2D)})$, and also in $C([0,T], C^{-j})$ for every $j \geq 2+2D$. We consider now the operators $\mathcal{L}(P_t), G(P_t)$ defined in (5.8), and we study their properties as operators on the functional spaces $C_b^j(\mathbb{R}^d), j \geq 0$.

Lemma 5.17. *Assume (K0), (K1) and that σ and b are in $C_b^j(\mathbb{R}^{2d})$, for a given integer $j \geq 2$. Then for every $t \geq 0$, $\mathcal{L}(P_t)$ operates from $C_b^{j+2}(\mathbb{R}^d)$ into $C_b^j(\mathbb{R}^d)$ and $G(P_t)$ operates from $C_b^j(\mathbb{R}^d)$ into itself. Moreover we have*

$$\|\mathcal{L}(P_t)\varphi\|_{C_b^j} \leq K\|\varphi\|_{C_b^{j+2}},$$
$$\|G(P_t)\varphi\|_{C_b^j} \leq K\|\varphi\|_{C_b^2} \leq K\|\varphi\|_{C_b^j}, \ \forall j \geq 2,$$

and the following continuity property:
if σ and b are in $C_b^{j+1}(\mathbb{R}^{2d})$ and φ is in $C_b^j(\mathbb{R}^d), j \geq 3$,

$$\|G(P_t)\varphi - G(P_s)\varphi\|_{C_b^j} \leq K \mid t-s \mid^{1/2} \|\varphi\|_{C_b^3} \leq K \mid t-s \mid^{1/2} \|\varphi\|_{C_b^j}.$$

The proof is easy and left to the reader.

Lemma 5.18. *Assume that σ and b belong to $C_b^j(\mathbb{R}^{2d})$, then*

$$\forall \varphi \in C_b^j(\mathbb{R}^d), \ \sup_{0\leq s\leq t\leq T} \| U(t,s)\varphi \|_{C_b^j} \leq K \| \varphi \|_{C_b^j}, \tag{5.22}$$
$$\forall \varphi \in C_b^{j+1}(\mathbb{R}^d), \| U(t,s)\varphi - U(t,r)\varphi \|_{C_b^j} \leq K \mid s-r \mid^{1/2} \| \varphi \|_{C_b^{j+1}} \tag{5.23}$$

The assertion (5.22) is fundamental since it exhibits a continuous linear operator on $C_b^j(\mathbb{R}^d), j \geq 0$.

Proof. As detailed in Appendix, we have

$$\| \varphi \|_{C_b^{2+2D}} = \sum_{k\leq 2+2D} \sup_{x\in\mathbb{R}^d} \mid D^k\varphi(x) \mid .$$

Observe that

$$\sup_x \mid E(\varphi(X_{st}(x))) \mid \leq \sup_x E(\mid \varphi(X_{st}(x)) \mid) \leq \sup_z \mid \varphi(z) \mid$$

and

$$\begin{aligned}
\sup_x |\nabla_x E(\varphi(X_{st}(x)))| &\le \sup_x \sum_{i=1}^d E(|\frac{\partial}{\partial x_i}(\varphi(X_{st}(x)))|) \\
&\le \sup_x \sum_{i=1}^d E(|\varphi'_{z_i}(X_{st}(x))\frac{\partial}{\partial x_i}(X_{st}(x))|) \\
&\le \sup_z \sum_{i=1}^d (|\varphi'_{z_i}(z)| \sup_x E(\frac{\partial}{\partial x_i}(X_{st}(x))|)) \\
&\le \sup_z |\nabla\varphi(z)|, \text{ owing to [13, Theorem 1, p.61]}.
\end{aligned}$$

For the other terms we obtain similar bounds.

The proof of the second assertion (continuity property) is similar. We just use the mean value theorem and Theorem 2.1 of Kunita [24] which asserts that

$$E(|X_{st}(x) - X_{rt}(x)|^p) \le K |s-t|^{p/2}$$

for any p greater than 2.

Thus, we deduce from Lemmas 5.17 and 5.18 that if φ belongs to $C_b^{6+2D}(\mathbb{R}^d)$, the functions $U(r,s)(\mathcal{L}(P_r)\varphi)$ and $\mathcal{L}(P_r)(U(t,r)\varphi)$ are Bochner integrable in the Banach space $C_b^{4+2D}(\mathbb{R}^d)$ and we obtain (5.20), (5.21) in $C_b^{4+2D}(\mathbb{R}^d)$:

$$\int_s^t U(r,s)(\mathcal{L}(P_r)\varphi)dr = U(t,s)\varphi - \varphi \tag{5.24}$$

$$-\int_s^t \mathcal{L}(P_r)(U(t,r)\varphi)dr = \varphi - U(t,s)\varphi. \tag{5.25}$$

Theorem 5.19. *Assume (K0), (K1) and take $\sigma, b \in C_b^{2+2D}(\mathbb{R}^{2d})$. Then the sequence (η^n) converges in law in $C([0,T], W_0^{-(2+2D),D})$ to a continuous process η which is the Ornstein-Uhlenbeck process, unique solution of the generalized Langevin equation (5.17) in $W_0^{-(4+2D),D}$.*

Proof. We only need to prove the uniqueness of the limit points. Let us consider η^1 and η^2 two limit points of η^n in $C([0,T], W_0^{-(2+2D),D})$ and then solutions of (5.17) in the space $W_0^{-(4+2D),D}$. We have

$$\tilde\eta_t = \eta_t^1 - \eta_t^2 = \int_0^t L(P_s)^* \tilde\eta_s ds,$$

and this equation makes also sense in $C^{-(4+2D)}$ since $W_0^{-(4+2D),D} \subset C^{-(4+2D)}$ (see Appendix).

We can not immediately conclude to the nullity of $\tilde\eta$ by Gronwall's inequality since the linear operator $L(P_s)^*$ is not bounded on $C_b^{4+2D}(\mathbb{R}^d)$. For $\varphi \in C_b^{6+2D}(\mathbb{R}^d)$, let us consider

$$\begin{aligned}\frac{d}{dr} < \tilde{\eta}_r, U(t,r)\varphi > &= < \frac{d}{dr}\tilde{\eta}_r, U(t,r)\varphi > + < \tilde{\eta}_r, \frac{d}{dr}U(t,r)\varphi > \\ &= < L(P_r)^*\tilde{\eta}_r, U(t,r)\varphi > - < \tilde{\eta}_r, \mathcal{L}(P_r)U(t,r)\varphi > \\ &= < \tilde{\eta}_r, G(P_r)U(t,r)\varphi > . \qquad (5.26)\end{aligned}$$

The first derivative is obtained by a differentiation theorem for the Bochner integral (cf. Yosida [53, Theorem 2, p.134]) and the second derivation is obtained owing to (5.25). The continuity of $r \to G(P_r)\varphi$ and $r \to U(t,r)\varphi$ in $C_b^{4+2D}(\mathbb{R}^d)$, and of $r \to \tilde{\eta}_r$ in $C^{-(4+2D)}$, proved in Lemmas 5.17, 5.18 and Proposition 5.10, implies that the term $< \tilde{\eta}_r, G(P_r)U(t,r)\varphi >$ is continuous in r for every φ in $C_b^{6+2D}(\mathbb{R}^d)$. We can then integrate (5.26). The equality being satisfied for every function $\varphi \in C_b^{6+2D}(\mathbb{R}^d)$, we have the equality of the operators in $C^{-(6+2D)}$:

$$\tilde{\eta}_t = \int_0^t U^*(t,r)G(P_r)^*\tilde{\eta}_r dr.$$

The integral term is well defined as a Bochner integral since:
for every function φ in $C_b^{6+2D}(\mathbb{R}^d)$, $\forall t, r \in [0,T]$,

$$\begin{aligned}&|< U^*(t,r)G(P_r)^*\tilde{\eta}_r, \varphi >| = |< \tilde{\eta}_r, G(P_r)U(t,r)\varphi > | \\ &\leq \| \tilde{\eta}_r \|_{-(2+2D),D} \| G(P_r)U(t,r)\varphi \|_{2+2D,D} \\ &\leq \| \tilde{\eta}_r \|_{-(2+2D),2D} \| G(P_r)U(t,r)\varphi \|_{C_b^{2+2D}} \quad \text{owing to (6.1)} \\ &\leq K \| \tilde{\eta}_r \|_{-(2+2D),2D} \| U(t,r)\varphi \|_{C_b^{2+2D}} \quad \text{owing to Theorem 5.17} \\ &\leq K \| \tilde{\eta}_r \|_{-(2+2D),2D} \| \varphi \|_{C_b^{2+2D}} \quad \text{owing to Lemma 5.18} \\ &\leq K \| \tilde{\eta}_r \|_{-(2+2D),2D} \| \varphi \|_{C_b^{6+2D}} .\end{aligned}$$

Remark 5.13 implies that at the limit, $\sup_{r \leq T} \| \tilde{\eta}_r \|_{-(2+2D),2D}$ is finite almost surely and $\| U^*(t,r)G(P_r)^*\tilde{\eta}_r \|_{C^{-(6+2D)}}$ is integrable. Yosida [53, Theorem 1, p.133] allows to conclude.

Since the operators $U^*(t,r)$ and $G(P_r)^*$ are linear continuous mappings from $C^{-(6+2D)}$ into $C^{-(6+2D)}$ (cf. Lemmas 5.17 and 5.18, we obtain by Gronwall's inequality that

$$\tilde{\eta}_t = 0, \quad \text{for } 0 \leq t \leq T.$$

We conclude to the uniqueness of the solutions to the generalized Langevin equation (5.17) in $C^{-(6+2D)}$ and then in $W_0^{-(2+2D),D}$. □

6. Appendix: Weighted Sobolev spaces

For every integer j, $\alpha \in \mathbb{R}_+$, let us consider the space of all real functions g defined on $\mathbb{R}^d$ with partial derivatives up to order j such that

$$\|g\|_{j,\alpha} = \left(\sum_{\bar{k} \leq j} \int_{\mathbb{R}^d} \frac{|D^k g(x)|^2}{1+|x|^{2\alpha}} dx \right)^{1/2} < +\infty$$

(where $|.|$ denotes the Euclidian norm on $\mathbb{R}^d$, and if $k = (k_1, k_2, ..., k_d)$, then $\bar{k} = \sum_{i=1}^d k_i$ and $D^k g = \frac{\partial^{\bar{k}} g}{\partial x_1^{k_1} ... \partial x_d^{k_d}}$). Let $W_0^{j,\alpha}$ be the closure of the set of functions of class C^∞ with compact support for this norm. $W_0^{j,\alpha}$ is a Hilbert space with norm $\|\cdot\|_{j,\alpha}$. We denote by $W_0^{-j,\alpha}$ its dual space.

Let $C^{j,\alpha}$ be the space of functions g with continuous partial derivatives up to order j and such that

$$\lim_{|x|\to\infty} \frac{|D^k g(x)|}{1+|x|^\alpha} = 0 \quad \text{for all } \bar{k} \leq j.$$

This space is normed with

$$\|g\|_{C^{j,\alpha}} = \sum_{\bar{k} \leq j} \sup_{x \in \mathbb{R}^d} \frac{|D^k g(x)|}{1+|x|^\alpha}$$

and $C^{j,0}$ is denoted by $C_b^j(\mathbb{R}^d)$. Let $C^{-j,\alpha}$ be the dual space of $C^{j,\alpha}$ and for $\alpha = 0$, C^{-j} the dual space of $C_b^j(\mathbb{R}^d)$.

We have the following embeddings (see Adams [1], in particular the proof of Theorem 6.53 can be adapted without difficulty for weighted Sobolev spaces):

$$W_0^{m+j,\alpha} \hookrightarrow C^{j,\alpha}, \quad m > \frac{d}{2},\ j \geq 0,\ \alpha \geq 0, \tag{6.1}$$

and

$$C_b^j(\mathbb{R}^d) \hookrightarrow W_0^{j,\alpha}, \quad \alpha > d/2, j \geq 0. \tag{6.2}$$

We have also

$$W_0^{m+j,\alpha} \hookrightarrow_{H.S.} W_0^{j,\alpha+\beta} \quad m > \frac{d}{2},\ j \geq 0,\ \alpha \geq 0,\ \beta > \frac{d}{2}, \tag{6.3}$$

where $H.S.$ means that the embedding is of Hilbert-Schmidt type. We deduce the following dual embeddings:

$$\begin{aligned} C^{-j,\alpha} &\hookrightarrow W_0^{-(m+j),\alpha}, \quad m > \frac{d}{2},\ j \geq 0,\ \alpha \geq 0, \\ W_0^{-j,\alpha} &\hookrightarrow C^{-j}, \quad \alpha > d/2,\ j \geq 0, \\ W_0^{-j,\alpha+\beta} &\hookrightarrow_{H.S.} W_0^{-(m+j),\alpha}, \quad m > \frac{d}{2},\ j \geq 0,\ \alpha \geq 0,\ \beta > \frac{d}{2}. \end{aligned}$$

In the following lemma, we consider some linear operators which appear in Section 5 and give estimates of their norms in some well-chosen weighted Sobolev spaces.

The parameter D is the first integer strictly greater than $d/2$, so $D = [d/2] + 1$. (The above inequalities justify the choice of D).

Lemma 6.1. *For every fixed $x, y \in \mathbb{R}^d$ the linear mappings D_{xy}, D_x, H_x : $W_0^{1+D,2D} \to \mathbb{R}$ defined by*

$$D_{xy}(\varphi) = \varphi(x) - \varphi(y); \quad D_x(\varphi) = \varphi(x); \quad H_x(\varphi) = \sum_{j=1}^{d} \frac{\partial \varphi}{\partial x_j}(x)$$

are continuous and

$$\|D_{xy}\|_{-(1+D),2D} \leq K_1|x-y|(1+|x|^{2D}+|y|^{2D}), \tag{6.4}$$

$$\|D_x\|_{-(1+D),2D} \leq K_2(1+|x|^{2D}), \tag{6.5}$$

$$\|H_x\|_{-(1+D),2D} \leq K_3(1+|x|^{2D}). \tag{6.6}$$

Proof. Let φ be a function of class C^∞ with compact support on $\mathbb{R}^d$, then

$$\begin{aligned} |\varphi(x) - \varphi(y)| &\leq |x-y| \sup_u |\sum_{j=1}^{d} \frac{\partial \varphi}{\partial x_j}(u)|, \quad \text{where } |u| \leq |x| + |y| \\ &\leq K|x-y|(1+|u|^{2D})\|\varphi\|_{(1+D),2D} \quad \text{from (6.1)} \\ &\leq K_1|x-y|(1+|x|^{2D}+|y|^{2D})\|\varphi\|_{(1+D),2D}. \end{aligned}$$

Then (6.4) follows from the definition of $\| \cdot \|_{-(1+D),2D}$ and by a density argument. The inequalities (6.5) and (6.6) are proved in a similar way. □

Remark 6.2. If $\{\varphi_p\}_{p\geq 1}$ is a complete orthonormal system in $W_0^{1+D,2D}$, Parseval's identity implies that for every fixed $x, y \in \mathbb{R}^d$,

$$\|D_{xy}\|^2_{-(1+D),2D} = \sum_{p\geq 1}(\varphi_p(x) - \varphi_p(y))^2 \leq K_1'|x-y|^2(1+|x|^{4D}+|y|^{4D})$$

$$\|D_x\|^2_{-(1+D),2D} = \sum_{p\geq 1}\varphi_p^2(x) \leq K_2'(1+|x|^{4D})$$

$$\|H_x\|^2_{-(1+D),2D} = \sum_{p\geq 1}\left(\sum_{j=1}^{d}\frac{\partial \varphi_p}{\partial x_j}(x)\right)^2 \leq K_3'(1+|x|^{4D}).$$

References

1. Adams, R. A., *Sobolev Spaces*, Academic Press (1978).
2. Aldous, D., "Stopping times and tightness", *Ann. Prob.* 6, 335-340 (1978).
3. Babovsky, H., Illner, R., "A convergence proof for Nanbu's simulation method for the full Boltzmann equation", *SIAM J. Numer. Anal.* 26, 45-65 (1994).
4. Billingsley, P., *Convergence of Probability Measures*, Wiley, (1969).
5. Bossy, M., *Vitesse de Convergence d'Algorithmes Particulaires Stochastiques et Application à l'Equation de Burgers*, Thèse de Doctorat de l'Université de Provence (1995).

6. Cercigniani, C., *The Boltzmann Equation and its Applications*, New York, Springer (1988).
7. Di Perna, R.J., Lions, P.L., "On the Cauchy problem for the Boltzmann equation: global existence and weak stability", *Ann. Math.* 130, 321-366 (1989).
8. Ethier, S.N., Kurtz, T.G., "Markov Processes, Characterization and Convergence", Wiley (1986).
9. Ferland, R., Fernique, X., Giroux, G., "Compactness of the fluctuations associated with some generalized nonlinear Boltzmann equations", *Can. J. Math.* 44, 1192-1205 (1992).
10. Fernandez, B., Méléard, S., "Multidimensional fluctuations on the McKean-Vlasov model: a Hilbertian approach", Prépublication 211 of the Laboratoire de Probabilités, University Paris 6 (1994).
11. Fernandez, B., Méléard, S., "Asymptotic behaviour for interacting diffusion processes with space-time random birth", Prépublication 295 du Laboratoire de Probabilités de l'Université Paris 6 (1995).
12. Gärtner, J., "On the McKean-Vlasov limit for interacting diffusions", *Math. Nachr.* 137, 197-248 (1988).
13. Gihman, I.I., Skorohod, A.V., *Stochastic Differential Equations*, Springer-Verlag (1972).
14. Graham, C., "Non linear diffusions with jumps", *Ann. I.H.P.* 28-3, 393-402 (1992).
15. Graham, C., "McKean-Vlasov Ito-Skorohod equations, and non linear diffusions with discrete jump sets", *Stoch. Proc. and Appl.* 40, 69-82 (1992).
16. Graham, C., Méléard, S. "Propagation of chaos for a fully connected loss network with alternate routing", *Stoch. Proc. and Appl.* 44, 159-180 (1993).
17. Graham, C., Méléard, S., "Chaos hypothesis for a system interacting through shared resources", *Prob. Theory and Rel. Fields* 100, 157-173 (1994).
18. Graham, C., Méléard, S., "Convergence rate on path space for particle approximations to the Boltzmann equation", Prépublication 249 du Laboratoire de Probabilités de l'Université Paris 6 (1994).
19. Hitsuda, M., Mitoma, I., "Tightness problem and stochastic evolution equations arising from fluctuation phenomena for interacting diffusions", *J. Mult. Anal.* 19, 311-328 (1986).
20. Illner, R., Neunzert, H., "On simulation methods for the Boltzmann equation", *Transport Theory Stat. Phys.* 16, 141-154 (1987).
21. Jacod, J., Shiryaev, A.N., *Limit Theorems for Stochastic Processes*, Springer (1987).
22. Joffe, A., Métivier, M., "Weak convergence of sequences of semimartingales with applications to multitype branching processes", *Adv. Appl. Prob.* 18, 20-65 (1986).
23. Kac, M., "Foundation of kinetic theory", Proc. Third Berkeley Symp. on Math. Stat. and Probab. 3, 171-197, Univ. of Calif. Press (1956).
24. Kunita, H., "Stochastic differential equations and stochastic flow of diffeomorphisms", *Ecole d'été de Probabilités de Saint-Flour XII - 1982*, *Lect. Notes in Math.* 1097, Springer (1984).
25. Léonard, C., "Une loi des grands nombres pour des systèmes de diffusions avec interaction et à coefficients non bornés", *Ann. I.H.P.* 22, 237-262 (1986).
26. Léonard, C., "Large deviations for long range interacting particle systems with jumps", *Ann. I.H.P.* 31-2, 289-323 (1995).
27. Lepeltier, J.P., Marchal, B., "Problème des martingales et équations différentielles stochastiques associées à un opérateur intégro-différentiel", *Ann. I.H.P.* 12-1, 43-100 (1976).

28. McKean, H.P., "Propagation of chaos for a class of non linear parabolic equations", in *Lecture Series in Differential Equations*, Vol. 7, 41-57 (1967).
29. Méléard, S., "Convergence of the fluctuations associated with generalized mollified Boltzmann equations", Prépublication 309 du Laboratoire de Probabilités de l'Université Paris 6 (1995).
30. Méléard, S., Roelly, S., "A propagation of chaos result for a system of particles with moderate interaction", *Stoch. Proc. and Appl.* 26, 317-332 (1987).
31. Méléard, S., Roelly, S., "Système de particules et mesures-martingales: un théorème de propagation du chaos", *Séminaire de Probabilités* 22, Lect. Notes in Math. 1321, Springer (1988).
32. Métivier, M., *Semimartingales, A Course On Stochastic Processes*, de Gruyter (1982).
33. Mitoma, I., "An ∞-dimensional inhomogeneous Langevin's equation", *J.F.A.* 61, 342-359 (1985).
34. Nanbu, K., "Interrelations between various direct simulation methods for solving the Boltzmann equation", *J. Phys. Soc. Japan* 52, 3382-3388 (1983).
35. Neunzert, H., Gropengeisser, F., Struckmeier, J., "Computational methods for the Boltzmann equation", *Applied and Indust. Maths*, (R. Spigler, ed.), Dordrecht: Kluwer Acad. Publ., 111-140 (1991).
36. Oelschläger, K., "A martingale approach to the law of large numbers for weakly interacting stochastic processes", *Annals of Prob.* 12, 458-479 (1984).
37. Oelschläger, K., "A law of large number for moderately interacting diffusion processes", *Z. Wahrsch. Verw. Geb.* 69, 279-322 (1985).
38. Perthame, B., "Introduction to the theory of random particle methods for Boltzmann equation", in *Progresses on Kinetic Theory*, World Scientific, Singapore (1994).
39. Perthame, B., Pulvirenti, M., "On some large systems of random particles which approximate scalar conservation laws", *Asympt. Anal.* (to appear).
40. Pollard, D., *Convergence of Stochastic Processes*, Springer, New York (1984).
41. Pulvirenti, M., Wagner, W., Zavelani Rossi, M.B., "Convergence of particle schemes for the Boltzmann equation", Preprint 49, Institut fur Angewandte Analysis und Stochastik, Berlin (1993).
42. Rachev, S.T., *Probability Metrics And The Stability Of Stochastic Models*, Chichester: John Wiley and Sons (1991).
43. Rebolledo, R., "La méthode des martingales appliquée à la convergence en loi des processus", *Mem. Soc. Math. France Suppl.*, 1-125 (1979).
44. Sznitman, A.S., "Equations de type Boltzmann spatialement homogènes", *Z. Wahrsch. Verw. Geb.* 66, 559-592 (1984).
45. Sznitman, A.S., "Nonlinear reflecting diffusion process, and the propagation of chaos and fluctuations associated", *J.F.A.* 56, 311-336 (1984).
46. Sznitman, A.S., "A propagation of chaos result for Burgers's equation", *Probability Theory and Rel. Fields* 71,581-613 (1986).
47. Sznitman, A.S., "Topics in propagation of chaos", *Ecole d'été de Probabilités de Saint-Flour XIX - 1989*, Lect. Notes in Math. 1464, Springer (1991).
48. Sznitman, A.S., "A fluctuation result for nonlinear diffusions", in *Infinite dimensional Analysis and Stochastic Processes*, (S. Albeverio, ed.), Pitman, 145-160 (1985).
49. Tanaka, H., "Limit theorems for certain diffusion processes with interaction", *Taniguchi Symp. on Stochastic Analysis*, Katata, 469-488 (1982).
50. Uchiyama, K., "Derivation of the Boltzmann equation from particle dynamics", *Hiroshima Math. J.* 18, 245–297 (1988).
51. Wagner, W., "A convergence proof for Bird's direct simulation method for the Boltzmann equation", *J. Stat Phys.* 66, 1011-1044 (1992).

52. Wagner, W., "A functional law of large numbers for Boltzmann type stochastic particle systems", Preprint 93, Institut fur Angewandte Analysis und Stochastik, Berlin, (1994).
53. Yosida, K., *Functional Analysis*, Fifth Edition, Springer-Verlag (1978).
54. Zolotarev, V.M., "Probability metrics", *Theory Prob. Appl.* 28-2, 278-302 (1983).

Kinetic limits for stochastic particle systems

Mario Pulvirenti

Dipartimento di Matematica, Università di Roma "La Sapienza", Roma, Italy

1. Introduction

The main objective of these lectures is to discuss some particle approximations to certain nonlinear partial differential equations of interest for the applications. The motivation is twofold: from one side these discretizations provide a background of ideas for practical numerical schemes; on the other we are led to exploit the passage from the "microscopic" to the "macroscopic" world and this is certainly interesting from the more fundamental point of view of the Mathematical Physics.

The equations which we are going to consider are of the following type.

1. *Conservation laws of mean field type.*

$$\partial_t f + div[(K * f)f] = 0 \tag{1.1}$$

where the unknown $f = f(x,t)$ is a time dependent probability density, t is the time, $x \in R^d$ and d is the dimension of the physical space. As usual we denote:

$$K * f(x) = \int_{R^d} f(y)K(x-y)dy \tag{1.2}$$

and $K : R^d \to R^d$ is a smooth vector kernel.

2. *Scalar conservation laws.*

$$\partial_t u + div F(u) = 0. \tag{1.3}$$

The unknown $u = u(x,t)$ is again a probability density and $F : R^+ \to R^d$ is a given function.

3. *Boltzmann − like equations:*

The unknown $f = f(x,v,t)$ is the probability density of finding a particle of a rarefied gas in the position x of the physical space, with velocity v. The evolution equation is of the form:

$$(\partial_t + v \cdot \partial_x) f(x,v,t) = Q(f,f)(x,v,t), \tag{1.4}$$

where the left hand side describes the free flow and the operator Q (bilinear whenever the interaction among the particles is binary) expresses the contribution to the evolution of f due to the interaction. Its form depends on the microscopic structure of the gas and we shall write it down explicitely later on for some particular cases.

For each of the three equations presented above there are various particle approximations (by which we mean a deterministic or stochastic dynamics involving a large number of particles) which should converge to the solutions of the equations when the number of particles diverges. We are interested in such convergence problems.

As we shall see these convergence problems are related to a deep property of the many particle systems which is the so called *propagation of chaos*, more or less asserting that, when the number of particles is very large, each particle behaves independently of the others. In other words, the physical states of a given pair of particles are distributed as independent random variables when the total number of particles is very large. In terms of equations, we start from a linear equation involving a large number of variables (the Liouville equation or the Master equation for stochastic system) and arrive to a nonlinear kinetic equation for a single variable.

In this analysis Eq.(1.1) is the easiest to deal with and we shall discuss it very briefly in this section.

The particle system which should approximate Eq.(1.1) will satisfy the following basic assumption:

each particle interacts with *many* others by means of a *small* individual interaction.

To be more precise, consider the N-particle system obeying the following ordinary differential equations:

$$\frac{dx_i}{dt} = \frac{1}{N}\sum_{j=1}^{N} K(x_i - x_j) \tag{1.5}$$

where $x_i \in R^d$ denotes the position of the ith-particle.

One can easily pass from a N-particle description, to a one-particle description by introducing a measure-valued function of the time t:

$$\mu^N(dx,t) = \frac{1}{N}\sum_{j=1}^{N} \delta(x - x_j(t)) \tag{1.6}$$

where $x_j(t), j = 1 \ldots N$ is the solution of Eq.(1.5) and δ denotes, as usual, the Dirac measure.The measure (1.6) is usually called empirical distribution.

Putting, for any smooth test function φ:

$$\langle \mu^N(t), \varphi \rangle = \int \mu^N(dx,t)\varphi(x), \tag{1.7}$$

we easily obtain that:

$$\frac{d}{dt}\langle \mu^N(t), \varphi \rangle = \langle \mu^N(t), K_t^N \cdot \nabla\varphi \rangle \tag{1.8}$$

where:

$$K_t^N(x) = K * \mu^N(t)(x) \equiv \int \mu^N(dy,t)K(x-y). \tag{1.9}$$

On the other hand integrating by parts Eq.(1.1) we obtain:

$$\frac{d}{dt}\langle f(t), \varphi\rangle = \langle f(t), K * f(t) \cdot \nabla\varphi\rangle \tag{1.10}$$

That is Eq.(1.10) is a weak form of Eq.(1.1). Actually Eq.(1.10) makes sense also if we replace f by a measure, and indeed Eq.(1.8) is exactly Eq.(1.10) which is nothing else than a generalization of the more familiar form (1.1). In other words the particle dynamics (1.5) induces through $\mu^N(t)$ a solution of Eq.(1.1) in a weak form.

The following natural question arises. Assume that at time zero:

$$\langle \mu^N(0), \varphi\rangle \to \langle f_0, \varphi\rangle \qquad \text{as} \quad N \to \infty \tag{1.11}$$

for any test function φ, then is it true that at time t the same convergence holds?

In other words we ask whether

$$\langle \mu^N(t), \varphi\rangle \to \langle f(t), \varphi\rangle \qquad \text{as} \quad N \to \infty \tag{1.12}$$

whenever $f = f(x,t)$ is the (smooth) solution of Eq.(1.1) with (smooth) initial datum $f_0 = f(x,0)$.

This convergence can indeed be proved and this property is nothing else that a continuity property of the solutions of Eq.(1.10) with respect to the weak convergece of the measures at time zero.

We do not give here the details of such a convergence, that is not very difficult to get in case of smoothness of K. We address the reader to the monography [15] where this convergence is discussed in the framework of the Euler equation for two-dimensional incompressible flows and vortex approximation. Rather, we shall discuss some consequences of (1.12).

Thanks to (1.12) we can say that Eq.(1.1) has been rigorously derived in the mean-field approximation, starting from the particle dynamics (1.5). In doing so we passed from a N-particle formalism (N here has to be considered as very large) to a one-particle description, but we paid a price: we replaced an ordinary differential system by a nonlinear partial differential equation.

Let us pass now to analyze another important feature of the Mean-Field limit that is the propagation of chaos. Suppose that the initial configuration of the above problem $(x_1 \dots x_N)$ is randomly distributed according to a probability measure ν^N which is a product measure, i.e.

$$\nu_0^N(x_1 \dots x_N) = f_0(x_1) \dots f_0(x_N) \tag{1.13}$$

Denote by $\nu^N(x_1 \dots x_N; t)$ the time evolved measure, defined by the identity:

$$\int \varphi(x_1 \dots x_N)\nu^N(x_1 \dots x_N; t)dx_1 \dots dx_N$$

$$= \int \varphi(x_1(t) \dots x_N(t))\nu_0^N(x_1 \dots x_N)dx_1 \dots dx_N \tag{1.14}$$

for any test function φ. Here $\{x_i(t)\}_{i=1}^N$ denotes the solution of Eq.(1.5) with initial datum $\{x_i\}_{i=1}^N$. $\nu^N(t)$ is the solution of the Liouville equation (a linear equation for which is not difficult to derive) with initial datum (1.13). Obviously at time t, $\nu^N(x_1 \dots x_N; t)$ is no longer factorized because the dynamics creates correlations and the random variables $(x_1(t) \dots x_N(t))$ are not independent anylonger even though they were at time zero. However, as a rather simple corollary of our previous discussion, it is possible to prove that the statistical independence we had at time zero, and we lost at time t, is recovered in the limit $N \to \infty$. Namely, defining the j-particle distribution functions f_j^N as:

$$f_j^N(x_1 \dots x_j; t) = \int dx_{j+1} \dots dx_N \nu^N(x_1 \dots x_N; t) \tag{1.15}$$

then:

$$\lim_{N\to\infty} f_j^N(x_1 \dots x_j; t) = f(x_1; t) \dots f(x_j; t) \tag{1.16}$$

where $f(x;t)$ is the solution of Eq.(1.1) with initial datum f_0 .The above limit must be interpreted in the sense of the weak convergence of the measures.

This important property is called propagation of chaos and can be viewed as a law of large numbers. Indeed

$$\mu^N(dx, t) = \frac{1}{N}\sum_{j=1}^N \delta(x - x_j(t)) \tag{1.17}$$

which is a measure valued stochastic process (being random the initial conditions $(x_1 J \dots x_N)$), converges to the certain (measure) variable $f(x;t)dx$ which is the limit of its expectation. In particular the limit does not fluctuate.

To prove (1.16) from the convergence (1.12), consider a test function $\Phi = \Phi(x_1 \dots x_j)$. Then the following identity holds:

$$\int f_j^N(t)\Phi dx_1 \dots dx_j = E_N \int \mu^N(dx_1, t) \dots \mu(dx_j, t)\Phi(x_1 \dots x_j) + O(\frac{1}{N}) \tag{1.18}$$

where the expectation E_N is done with respect to ν_0^N. In fact:

$$E_N \int \mu^N(dx_1, t) \dots \mu(dx_j, t)\Phi(x_1 \dots x_j)$$

$$= \int \nu_0^N(x_1 \dots x_N)\frac{1}{N^j}\sum_{i_1 \dots i_j} \Phi(x_{i_1}(t) \dots x_{i_j}(t))$$

$$= \frac{N(N-1)\dots(N-j+1)}{N^j} \int \nu_0^N(x_1 \dots x_N)\Phi(x_1(t)\dots x_j(t)) + O(\frac{1}{N})$$

$$= \frac{N(N-1)\dots(N-j+1)}{N^j} \int (f_j^N(t)\Phi)(x_1\dots x_j)dx_1\dots dx_j + O(\frac{1}{N}) \quad (1.19)$$

The third step in the above computation follows by separating in the sum $\sum_{i_1 \dots i_j}$ the contribution in which all the i_k are different and the rest, which is $O(\frac{1}{N})$. Also the symmetry of the measure has been used.

(1.16) is now a consequence of (1.18), the law of large numbers at time zero and the continuity of the dynamics with respect to the initial condition as expressed by (1.11) and (1.12). To summarize the law of large numbers at time zero plus the weak continuity with respect to the initial conditions imply the law of large numbers at time t.

How was it possible that the correlations created by the dynamics disappear in the limit? This is not difficult to understand. If we focus our attention to particle 1 and particle 2 and compare the dynamical behavior of particle 1 with that when the particle 2 is absent, we find:

$$\frac{dx_1}{dt} = \frac{1}{N}\sum_{j=1}^{N} K(x_1 - x_j) \quad (1.20)$$

and:

$$\frac{dy_1}{dt} = \frac{1}{N} \sum_{\substack{j=1 \\ j \neq 2}}^{N} K(x_1 - x_j). \quad (1.21)$$

The right hand sides of (1.20) and (1.21) are different by a factor $O(\frac{1}{N})$ so that $x_1(t) \approx y_1(t)$. The same argument holds for particle 2 with respect to particle 1 so that we expect the two particles to be independent in the limit. Actually it is so as the propagation of chaos shows.

In conclusion we can say that the mean-field limit is conceptually well understood even though some problems, essentially related to possible singularities of the kernel K, (as for instance the Vlasov-Poisson case) are still open. Technically speaking, this kind of limit is a continuity property of the dynamics with respect to the initial condition.

The analysis we have developed so far, applies to a Hamiltonian particle systems as well. The limit equation, in this case, is the Vlasov equation which is nothing else than the Newton equation for a continuum system of particles of infinitesimal charge.

We conclude this section by giving some references. In [14] McKean introduces the general ideas we have rewied in this section in a more general context, including diffusion processes. More details and references are given in the S. Meleard lecture in this volume. The Vlasov equation has been treated by Braun and Hepp [2], Dobrushin [8] and Neunzert [16]. Similar ideas can also be used for approximation problems in two-dimensional incompressible fluid dynamics. See the monograph [15] and the references quoted therein.

When the dynamics is not of mean field type all this framework becomes more involved. As we shall see, there is not a leading thecniques allowing us to treat all the cases of interest. We shall approach in the next sections Eq.(1.3) (or better one kinetic approximation of it) by a coupling method. Moreover the particle approximation for the Boltzmann-like equations of the type given by Eq.(1.4) will be analyzed by the correlation function thecnique and by the cluster expansion analysis. This three methods will constitute the main thecnical content of these lectures.

2. Scalar Conservation Laws

Eq.(1.3) can be recovered in terms of a suitable kinetic picture which we are going to illustrate. Consider the following evolution equation for a probability density $f = f(x, v, t)$:

$$(\partial_t + a(v) \cdot \partial_x) f(x, v, t) = \frac{1}{\varepsilon}(\chi_u - f)(x, v, t), \tag{2.1}$$

where $v \in R^+$ and $x \in R^d$. $u = u(x, t)$ is the spatial density associated to f:

$$u(x, t) = \int_{R^d} f(x, v, t) dv$$

and:

$$\chi_u(v) = 1_{\{v \leq u(x,t)\}}.$$

Finally $a = a(v) \in R^d$ is a C^1 given vector valued function.

Eq.(2.1) describes the evolution for the probability density of the following stochastic process $(x(t), v(t))$. $x(t)$ evolves accordingly to the free motion:

$$dx(t) = a(v)dt.$$

Moreover $v(t)$ performs jumps according to a Poisson process of intensity $\frac{1}{\varepsilon}$. The outgoing velocity after a jump is uniformly choosen in the interval $[0, u]$. Notice that the process is defined in terms of the probability distribution of the process itself. In this sense the process is said nonlinear as the evolution equation (2.1). We now pass to analyze the limit behavior when $\varepsilon \to 0$. This is a singular limit which can be handled in the following formal way. If $f_\varepsilon \to f$ (here we are expliciting the ε- dependence of the solution of (2.1)), then it must be:

$$f = \chi_u \tag{2.2}$$

On the other hand integrating Eq.(2.1) with respect to v, we obtain:

$$\partial_t u_\varepsilon(x,t) + \int dva(v) \cdot \partial_x f_\varepsilon(x,v,t) = 0 \tag{2.3}$$

Therefore:

$$\partial_t u(x,t) + \int dva(v) \cdot \partial_x \chi_{u(x,t)} = \partial_t u(x,t) + div \int_0^u dva(v) = 0. \tag{2.4}$$

Thus we expect that the limit $u = u(x,t)$ satisfies Eq.(1.3), provided that $F_i' = a_i, i = 1 \dots d$.

The limit $\varepsilon \longrightarrow 0$ we have discussed above can be proven rigorously (see [18] and [13]). In addition the solution of the scalar conservation law choosen by the kinetic equation (2.1) is the (unique) entropy solution.

The above procedure is usually called hydrodynamic limit (for the kinetic equation (2.1)) exactly as the Euler or Navier-Stokes equations, describing real fluids, are formally or, some times rigorously, derived from the Boltzmann equation in the same kind of limit. The kinetic equations are an intermediate (mezoscopic) description between the microscopic world (described by fundamental laws like the Newton equations) and the macroscopic picture given by fluid-dynamical equations like the Euler or Navier-Stokes equtions. For realistic system the kinetic picture is usually given by the Boltzmann equation. In our case (we are considering only a mathematical model) the kinetic picture is given by Eq.2.1.

The next step is to construct a particle model describing the underlying microscopic world, which also provide a reasonable scheme for Eq.(2.1) and hence Eq.(1.3). From now on we shall assume ε fixed so that we choose the unities in such a way that $\varepsilon = 1$.

We begin by giving a sense to the solutions to Eq.(2.1) and the corresponding stochastic process. This is very easy. Indeed by the obvious inequality:

$$||\chi_{u_1} - \chi_{u_2}||_{L_1} \leq ||f_1 - f_2||_{L_1} \tag{2.5}$$

where $u_i(x,t) = \int dv f_i(x,v)$ for $i = 1,2$, we get, by the usual contraction mapping principle, a unique positive solution $f \in C^1([0,T]; L_1(dx,dv))$. Once we know the solution we can also construct the stochastic proces $(x(t), v(t))$ defined above. These steps are trivial and they will be considered as achieved.

We now introduce our particle scheme. The physical domain under conideration is $[0,1]^d$. This domain will be thought covered by cubic, uniform disjoint cells $\Delta \in C$ (C is the set of all the cells) of volume $|\Delta|$ (so that $\sum_{\Delta \in C} |\Delta| = 1$). We consider a particle system whose positions are denoted by the vector $X_N = (x_1, x_2 \dots x_N) \in [0,1]^d$. Moreover we also denote by Δ_i the cell to which x_i belongs.

We now count the density number of coordinates of X_N which belong to the same cell than x_i:

$$\rho_i(X_N) = \frac{1}{N|\Delta|} \sum_{j=1}^N \chi_{\Delta_i}(x_j) \tag{2.6}$$

where χ_A is the indicator function of the set A. Finally we set $V_N = (v_1, v_2, \dots v_N)$.

The motion of the system is defined in this way. Each particle moves freely between two collisions. The free motion is given by:

$$dx_i(t) = a(v_i(t))dt. \tag{2.7}$$

The boundary conditions are choosen periodic to avoid inessential complications.

Notice that the real velocity of each particle is $a(v_i)$ while v_i is just a parameter (sometimes called velocity paramter).

Collisions occurr according N independent Poisson stopping times of intensiry one. When the particle i suffers a collision its velocity parameter v_i jumps from v_i to v_i' with a transition probability $k(v_i \to v_i')$ given by:

$$k(v_i \to v_i') = \frac{1}{\rho_i(X_N(t))}\chi_{\rho_i(X_N(t))}(v_i') \tag{2.8}$$

(remind that $\chi_b(v)$ is the indicator function of the set $[0, b]$). In other words the outgoing velocity parameters are sampled uniformly from zero to the empirical spatial density of the particles in the cell Δ_i independently of the initial velocity parameter. In this way the process $(X_N(t), V_N(t))$ is perfectly defined and, as usual, a probability density $\mu_N = \mu_N(X_N, V_N, t)$ evolves according to the formula:

$$\int \mu_N(X_N, V_N, t)\varphi(X_N, V_N)dX_N dV_N$$
$$= E[\int \mu_N^0(X_N, V_N)\varphi(X_N(t), V_N(t))dX_N dV_N] \tag{2.9}$$

where (X_N, V_N) is the initial condition for the process, φ is any test function, E denotes the expectation with respect to the underlying probabiltiy space and $\mu_N^0(X_N, V_N) = \mu_N(X_N, V_N, 0)$ is the initial distribution density which is a given positive, normalized function, symmetric in the exchange of the particles and periodic in $[0,1]^{dN}$ for any $V_N \in R_+^N$.

We expect that the particle process we have defined is well suited to approximate the solutions of Eq. (2.1) or the process $(x(t), v(t))$. Indeed $(x_1(t), v_1(t))$, the stochastic process associated to the first particle (as well as the stochastic process associated to any other particle), has the same dynamics of the limiting process, the only difference being that the transition probability on the velocity parameter is in the two cases $\rho_1(X_N(t))$ and $u(t)$ respectively. If a law of large number is going to be satisfied in the limit $N \to \infty$, we expect that such two quantities are close for small $|\Delta|$, namely that the empirical density converge to the expected density. This would also imply the convergence of the processes.

To make rigorous the above argument it is convenient to introduce the equation (sometimes called Master equation) associated to the evolution of a probability measure. By standard arguments we argue that if $\mu_N = \mu_N(x_1, v_1 \dots, x_N, v_N, t)$ is a time dpendent probability density on the states of the system, then:

$$\partial_t \mu_N + \sum_{i=1}^{N} a(v_i) \cdot \partial_{x_i} \mu_N + N\mu_N = K_N[\mu_N] \tag{2.10}$$

where:

$$K_N[\mu_N](X_N, V_N, t) = \sum_{i=1}^{N} k(v_i) \int_{R+} \mu_N(X_N, \bar{V}_N^i, t) d\bar{v}_i, \tag{2.11}$$

$$k(v_i') = \frac{1}{\rho_i(X_N)} \chi_{\rho_i(X_N)}(v_i') \tag{2.12}$$

and, for a given velocity vector $V_N = (v_1, \dots v_N)$, we put:

$$\bar{V}_N^i = (v_1 \dots, v_{i-1}, \bar{v}_i, v_{i+1}, \dots, v_N) \tag{2.13}$$

The j-particle distribution functions are defined, as usual, by:

$$f_j^N(X_j, V_j, t) = \int \mu_N(X_N, V_N, t) dx_{j+1} dv_{j+1} \dots dx_N dv_N \tag{2.14}$$

To investigate the asymptotic behavior of the particle process in the limit $N \to \infty$ it is convenient to introduce a new process, denoted by $(X_N, Y_N; V_N, W_N)$, with $2N$ particles, defined as follows. Initially $X_N = Y_N$ and $V_N = W_N$. In addition to the free stream there is a simultaneous transition $(v_i, w_i) \to (v_i', w_i')$ at random Poisson times. The outgoing velocities are choosen as close as possible in such a way that this new process is a pairing between our particle process and n independent copies of the limiting process. More precisely, defining $g_N(X_N, Y_N; V_N, W_N; t)$ a probability density, we require that g_N satisfies the following evolution equation:

$$\partial_t g_N + \sum_{i=1}^{N} (a(v_i) \cdot \partial_{x_i} g_N + a(w_i) \cdot \partial_{y_i} g_N) + N g_N$$

$$= \sum_{i=1}^{N} K_N^i(X_N, y_i; v_i, w_i; t) \int_{(R+)^2} g_N(X_N \bar{V}_N^i; Y_N, \bar{W}_N^i; t) d\bar{v}_i d\bar{w}_i \tag{2.15}$$

where:

$$K_N^i = \lambda_N^i \delta(v_i - w_i) + \frac{(k_N^i - \lambda_N^i(v_i))(h_N^i - \lambda_N^i(w_i))}{1 - \int \lambda_N^i(v) dv} \tag{2.16}$$

and

$$\lambda_N^i = min(k_N^i, h_N^i)$$

with

$$h_N^i = h_N^i(y_i, v, t) = \frac{\chi(v)_{u(y_i,t)}}{u(y_i,t)},$$

$$k_N^i = k_N^i(X_N, v) = \frac{\chi(v)_{\rho_i(X_N))}}{\rho_i(X_N)}.$$

As initial condition we choose:

$$g_N(X_N, V_N; Y_N, W_N, 0) = \prod_{j=1}^{N} \delta(v_j - w_j)\delta(x_j - y_j) f_0(x_j, v_j), \qquad (2.17)$$

where f_0 is a given probability density.

Notice that according to the comment before Eq.(2.15) the measure g_N evolves with the tendence to concentrate as much as possible on the diagonal $v_i = w_i$, which depends on how far $u(y,t)$ is from $\rho_i(X_N)$. Therefore the mass outside such a diagonal is also a measure of the distance between $\mu_N(t)$ and $\bigotimes f(x_i, v_i, t)$, where $f(x, v, t)$ is the solution of Eq.(2.1) with initial datum f_0. To be more precise we first show that g_N is a joint representation of $\mu_N(t)$ and $\bigotimes f(x_i, v_i, t)$. Indeed as consequence of the identities:

$$\int K_N^i dv_i = h^i; \qquad \int K_N^i dw_i = k_N^i$$

we get:

$$\int g_N(t) dY_N dW_N = \mu_N(t) \qquad (2.18)$$

and

$$\int g_N(t) dX_N dV_N = \prod_{j=1}^{N} f(y_i, w_i, t) \qquad (2.19)$$

To control the mass outside the diagonal we introduce the quantity:

$$Q(t) = \int g_N(t)[d(x_1, y_1) + (v_1, w_1)] dX_N dY_N dV_N dW_N \qquad (2.20)$$

where $d(x, y) = 1$ if $x \neq y$ and $d(x, y) = 0$ if $x = y$.

We readly realize that a control of Q implies a control on the L_1-norm of the differences of the $j-$ particle distributions. In facts:

$$\int |f_1^N - f|(x, v, t) dx dv = \int (f_1^N - f) sgn(f_1^N - f)(x, v, t) dx dv =$$

$$\int g_N(X_N, V_N; Y_N, W_N; t)[sgn(f_1^N - f)(x_1, v_1, t)$$

$$- sgn(f_1^N - f)(y_1, w_1, t)] dx_N dY_N dV_N dW_N \leq 2Q(t) \qquad (2.21)$$

Here we have used the obvious inequality:

$$sgn(f_1^N - f)(x,v,t) - sgn(f_1^N - f)(y,w,t) \leq 2[d(x,y) + d(v,w)].$$

By the same argument (using the symmetry of the measures) we also show that:

$$\int |f_j^N(x_1, v_1 \dots x_j, v_j, t) - \prod_{i=1}^{j} f(x_i, v_i, t)| dx_1 dv_1 J \dots dx_j dv_j \leq 2jQ(t) \tag{2.22}$$

Thus our goal is to to control $Q(t)$. By Eq.(2.15) we have (neglecting sometimes the index N for notational simplicity):

$$g_N(X,V;Y,W;t) - g_N(X - A(V)t, V; Y - A(W)t, W; 0)$$
$$= \sum_{i=1}^{N} \int_0^t ds \{ \int d\bar{v}_i d\bar{w}_i K_N^i (X - A(V)s, y_i - a(w_i)s; v_i, w_i; t-s)$$
$$g_N(X - A(V)s, \bar{V}^i; Y - A(W)s, \bar{W}^i; t-s)$$
$$g_N(X - A(V)s, V; Y - A(W)s, W; t-s)\} \tag{2.23}$$

where $A(V) = \{a(v_i)\}_{i=1}^N$. Now, integrating $d(x_1, y_1) + d(v_1, w_1)$ with respect to $g_N(X,V;Y,W;t)dXdVdYdW$, we observe that all terms in the above sum, with $i \neq 1$, vanish. In facts by the identity:

$$\int dv dw K_N^i(X, y_i; v, w; t-s) = 1, \tag{2.24}$$

we have that $(i \neq 1)$:

$$\int dXdYdVdWd\bar{v}_i d\bar{w}_i K_N^i(X - A(V)s, y_i - a(w_i)s, v_i, w_i; t-s)$$
$$[d(x_1, y_1) + d(v_1, w_1)] g_N(X - A(V)s, \bar{V}^i; Y - A(W)s, \bar{W}^i; t-s)$$
$$= \int dXdYdVdWd\bar{v}_i d\bar{w}_i K_N^i(X, y_i; v_i, w_i; t-s)$$
$$[d(x_1 + a(v_1)s, y_1 + a(w_1)s) + d(v_1, w_1)] g_N(X, \bar{V}^i; Y, \bar{W}^i; t-s)$$
$$= \int dXdYdVdW [d(x_1 + a(v_1)s, y_1 + a(w_1)s) + d(v_1, w_1)]$$
$$g_N(X,V;Y,W;t-s). \tag{2.25}$$

The last term is exactly what we get from the loss term by the same changment of variable. So we are reduced to estimate the term $i = 1$ which is, after a changment of variables and putting $g_N(X,V;Y,W;t)dXdVdYdW = dg_N(t)$:

$$\int dg_N(t-s) \{ \int d\bar{v}_1 d\bar{w}_1 K_N^i(X, y_i, \bar{v}_i, \bar{w}_i; t-s)$$
$$d(x_1 + a(\bar{v}_1)s, y_1 + a(\bar{w}_1)s) + d(\bar{v}_1, \bar{w}_1) - d(x_1 + a(v_1)s, y_1 + a(w_1)s) - d(v_1, w_1)\}$$

$$\leq \int dg_N(t-s)\{\int d\bar{v}_1 d\bar{w}_1 K_N^i(X, y_i, \bar{v}_i, \bar{w}_i; t-s) d(x_1, y_1) + 2d(\bar{v}_1, \bar{w}_1) - d(x_1, y_1)\}$$

$$= 2\int dg_N(t-s) d\bar{v}_1 d\bar{w}_1 K_N^i(X, y_i, \bar{v}_i, \bar{w}_i; t-s) d(\bar{v}_1, \bar{w}_1) \qquad (2.26)$$

Again here we used Eq.(2.24).

By the choice of the initial condition for g_N, yielding

$$\int g_N(X - A(V)t, V; Y - A(W)t, W; 0)(d(x_1, y_1) + d(v_1, w_1)) = 0 \qquad (2.27)$$

we arrive to:

$$Q(t) \leq \int_0^t ds \int dg_N(s) d\bar{v}_1 d\bar{w}_1 K_N^i(X, y_i, \bar{v}_i, \bar{w}_i; s) d(\bar{v}_1, \bar{w}_1) \qquad (2.28)$$

On the other hand:

$$\int dg_N(s) d\bar{v}_1 d\bar{w}_1 K_N^i(X, y_i, \bar{v}_i, \bar{w}_i; s) d(\bar{v}_1, \bar{w}_1)$$

$$= \frac{1}{1 - \int dv \lambda_N^1(v)} \int (k_N^1 - \lambda_N^1(\bar{v}_1)(h^1 - \lambda_N^1(\bar{w}_1)) d\bar{v}_1 d\bar{w}_1$$

$$= 1 - \int \lambda_N^1(v) dv$$

$$= \frac{|u(y_1, s) - \rho_1(X)|}{Max(u(y_1, s), \rho_1(X))} \qquad (2.29)$$

Here we used the identiy:

$$1 - \int Min(\frac{\chi_a(v)}{a}, \frac{\chi_b(v)}{b}) dv == \frac{|b-a|}{Max(a,b)}$$

which is easy to check. In conclusion:

$$Q(t) \leq 2\int_0^t ds \int dg_N(s) \frac{|u(y_1, s) - \rho_1(Y)|}{u(y_1, s)}$$

$$- \frac{|\rho_1(Y) - \rho_1(X)|}{Max(\rho_1(Y), \rho_1(X)} \qquad (2.30)$$

Observe now that the first term in the right hand side of Eq. (2.30) involves only the limit process. This term is expected to be small (for the law of large numbers for independent random variables), whenever the solution $u(y,t)$ is sufficiently smooth, bounded away from zero, N is large and $|\Delta|$ is small. Indeed it can be proved, under suitable assumptions on the initial datum (see Theorem 2.1 below) that such a term can be bounded by:

$$Const(\frac{1}{N}|\Delta| + |\Delta|^{1/d})^{1/2}$$

Moreover the second term in the right hand side of Eq.(2.30) can be bounded in terms of $Q(s)$. This is not surprising since such a term gives a measure of the distance between the two processes. To prove this we consider separately the cases when x_1 and y_1 do, or do not belong to the same cell and define:

$$n_\Delta(X) = \sum_{j=1}^{N} \chi_\Delta(x_j)$$

and:

$$\rho_\Delta(X) = \frac{n_\Delta(X)}{N|\Delta|}.$$

Then the contribution due to the event when x_1 and y_1 are in different cells is estimated by:

$$\sum_\Delta \int_{y_1 \notin \Delta} \chi_\Delta(x_1) \frac{|\rho_1(Y) - \rho_\Delta(X)|}{Max(\rho_1(Y), \rho_1(X))} dg_N(s) \leq$$

$$\sum_\Delta \int \chi_\Delta(x_1) d(x_1, y_1) dg_N(t) \leq Q(t) \tag{2.31}$$

The other contribution is bounded by:

$$\begin{aligned}
&\sum_\Delta \int \chi_\Delta(x_1) dg_N(s) \frac{|\rho_\Delta(Y) - \rho_\Delta(X)|}{\rho_\Delta(X)} \\
&= \frac{1}{N} \sum_\Delta \sum_{j=1}^{N} \int \frac{|n_\Delta(Y) - n_\Delta(X)|}{n_\Delta(X)} \chi_\Delta(x_j) dg_N(s) \\
&= \frac{1}{N} \sum_\Delta \int |n_\Delta(Y) - n_\Delta(X)| dg_N(s) \\
&= \frac{1}{N} \sum_\Delta \int |\sum_{k=1}^{N} (\chi_\Delta(x_k) - \chi_\Delta(y_k)| dg_N(s) \\
&\leq \frac{1}{N} \sum_{k=1}^{N} \sum_\Delta \int |\chi_\Delta(x_k) - \chi_\Delta(y_k)| dg_N(s) \\
&\leq \frac{1}{N} \sum_{k=1}^{N} \sum_\Delta \int d(x_k, y_k)(\chi_\Delta(x_k) + \chi_\Delta(y_k)) dg_N(s) \\
&\leq \frac{2}{N} \sum_{k=1}^{N} \int d(x_k, y_k) dg_N(s) \\
&= 2Q(s)
\end{aligned} \tag{2.32}$$

Combining (2.32) and (2.31) we arrive finally to:

$$Q(t) \leq 6 \int_0^t Q(s) + C(\frac{1}{N}|\Delta| + |\Delta|^{1/d})^{1/2} \tag{2.33}$$

which allows us, via the Gronwall lemma, to prove the following theorem.

Theorem 2.1. *Assume that for $\delta > 0, K > 0$ and $K_1 > 0$:*

$$\chi_K(v) \geq f_0(x,v) \geq \chi_\delta(v),$$

$$\int dvdx\, f_0(x,v) = 1,$$

and

$$\int \sum_{k=1}^{d} |\partial_{x_k} f_0(x,v)| \leq K_1.$$

Consider the initial distribution for the particle system given by:

$$\mu_N^0(X_N, V_N) = \prod_{i=1}^{N} f_0(x_i, v_i)$$

and the time evolved j-particle distribution functions $f_j^N(X_j, V_j, t)$. Denote by $f = f(x,v,t)$ the solution of Eq.(2.1) with initial datum f_0. Then:

$$\int |f_j^N(t) - \bigotimes_{i=1}^{j} f(t)| dx_1 dv_1 \ldots dx_j dv_j \leq Cj(\frac{1}{N}|\Delta| + |\Delta|^{1/d})^{1/2} e^{6t}. \tag{2.34}$$

In particular if $|\Delta| \to 0$ as N^{-1+a} with $1 > a > 0$ we have by (2.33) convergence and propagation of chaos.

Some comment are in order. We have showed how to control the dynamics of our particle system in terms of a deviation between the spatial density and the empirical density obtained by many identical copies of the limiting process, smeared on the scale of Δ. As we have shown this term is small provided that $|\Delta|$ goes to zero gently with respect to N^{-1}. This is essential because we need a sufficiently large number of particles for any cell in order that the law of large number could work.

The results of this section are contained in a joint work of B. Perthame and the author [17].

3. Kinetic Limits

Consider a system of hard spheres, of diameter d in the whole space (or in a bounded domain $\Lambda \subset R^3$) with the usual collision law:

$$v_1' = v_1 - n[n \cdot (v_1 - v_2)]$$
$$v_2' = v_2 + n[n \cdot (v_1 - v_2)].$$

v_1' and v_2' denote the outgoing velocities after a collision between two particles with ingoing velocities given by v_1 and v_2 and $n = \frac{x_1 - x_2}{|x_1 - x_2|}$ is the unit vector directed along the line joining the centers of the two spheres (x_1 and x_2) at the moment of the impact. Then, if we denote by $f(x, v, t)$ the probability density of finding a particle centered in x with velocity v, we find (following the usual arguments given in the textbooks) that:

$$(\partial_t + v \cdot \partial_x) f(x, v, t) =$$
$$Nd^2 \int dv_1 \int_{S+} dn\, n \cdot (v - v_1)[f(x, v')f(x, v_1') - f(x, v)f(x, v_1)] \tag{3.1}$$

where N is the total number of particles, n is the unit vector and S^+ denotes the half unit sphere $\{n | n \cdot (v - v_1) \leq 0\}$.

Eq. (3.1) is the famous Boltzmann equation. In the heuristic derivation of Eq.(3.1) an important hypothesis has been done: it is *assumed* that the two-particle distribution function of the system factorizes into the product of the two one-particle distributions. In other words some statistical independence has been achieved by the system.

The fact that Eq.(3.1) presents some foundational problems, was clear immediately after the equation was written by L.Boltzmann in 1872, because of the well known paradoxes arising by assuming its complete equivalence with the Newton laws. It is also true that Boltzmann himself had clear in mind how his equation could only be understood in a statistical sense. More recently H.Grad clarified that Eq.(3.1) must be interpreted as a description of an asymptotic situation, which is exactly that of a very dilute gas, namely when $N \to \infty$, $d \to 0$ keeping the product $Nd^2 = \lambda$ constant (see [10]). This limit, which is called the Boltzmann-Grad limit for obvious reasons, is really a low density limit. In contrast with the hydrodinamical limit in which each tagged particle has, in the limit, an infinite number of collisions per unit time, here the average number of collisions suffered by any particle is controlled by the parameter λ. In other words, even if the total number of particles is diverging, the diameter d of the particles is suitably rescaled in such a way that the expected number of collisions of each particle is bounded.

A rigorous derivation of the Boltzmann equation for hard-spheres (that is a mathematical control of the Boltzmann-Grad limit) was given, for short times, by O. Lanford (see [12]).

The problem of a rigorous derivation, global in time, of the Boltzmann equation from a system of hard spheres is a challenging, hard problem which we shall not discuss in this lecture. We address the reader to Chapt 4 of the monography [5] for a more detailed analysis of the problem. In this section we shall deal with a much simpler problem, that with the convergence of certain stochastic particle schemes (essentially those really used in the applications) to the solutions of eq.(3.1). As a first step we shall not consider eq.(3.1) but a regularized one, whose solutions would converge to those of Eq.(3.1), whenever such solutions exist and are sufficiently regular. The problem concerning the existence and regularity for large data of the Boltzmann equation, is still poorely understood, although recently important steps forward have been achieved by Di Perna and Lions [9].

The regularization we consider is done in the following way. Consider again a grid in our domain assumed to be $[0,1]^d$ with periodic boundary conditions. This domain will be thought covered by cubic, uniform disjoint cells $\Delta \in C$ (C is the set of all the cells) of volume $|\Delta|$ (so that $\sum_{\Delta \in C} |\Delta| = 1$).

Consider the equation:

$$(\partial_t + v \cdot \partial_x) f(x,v,t) = Q_\delta(f,f)(x,v,t) \tag{3.2}$$

with the collision operator given by the following expression:

$$Q_\delta(f,f)(x,v,t) = \int dy \int dv_1 \int_{S^{d-1}} dn h_\delta(x,y) q_\delta(v,v_1,n)$$
$$[f(x,v')f(y,v_1') - f(x,v)f(y,v_1)] \tag{3.3}$$

where (for $\delta > 0$):

$$h_\delta(x,y) = \sum_\Delta \chi(x \in \Delta)\chi(y \in \Delta)\frac{1}{\delta^d} \tag{3.4}$$

and

$$q_\delta(v,v_1,n) = n \cdot (v - v_1)\chi(|v_1 - v| \leq \frac{1}{\delta})\chi(n \cdot (v - v_1) \geq 0) \tag{3.5}$$

Notice that $\int dx h_\delta(x,y) = \int dy h_\delta(x,y) = 1$ so that $h_\delta(x,y)$ is an approximation of the delta-function.

The physical meaning of Eq.(3.2) is that we are delocalizing the interaction between two particles and also that we are neglecting collisions with relative velocity larger than $\frac{1}{\delta}$. We shall discuss later on the physical meaning of Eq.(3.2) in some detail. For the moment we observe that Eq.s (3.1) and (3.2) are the same in the limit $\delta \to 0$. On the other hand Eq.(3.2) is much easier to handle. Indeed, introducing the symmetrized collision operator:

$$Q_\delta(f,g)(x,v,t) = \frac{1}{2} \int dv_1 \int_{S^{d-1}} dn h_\delta(x,y) q_\delta(v,v_1,n)$$

$$[f(x,v')g(y,v_1')+g(x,v')f(y,v_1')-f(x,v)g(y,v_1)-g(x,v)f(y,v_1)], \quad (3.6)$$

we have the following Lipschitz estimate:

$$\begin{aligned}\|Q_\delta(f,f)-Q_\delta(g,g)\|_{L_1} &= \|Q_\delta(f+g,f-g)\|_{L_1} \\ &\le C(\delta)\|f+g\|_{L_1}\|f-g\|_{L_1} \end{aligned} \quad (3.7)$$

By (3.7) it is not hard to obtain a unique positive solution, $C^1([0,T];L_1(x,v))$ to Eq.(3.2). We do not insist further on this point (which will be incidentally showed later on), but we start to introduce the approximating particle system.

We denote by $Z_N=(X_N,V_N)=(x_1,\dots x_N,v_1,\dots v_N)$ the set of positions and velocities of N particles. They are moving freely $dx_i=v_idt$ and perform binary collisions at random times described by $\frac{N(N-1)}{2}$ independent Poisson processes (how many are the pairs) depending on the state of the system. According to the collision rules of the hard spheres, the outgoing velocities are given by the incoming velocities v, v_1 and an impact parameter n which is random and uniformly choosen in the half–sphere $n\cdot(v-v_1)\ge 0$.

We do not specify further the process, but introduce its generator (more precisely its adjoint). Define:

$$A_N\mu_N(Z_N)=\frac{1}{N}\sum_{i<j}\int_{S^{d-1}} dn h_\delta(x_i,x_j)q_\delta(v_i,v_j,n)$$

$$[\mu_N(Z_N'(i,j))-\mu_N(Z_N)] \quad (3.8)$$

where

$$Z_N'(i,j)=(x_1,v_1\dots x_i,v_i'\dots x_j,v_j'\dots x_N,v_N)$$

if

$$Z_N=x_1,v_1\dots x_i,v_i\dots x_j,v_j\dots x_N,v_N).$$

In other words $Z_N'(i,j)$ is the state just after a collision between the particle i and j, with impact parameter n.

According to the form of the generator, we have cutoffed the interaction: two particle collide if and only if they are in the same cell. Moreover if their relative velocity is too large, they go ahead.

It is possible to characterize the above process in terms of Poisson processes of constant intensity by introducing the "ficticious collisions". This is not important as regards the convergence problem, which will be treated only at level of distribution functions, however this allows to describe precisely the process so that we spend some time in doing this.

Define:

$$\tilde\alpha(Z_N,i,j,n)=h_\delta(x_i,x_j)q_\delta(v_i,v_j,n),$$

$$\bar\alpha(n)=\sup_{Z_N,i,j}\tilde\alpha(Z_N,i,j,n),$$

$$\alpha=\int dn\bar\alpha(n).$$

Now consider a Poisson process of intensity $\frac{\alpha(N-1)}{2}$. For each exponential time choose a pair of particles , say i and j, a random parameter η uniformly distributed in $[0,1]$ and an impact parameter n on the half–sphere. Then if $\eta < \frac{\tilde{\alpha}(Z_N,i,j,n)}{\bar{\alpha}(n)}$ preform the collision, otherwise not. In this last case the collision is said ficticious. Notice that, by the above definitions, a collision between two particles can occurr if and only if they are in the same cell with moderate relative velocity.

Denoting by $\mu_N(Z_N,t) = \mu_N(x_1, v_1 \dots, x_N, v_N, t)$ the evolution of a probability density according to the above process, we have the following equation:

$$(\partial_t + \sum_{i=1}^{N} v_i \cdot \partial_{x_i})\mu_N = \bar{A}_N \mu_N \tag{3.9}$$

where:

$$\bar{A}_N\mu_N(Z_N) = \frac{1}{N}\sum_{i<j}\int_{S^{d-1}} dn \int_0^1 d\eta\bar{\alpha}(n)$$
$$[\mu_N(\bar{Z}'_N(i,j)) - \mu_N(Z_N)] \tag{3.10}$$

and where

$$\bar{Z}'_N(i,j)) = Z'_N(i,j) \qquad if \qquad \eta < \frac{\tilde{\alpha}(Z_N,i,j,n)}{\bar{\alpha}(n)}$$

and

$$\bar{Z}'_N(i,j)) = Z_N$$

otherwise.

It is just matter of easy computations to show that $\bar{A}_N = A_N$. For the sequel we choose the form (3.8) for the generator A_N only for technical reasons.

By the simple estimate:

$$||A_N\mu_N||_{L_1} \leq NC(\delta)||\mu_N||_{L_1} \tag{3.11}$$

we construct a strongly continuous (in L_1) Markov semigroup $P^t\mu_N^0$ which solves Eq.(3.10) with initial datum μ_N^0. This is the solution of the following integral equation:

$$P^t\mu_N^0(t) = S(t)\mu_N^0 + \int_0^t dsS(t-s)A_N P^s\mu_N^0 \tag{3.12}$$

where $(S(t)\mu_N(x_1, v_1 \dots x_N, v_N) = \mu_N)(x_1 - v_1t, v_1 \dots x_N - v_Nt, v_N)$ is the free-stream operator.

Obviously, Eq.(3.11) can be solved by a series expansion as it is usual in the perturbation theory of linear semigroups.

Introducing the j–particle distribution function f_j^N (see Eq. (2.14)) we obtain from Eq. (3.9), the following hierarchy of equation:

$$(\partial_t + \sum_{i=1}^{j} v_i \cdot \partial_{x_i}) f_j^N = \frac{j}{N} A_j f_j^N + \frac{N-j}{N} C_{j,j+1} f_{j+1}^N \tag{3.13}$$

where

$$\begin{aligned} &C_{j,j+1} f_{j+1}(x_1, v_1 \dots x_j v_j) \\ &= \sum_{i=1}^{j} \int dn \int dv_{j+1} \int dx_{j+1} h_\delta(x_i, x_{j+1}) q_\delta(v_i, v_{j+1}, n) \\ &\qquad [f_{j+1}^N(x_1, v_1 \dots x_i, v_i', \dots x_{j+1} v_{j+1}') - f_{j+1}^N(x_1, v_1 \dots x_{j+1} v_{j+1})] \end{aligned}$$

Eq.(3.13) has been formally obtained from Eq.(3.9) by integrating both sides over all the variables $(x_{j+1}, v_{j+1} \dots x_N, v_N)$ and using the symmetry. Notice that the last equation of the hierarchy (3.13) is just Eq.(3.9).

The reason why we consider such a hierarchy is that we are interested in the asymptotic behavior (as N diverges) of f_j^N for any fixed j. From Eq.(3.13) we argue that the time variation of f_j^N is due to the interaction among the group of the first j–particles (which is negligible because of the factor $\frac{j}{N}$ in front to A_j) plus the interaction with the remaining particles, which is the relevant part in the limit.

Consider now a solution to the Cauchy problem associated to Eq.(3.2) with an initial datum $f_0 \in L_1(x, v)$. Denote it by $f = f(x, v, t)$ and consider also the tensor products:

$$f_j(x_1, v_1 \dots x_j, v_j, t) = \prod_{i=1}^{j} f(x_i, v_i, t) \tag{3.14}$$

Then a simple algebraic computation shows that

$$(\partial_t + \sum_{i=1}^{j} v_i \cdot \partial_{x_i}) f_j = C_{j,j+1} f_{j+1} \tag{3.15}$$

Notice that the hierarchy (3.15) is unbounded, in the sense that it is not truncated at some j as for the hierarchy (3.12).

We are now in position to prove the main result of this section.

Theorem 3.1. *Consider an initial probability measure*

$$\mu_N^0 = \prod_{i=1}^{j} f_0(x_N, v_N)$$

and its time evolved $\mu_N(t)$ according to Eq.(3.9). Denote by $f_j^N(t)$ the j–particle distribution functions. Then:

$$\lim_{N\to\infty} \|f_j^N(t) - f_j(t)\|_{L_1(x,v)} = 0 \tag{3.16}$$

for all positive t.

Theorem 3.1 shows that the particle system converge to the unique solution of Eq.(3.2) in the sense of the L_1 convergence of the j–particle distribution functions. Moreover the propagation of chaos is also showed. Indeed by (3.14) and (3.16) the correlations created by the particle dynamics are going to disappear in the limit.

Proof. The proof of Theorem 3.1 is simple and direct. Indeed we express f_j^N and f_j in terms of series expansions:

$$f_j^N(t) = \sum_{n\geq 0} \frac{(N-j)\dots(N-j+n)}{N^n} \int_0^t dt_1 \int_0^{t_1} dt_2 \dots \int_0^{t_{n-1}} dt_n$$

$$P_j^N(t-t_1)C_{j,j+1}\dots P_{j+n-1}^N(t_{n-1}-t_n)C_{j+n-1,j+n}P_{j+n}^N(t_n)f_0^{\bigotimes n+j} \tag{3.17}$$

where $P_j^N(t)$ is the Markov semigroup generated by the free stream and $\frac{j}{N}A_j$.

Analogously:

$$f_j(t) = \sum_{n\geq 0} \int_0^t dt_1 \int_0^{t_1} dt_2 \dots \int_0^{t}{}_{n-1} dt_n$$

$$S(t-t_1)C_{j,j+1}\dots S(t_{n-1}-t_n)C_{j+n-1,j+n}S(t_n)f_0^{\bigotimes n+j} \tag{3.18}$$

We now show that the two series (3.17) and (3.18) are absolutely convergent in L_1 for sufficiently small time t. Indeed by the estimate:

$$\|C_{j,j+1}f_{j+1}\|_{L_1} \leq jC(\delta)\|f_{j+1}\|_{L_1}, \tag{3.19}$$

where we have used the fact that the Jacobian of the tranformation $(v_i, v_j) \to (v_i', v_j')$ is the unity, for any value of the impact parameter n, and the obvious relations:

$$\|P^N(t)_j f_j\|_{L_1} = \|f_j\|_{L_1}, \qquad \|S(t)f_j\|_{L_1} = \|f_j\|_{L_1} \tag{3.20}$$

we have that the series of the L_1 norms relative to (3.18) and (3.19) are both bounded by:

$$\sum_{n\geq 0} j(j+1)\dots(j+n-1)C(\delta)^n \int_0^t dt_1 \int_0^{t_1} dt_2 \dots \int_0^{t_{n-1}} dt_n (\|f_0\|_{L_1})^n$$

$$\leq \sum_{n\geq 0} \frac{j(j+1)\dots(j+n-1)}{n!}[tC(\delta)]^n$$

$$\leq \sum_{n\geq 0} 2^j[tC(\delta)]^n \tag{3.21}$$

On the other hand it is easy to show that:

$$P_j^N(t-t_1)C_{j,j+1}\dots P_{j+n-1}^N(t_{n-1}-t_n)C_{j+n-1,j+n}P_{j+n}^N(t_n)f_0^{\bigotimes n+j} \rightarrow$$

$$S(t-t_1)C_{j,j+1}\dots S(t_{n-1}-t_n)C_{j+n-1,j+n}S(t_n)f_0^{\bigotimes n+j} \qquad (3.22)$$

as $N \to \infty$ in the L_1 sense for any fixed j. This follows by:

$$\|[P_j^N(t)-S(t)]f_j\|_{L_1} \to 0 \qquad (3.23)$$

as $N \to \infty$, for any fixed j.

Since the series (3.18) converges to the series (3.19) term by term and both are unformly bounded by a series with positive terms independent of N, we conclude, by the dominated convergence theorem, that the series (3.18) converge to the series (3.19) for a small time t_0. The same result could be obtained as well if the initial distribution $\mu_N = \prod_{i=1}^N f_0(x_i, v_i)$ would be replaced by ν_N, whose j-particle distribution functions $g_j(x_1, v_1 \dots x_j, v_j)$ would coverge to $\prod_{i=1}^N f_0(x_i, v_i)$ in the L_1 sense. This allows us to extend the convergence from $[0, t_0]$ to $[0, 2t_0]$ and so on. Therefore the proof of Theorem 3.1 is complete. □

By Theorem 3.1 we can get informations on the particle process. In particular, by the arguments given in Section 1, we can conclude that the empirical distribution of the particle process, namely $\frac{1}{N}\sum_{i=1}^N \delta(x - x_i(t))\delta(v - v_i(t))$, which is a measure valued random process (here the initial data (x_i, v_i) are supposed distributed according to $\prod_{i=1}^N f_0(x_i, v_i)$), is weakly convergent, in the sense of the probability measures, to the measure $f(x, v, t)$ solution of Eq.(3.2) with initial datum f_0.

The method illustrated above, inspired to the classical papers of Kac [11] and Lanford [12], can be used to obtain an explicit rate of convergence for the time discretization of the particle process (see [19]), that is a numerical scheme really used in practice for the numerical simulation of rarefied flows (see, for instance, [5, Chapter 10]).

Notice that the estimates presented in the present Section are dependent on the parameter δ. The cutoff on the velocities is somehaw innncent: it could be removed with some technical effort. On the contrary the spatial cutoff is very delicate to remove. In the next section we shall study a one–dimensional model with a stricly local interction for which a different and more sofisticated technique is necessary.

4. Cluster Expansion

In this section we shall consider a one-dimensional particle system without any spatial cutoff and the corresponding one–dimensional Boltzmann equation. For such a system we shall prove convergence and propagation of chaos.

Let us consider a system of N identical particles in the line. Denote by $Z_N = (X_N, V_N) = \{x_1v_1, \dots, x_Nv_N\}$ a state of the system, where x_i and v_i are the position and the velocity of the i-th particle, $x_i \in R$ and $v_i \in [-1, 1]$. A collision between two particles is defined in the following way. Denoting by v_1, v_2 and by v_1', v_2' the ingoing and the outgoing velocities respectively, then:

$$v_1' = -v_2, \quad v_2' = -v_1 \quad if \quad sgn(v_1) \neq sgn(v_2) \tag{4.1-1}$$

$$v_1' = v_1, \quad v_2' = v_2 \quad if \quad sgn(v_1) = sgn(v_2) \tag{4.1-2}$$

Notice that, due to the fact that the particles are identical, the above collision rule is equivalent to a specular reflection of each colliding particle. However we prefer a labelling for which the direction of the velocity does not change in the collision because in this way a given pair of colliding particles cannot interact anymore in the future. Moreover, according to (4.1), collisions between particles travelling in the same direction are not considered.

Notice also that the relative velocity is preserved during the collision:

$$|v_i - v_j| = |v_i' - v_j'|. \tag{4.2}$$

The dynamics of the system is assumed stochastic. The particles move freely up to the first time in which two of them arrive at the same point. Then they collide (independently) with probability ε and go ahead with probability $1 - \varepsilon$.

The stochastic process introduced above defines a semigroup P_N^t in the usual way. Let $\nu_N^0 = \nu_N^0(Z_N)$ be a symmetric probability density, then $\nu_N(Z_N, t) = P_N^t \nu_N^0(Z_N)$ is the time evolved measure defined by

$$\int \nu_N(Z_N, t)\varphi(Z_N)dZ_N = \int \nu_N^0(Z_N)E\varphi(T^t Z_N)dZ_N, \tag{4.3}$$

where φ is any smooth test function, $T^t Z_N$ is the N-particle stochastic process starting almost surely from Z_N and E denotes the expectation. Formally $\nu_N(t)$ satisfies:

$$(\partial_t + \sum_{i=1}^{N} v_i \partial_i)\nu_N(Z_N, t) =$$

$$\frac{1}{2}\varepsilon \sum_{i \neq j} \delta(x_i - x_j)\chi(i,j)|v_i - v_j|\{\nu_N(Z_N'(i,j), t) - \nu_N(Z_N, t)\}, \tag{4.4}$$

where ∂_i is the derivative with respect to x_i, $\delta(\cdot)$ is the δ -function centered in zero, $\chi(i,j) = 1$ if $sgn(v_1) \neq sgn(v_j)$ and 0 otherwise,

$$Z_N'(i,j) = (x_1v_1, \dots x_iv_i', \dots x_jv_j', \dots x_Nv_N)$$

if

$$Z_N = (x_1 v_1, \dots x_i v_i, \dots x_j v_j, \dots x_N v_N).$$

Eq.(4.4) must be complemented by the initial conditions

$$\nu(Z_N, 0) = \nu_N^0(Z_N).$$

The equivalent of the Boltzmann-Grad limit is, in this context, $N \to \infty$, $\varepsilon \to 0$ in such a way that the combination $\varepsilon N = \lambda$ is constant. The espected result is that the measure μ^N is going to factorize in this limit (propagation of chaos) and that the one-particle distribution function approximates the solution of the Boltzmann equation relative to the model under consideration. This is given by:

$$(\partial_t + v\partial_x) f(x,v,t) = \lambda \int_{vv_1<0} |v-v_1| \{ f(x,v',t) f(x,v_1',t) - f(x,v,t) f(x,v_1,t) \} \tag{4.5}$$

Eq. (4.5) can be formally obtained from Eq. (4.4) by introducing a hierarchy of equations for the j-particle distribution functions f_j^N and integrating with respect to the last variables, as we did in the previous section. We have:

$$(\partial_t + \sum_{i=1}^{j} v_i \partial_i) f_j^N = A_j f_j^N + \varepsilon(N-j) C_{j,j+1} f_{j+1}^N \tag{4.6}$$

where $-\sum_{i=1}^{j} v_i \partial_i + A_j f_j^N$ is the generator of the j-particle stochastic process (that is the generator of the semigroup P_j^t) and:

$$C_{j,j+1} f_{j+1}(x_1, v_1 \dots x_j v_j)$$

$$= \sum_{i=1}^{j} \int dv_{j+1} \chi(i, j+1) |v_i - v_{j+1}|$$

$$[f_{j+1}^N(x_1, v_1 \dots x_i, v_i', \dots x_i v_{j+1}') - f_{j+1}^N(x_1, v_1 \dots x_i v_{j+1})] \tag{4.7}$$

By Eq.(4.6) we see that the right scaling is indeed $\varepsilon \to 0$ and $N \to \infty$ in such a way that $\varepsilon N = \lambda$ where λ is a finite number. In this way the generator of the j-particle process converge formally to that of the free stream, while the interaction of the group of the first j particles with the rest stays bounded.

The Boltzmann equation is then formally derived by Eq.(4.7) for $j = 1$, assuming the propagation of chaos and taking the limit $N \to \infty$.

In spite of a formal similarity, the problem we takle here is much more difficult than that solved in Sect.3. Here the operator $C_{j,j+1}$, from $(L_1)^{j+1}$ to $(L_1)^j$, is not continuous anylonger. However it is continuous in L_∞:

$$|C_{j,j+1} f_{j+1}| \leq Const \quad j \| f_{j+1} \|_\infty \tag{4.8}$$

so that we can hope to follow the same strategy of the series expansion technique explained in Sect.3, working in L_∞. Indeed it can be done and the result is that the series expansion associated to Eq.(4.7) is absolutely convergent in L_∞ for a short time $t_0 = \frac{1}{CM\lambda}$ where $M = \|f_0\|_{L_\infty}$. The same is also true for the limiting hierarchy associated to the products of the solutions of Eq.(4.5). For Eq.(4.5) it is possible to prove that there exists a unique solution $f \in C([0,T]; L_1)$ which is also bounded for all t, provided that $M < +\infty$ (see for instance [1] and [3]). Then we can prove the following Theorem.

Theorem 4.1. *Consider an initial probability measure*

$$\nu_N^0 = \prod_{i=1}^{j} f_0(x_i, v_i)$$

with $\|f_0\|_{L_\infty} = M < +\infty$. Let $\mu_N(t)$ be the time evolved measure according to the stochastic process defined above. Denote by $f_j^N(t)$ its j-particle distribution functions. Then for $t < t_0 = \frac{1}{CM\lambda}$:

$$\lim_{N\to\infty} \|f_j^N(t) - f_j(t)\|_{L_1(x,v)} = 0. \tag{4.9}$$

Here $f_j(t) = f^{\bigotimes j}(t)$, $f(t)$ being the unique bounded solution to Eq.(4.5) with initial datum f_0

We do not give the detais of the proof of the above theorem. This can be done by using the arguments in [11] and [12], namely to use the series expansions technique explained in the previous Section, in a L_∞ setup. Thm 4.1 is satisfactory, however it works for short times only and it is not obvious how to extend it for arbitray times. Actually we would need an estimate of the type $f_j^N < K^j$ for a fixed, but arbitrary time interval $[0,T]$ to apply the Theorem 4.1 in time intervals of size $1/C\lambda K$. Unfortunately such estimate on the distribution functions is difficult to obtain globally in time (we mean for an arbitrary time interval) so that we need to consider different strategies exploiting more carefully the properties of the process. The property we really need is expressed by the following theorem.

Theorem 4.2. *Consider two initial probability measures μ_N^0 and ν_N^0 and their time evolved $\mu_N(t)$ and $\nu_N(t)$. Denote by $g_j^N(t)$ and $f_j^N(t)$ their j-particle distribution functions respectively. Then, if λ is sufficiently small and if, for all j:*

$$\lim_{N\to\infty} \|f_j^N(0) - g_j^N(0)\|_{L_1(x,v)} = 0, \tag{4.10}$$

the following holds:

$$\lim_{N\to\infty} \|f_j^N(t) - g_j^N(t)\|_{L_1(x,v)} = 0 \tag{4.11}$$

for all j and all positive t .

Theorem 4.2 expresses a continuity property with respect to the initial condition which is enough to extend Theorem 4.1 for arbitrary times. Indeed let:

$$K = \sup_{t \in [0,T]} f(x,v,t) \tag{4.12}$$

and $t_0 = \frac{1}{C\lambda K}$. For $t \in [t_0, 2t_0]$ consider the difference:

$$\|f_j^N(t) - f_j(t)\|_{L_1(x,v)}$$
$$\leq \|f_j^N(t) - g_j^N(t)\|_{L_1(x,v)} + \|f_j(t) - g_j^N(t)\|_{L_1(x,v)} \tag{4.13}$$

where $g_j^N(t)$ are the j–particle distribution functions associated to the N–particle dynamics with initial measure $f(t_0)^{\bigotimes N}$. Then by Theorem 4.1 (applied to the initial datum $f_j(t_0)$), the second term of the right hand side of (4.13) is vanishing. Moreover by Theorem 4.2 also the first term is vanishing because we have convergence at the initial time t_0 by Theorem 4.1. This proves convergence up the time $2t_0$. The argument can be extended to arbitrary times so that we achieve convergence and propagation of chaos globally in time even though for λ small. How to pass from small to large λ will be discussed later.

It remains to prove Theorem 4.2. To do this it is convenient to give an explicit representation of the stochastic process T^t in a suitable sample space. Define:

$$\Omega = \{0,1\}^{\frac{N(N-1)}{2}}. \tag{4.14}$$

$\omega \in \Omega$, $\omega = \omega(i,j) = \omega(j,i;\alpha)$, $i \neq j$, is a function defined on the set of all pairs of particles, taking values 0 and 1. Define the flow T_ω^t in the following way. $T_\omega^t Z_N$ is the free motion unless at time t two particles, say i and j, are at the same point with opposite velocities. Then if $\omega(i,j) = 1$ the two particles collide according to the law (4.1), otherwise they keep their free motion. In other words T_ω^t is a completely deterministic flow.

We define the probability of a single event $\omega \in \Omega$ by

$$p(\omega) = \prod_{i,j} \varepsilon^{\omega(i,j)} (1-\varepsilon)^{1-\omega(i,j)} \tag{4.15}$$

where $\prod_{i,j}$ denotes the product on all pairs. $\{\Omega, p\}$ is a probability space and we can write:

$$E(\varphi(T^t Z_N)) = E(\varphi(T_\omega^t Z_N)), \tag{4.16}$$

where φ is a test function, E denotes the expectation with respect to $\{\Omega, p\}$.

Notice that this explicit representation for our process is possible because we decided to choose a representation of the two equivalent collision laws for which two particle that have interacted once, cannot interact anymore.

In what follows we shall use the notation I_j, $j = 1 \dots N$ for the set of the first j integers.

Given $\omega \in \Omega$, we denote by $ch(i)$ ("chain" starting from i), any set $I \subset I_N/i$ of indices such that

$$\omega(i, i_1; \alpha_1)\omega(i_1, i_2; \alpha_2)\dots\omega(i_{k-1}, i_k; \alpha_k) = 1$$

for some ordering $i_1, \dots, i_k$ of the set I and some sequence $\alpha_1 \dots \alpha_k, \alpha_i = 1 \dots r$. The union of all chains starting from the particle i will be called "cluster of i" and denoted by

$$cl(i) = \bigcup ch(i). \tag{4.17}$$

Notice that the index i is not included in $cl(i)$. Moreover we denote by:

$$cl(I) = \bigcup_{i \in I} cl(i)/I, \tag{4.18}$$

the cluster of the set I. Notice that $cl(I)$ is the set of all particles which can influence the dynamics of the set I (excluded the set I itself).

With these definitions we have:

$$\langle f_j^N(t), \varphi_j \rangle = E\Big(\int \nu^0(Z_N)\varphi_j((T^t Z_N)_j) dZ_N\Big)$$

(where φ_j is any suitable test function depending on the first j components, and $(T^t Z_N)_j$ denotes the first j components of $T^t Z_N$)

$$= \sum_{S \subseteq I_N/I_j} E\Big[(\chi(cl(I_j) = S) \int \nu^0(Z_N)\varphi_j((T_\omega^t Z_N)_j) dZ_N\Big] \tag{4.19}$$

In the expansion (4.19) we have decomposed the events according to the clusters of I_j. The advantage of this decomposition becomes clear if we consider that, if $\chi(cl(I_j)) = 1$, then:

$$\int \nu^0(Z_N)\varphi_j((T_\omega^t Z_N)_j) dZ_N = \int \nu^0(Z_N)\varphi_j(T_\omega^{-t} Z_{I_j \cup S_j}) dZ_N$$

$$= \int f^N_{I_j \cup S)}(Z_{I_j \cup S}, 0)\varphi_j((T_\omega^{-t} Z_{I_j \cup S})_j) dZ_{I_j \cup S} \tag{4.20}$$

This is because the dynamics of the particles of the group $I_j \cup S$ is independent of that of the group $I_N/(I_j \cup S)$. by definition of cluster. Here $f_S^N(0)$ denotes the distribution of the particles S at time zero. Therefore:

$$\langle (f_j^N(t) - g_j^N(t)), \varphi_j \rangle = \sum_{S \subseteq I_N/I_j} E[(\chi(cl(I_j) = S)$$

$$\cdot \int (f^N_{I_j \cup S} - g^N_{I_j \cup S})(Z_{I_j \cup S}, 0)\varphi_j((T_\omega^{-t} Z_j)_j) dZ_{I_j \cup S} \tag{4.21}$$

Notice now that the above sum $\sum_S$ can be split into two contributions: the first one is the event in which all the clusters of each particle of the set I_j are disjoint; the second its complement. Denoting the first contribution by Σ and the remaining term by T, we have:

$$\Sigma = \sum_{\substack{S\subseteq I_N\\ S\cap I=\emptyset}} \sum_{\substack{S_1\dots S_j:\\ \bigcup_i S_i=S\\ S_r\cap S_k=\emptyset, r\neq k}} E[(\prod_{i\in I_j}\chi(cl(i)=S_i)$$

$$\cdot \int (f^N_{I_j\cup S} - g^N_{I_j\cup S})(Z_{I_j\cup S},0)\varphi_j((T_\omega^{-t}Z_{I_j\cup S})_j)dZ_{I_j\cup S}] \qquad (4.22)$$

from which:

$$|\Sigma| \le \|\varphi\|_{L_\infty} \sum_{k\ge 0} \|f^N_{j+k}(0) - g^N_{j+k}(0)\|_{L_1}$$

$$\cdot \sum_{\substack{S\subseteq I_N\\ S\cap I=\emptyset}} \sum_{S:|S|=k} \sum_{\substack{S_1\dots S_j:\\ \bigcup_i S_i=S\\ S_r\cap S_k=\emptyset, r\neq k}} E(\prod_{i\in I_j}\chi(cl(i)=S_i) \qquad (4.23)$$

Then we are reduced to estimate independent events. We have:

$$E(\chi(cl(1)=S)) = \sum_{h\ge k} E(\chi(cl(1)=S)\chi(n=h))$$

where $k = |S|$=cardinality of S and n is the number of active links in the group $S\cup\{1\}$, that is $\sum_{i,j\in S\cup\{1\}}\omega(i,j)$.

If $n=k$, that is the cluster is created with a minimum number of links, we have:

$$E(\chi(cl(1)=S)\chi(n=k)) \le \varepsilon^k k! e^{Ck}$$

Indeed it is not difficult to estimate the number of ways to realize a clusters of S particles by $k!e^{Ck}$. See [6] for details. The factor ε^k arises from the probability of a single event. If $h>k$ we have extra $h-k$ links to put. This can be done in

$$\binom{H}{h-k}$$

different ways, being $H = \frac{(k+1)k}{2}$ the number of pairs. Hence (using that $\varepsilon H \le \lambda k$):

$$E(\chi(cl(1)=S)) = \sum_{h\ge k} E(\chi(cl(1)=S)\chi(n=h)) \le \sum_{h\ge k} \varepsilon^h k!\, e^{Ck}\binom{H}{h-k}$$

Hence we get

$$E(\chi(cl(1)=S)) = k!\, e^{Ck}\, \varepsilon^k \sum_{r=0}^{H} \binom{H}{r} \varepsilon^r \le k!\, e^{Ck}\, \varepsilon^k\, e^{\lambda k} \tag{4.24}$$

Inserting estimate (4.24) in (4.23), we arrive to:

$$|\Sigma| \le \|\varphi\|_{L_\infty} \sum_{k\ge 0} \binom{N}{k} \|f^N_{j+k}(0) - g^N_{j+k}(0)\|_{L_1}$$

$$\cdot \sum_{\substack{k_1\dots k_j: \\ \sum k_i = k}} k_1! \dots k_j!\, e^{Ck} \left(\frac{\lambda}{N}\right)^k \binom{k}{k_1}\binom{k-k_1}{k_2}\dots\binom{k-\sum_{i=1}^{j-1} k_i}{k_j}$$

$$\le \|\varphi\|_{L_\infty} \sum_{k\ge 0} \|f^N_{j+k}(0) - g^N_{j+k}(0)\|_{L_1} \sum_{\substack{k_1\cdots k_j: \\ \sum k_i = k}} \frac{N!}{k!N^k} \frac{k!}{k_1!\cdots k_j!} k_1! \dots k_j!\, e^{Ck} \lambda^k$$

$$\le \|\varphi\|_{L_\infty} \sum_{k\ge 0} \|f^N_{j+k}(0) - g^N_{j+k}(0)\|_{L_1} e^{Ck} \lambda^k. \tag{4.25}$$

The last step is due to the inequality:

$$\sum_{\substack{k_1\cdots k_j: \\ \sum k_i = k}} 1 \le C^{k+j}$$

Therefore Σ goes to zero as $N \to \infty$ by the dominated convergence theorem.

It remains to control T whose expresion is:

$$T = \sum_{\substack{S \subseteq I_N \\ S\cap I = \emptyset}} \sum^{*}_{\substack{S_1\dots S_j: \\ \bigcup_i S_i = S}} E[(\prod_{i\in I_j} \chi(cl(i) = S_i)$$

$$\int (f^N_{I_j \cup S} - g^N_{I_j\cup S})(Z_{I_j\cup S}, 0)\varphi_j((T_\omega^{-t} Z_{I_j\cup S})_j) dZ_{I_j\cup S}] \tag{4.26}$$

where $\sum^*$ means that we are summing over those sets $S_1 \dots S_j$ such that, at least two of them have non-empty intersection. Hence:

$$|T| \le 2\|\varphi\|_{L_\infty} \sum_{k\ge 0} \sum_{\substack{k_1\dots k_j: \\ \sum k_i = k}} \sum_{\substack{S: \\ |S| = k}} \sum^{*}_{\substack{S_1\dots S_J: \\ \bigcup_i S_i = S}} E(\prod_{i\in I_j} \chi(cl(i) = S_i) \tag{4.27}$$

Such a contribution can be handled as above the only difference is that here we have at least an extra links connecting two clusters. This allows us to gain an ε-factor. Therefore:

$$|T| = O(\frac{1}{N}) \tag{4.28}$$

In conclusion:

Theorem 4.3. *Theorem 4.1 holds for arbitrary times provided that λ is sufficiently small.*

The next problem is now to extend the result to arbitrary λ's. Here the proof is rather technical and will be only sketched (see [6] for details).

Consider the interval $\Lambda = [b_1, b_2]$. If $t \leq a$, due to the bound $|v| \leq 1$, all the particles which are in Λ at time t are influenced only by those particles sitting, at time zero, in the region $\Lambda_a = \Lambda \cup [b_1, -a]J \cup [b_2, +a]$ of total measure $meas\Lambda_a = meas\Lambda + 2a$. The average number of particles at time zero on such a region, say $\bar{N}_a$ is then:

$$\bar{N}_a = N \int_{\Lambda_a} dx \int dv f_0(x, v) \leq NM meas\Lambda_a \tag{4.29}$$

Suppose, for the moment, that the real number of particles in this region is exactly its average values. Since the solution to the Boltzmann equation at time t, when retricted to Λ, depends only on f_0 restricted in Λ_a, if we restrict ourselvelves to this region, everything goes on as if $\lambda \to N\varepsilon meas\Lambda_a$ which can be made as small as we want, provided that $|b_2 - b_1| + a$ is made sufficiently small. Therefore our problem is to apply the law of large numbers at time zero and to control the fluctuations. This can indeed be done in a rather stightforward way. At this point, using also the fact that $\sup_{t \leq T} f(x, v, t) \leq M(T)$, it is not difficult to extend the argument from a to $2a$ and so on. The exact statment which can be proved is summarized below.

Theorem 4.4. *Under the hypotheses of Theorem 4.1, for any $\lambda > 0$, $t > 0$ and $j > 0$:*

$$\lim_{N \to \infty} \langle f_j^N(t), \varphi_j \rangle = \langle f_j(t), \varphi_j \rangle \tag{4.30}$$

for any bounded continuous function φ_j.

5. Concluding remarks

In this lectures we have discussed the transition from some particle systems to the reduced description given by suitable kinetic non linear partial differential equations. In doing this we have analyzed different strategies. Of course our presentation does not exhaust the existing literature in the field. We have discussed the mean-field limit only marginally: the S.Meleard's lectures in this volume, contain many arguments and references to which we address the interested readers.

For the so called hydrodynamic limits which have just mentioned in the present paper, we suggest the S.R.S. Varadhan review [21] and the references quoted therein.

As regards the genuine kinetic limits (of Boltzmann-Grad type), beyond the paper discussed and quoted in Chapt.4 of [5], there are a couple of other results concerning stochastic particle systems on lattices, which should be mentioned. In [20] the author derives one-dimensional kinetic equations by mimicing the same strategy used for the existence theory of the solutions of the limiting equation, for the corresponding discretized quantities at the microscopic level.

In [4] the two-dimensional Brodwell equation is derived, assuming the existence of a smooth solution for the Cauchy problem relative to the initial datum under consideration. Such a solution allows to conclude that, in some sense, the particle system behaves correctly as regards the scaling limit under consideration.

We conclude this section by criticizing the results we have presented here. In most of the practical numerical simulations of real rarefied gases, we are interested to compute stationary non-equilibrium profiles. For instance, in the classical simulation problem of a flow past an obstacle, after an inessential transient, the system stabilizes and exhibit a stationary temperature, density and momentum profiles whose numerical calculation is the scope of the simulation. The time dependent analysis we have developed so far describes only the initial layer while the physical interesting features are eluded. As a consequence it would be very interesting to analyze such kind of scaling limits in a stationary regime. In other words we would like to have convergence results of phyically relevant stationary measures for the particle system, towards stationary solutions of the kinetic equations. The analysis for the time dependent case does not help so much. We have obtined results only for a finite time interval, and they are not expected to hold uniformly in time. Therefore other methods must be exploited. We believe that this is an interesting field on which very little is known: the only result of which we are aware, is obtained for the one-dimensional system of Sect 4, localized in an interval with diffusive boundary condition and for a moderate value of the coupling contant λ (see [7]).

References

1. Arkeryd, L., *Arch Rat Mech and Anal* 103, 139-149 (1988).
2. Braun, W., Hepp, K., *Comm. Math. Phys.* 56, 101-120 (1977).
3. Cercignani, C., "Recent results in kinetic theory of gases", *Rendiconti Sem. Mat. Univ. Torino*, 47-64 (1990).
4. Caprino, S., De Masi, A., Presutti, E., Pulvirenti, M., *Comm. Mat. Phys* 135, 443-465 (1991).
5. Cercignani, C., Illner, R., Pulvirenti, M., "The mathematical theories of diluite gases", Springer series in Appl. Math. 106 (1994).
6. Caprino, S., Pulvirenti, M., *Comm. Math. Phys.* 166, 603-621 (1994).
7. Caprino, S., Pulvirenti, M., "The Boltzmann-Grad limit for a one-dimensional Boltzmann equation in a stationary state", preprint (1994).
8. Dobrushin, R.L., *Sov. J. Funct. Anal.* 13, 115-119 (1979).

9. Di Perna, R.J., Lions, P.L., "On the Cauchy problem for Boltzmann equations: global existence and stability", *Ann. Math.* 130, 321-366 (1989).
10. Grad, H., *Comm. Pure Appl. Math.* 2, 331-407 (1949).
11. Kac, M., *Probability and Related Topics*, Interscience, New York (1959).
12. Lanford III, O., "The evolution of large classical system", *Lect. Notes in Physics* 35, (J. Moser, ed.), Springer (1975).
13. Lions, P.L., Perthame, B., Tadmor, E., "A kinetic formulation of multidimensional scalar conservation laws and related eqautions", *J.A.M.S.* 7-1, 169-191 (1992).
14. McKean, H.P., "Lectures in differential equations", (Aziz, ed.), 2, 177 (1969).
15. Marchioro, C., Pulvirenti, M., *Mathematical Theory of Incompressible Nonviscous Fluids*, Springer series in Appl. Math. 96 (1994).
16. Neunzert, H., *Lect. Notes Math.* 1048, (C. Cercignani, ed.), 60-110, Springer (1984).
17. Perthame, B., Pulvirenti, M., *Asympt. Anal.* (1995).
18. Perthame, B., Tadmor, E., *Comm. Math. Phys.* 136, 501-517 (1991).
19. Pulvirenti, M., Wagner, W., Zavelani, M.B., *Eur. J. Mech., B/Fluids* 3, 339-351 (1994).
20. Rezakhanlou, F., "Kinetic limits for a class of interacting particle systems", *Probab Theory Related Fields*, to appear (1995).
21. Varadhan, S.R.S., *Lect. Notes Math.* 1551, (C. Cercignani and M .Pulvirenti, eds.), Springer (1994).

A statistical physics approach to large networks

Carl Graham

CMAP, URA CNRS 756, École Polytechnique, 91128 Palaiseau, France

Introduction

A communication or computer network is a large complex system constituted of different resources such as circuits, queues, multiplexers, memory, screens, printers, and processors. It can be usually given a Markovian description only by suitably enlarging the state space with respect to the variables which are of practical interest, thus introducing auxiliary bothersome variables. Natural networks have little or no symmetry.

Such networks can be considered as large multitype strongly interacting particle systems, the interaction coming from the competition for limited resources. It is tempting to consider this from a statistical mechanic point of view in order to obtain a simple limit for the macroscopic variables of interest, when the size of the network increases. The auxiliary variables may vanish at this limit.

We would consider then a limit situation, often nonlinear but in low dimensions, instead of a huge linear system. Computation of different transitory and equilibrium quantities such as blocking probabilities is often much simpler in the limit nonlinear model. If the limit links act as if they were independent, we may use product-form solutions or fixed-point approximations.

We are thus interested in the chaos hypothesis (Boltzmann's *stosszahlansatz*) according to which particles become independent as their number grows. We give two examples of networks, one exchangeable and the other with a graph structure. We define a wide class of systems including these examples, in which the interaction may be strong and local and there is no symmetry, for which the chaos hypothesis holds under simple assumptions which depend little on the specifics of the systems. The results do not depend on the law of the holding times, a property called insensitivity.

We use interaction graphs in order to construct the processes in reverse time with the least knowledge of the "past history" of the particles. We prove the chaos hypothesis by coupling graphs. This result gives asymptotics for empirical measures over certain sets of k-tuples. The graph for one particle is similarly coupled with a Boltzmann tree in which all the particles are independent.

Often the total effect of the system on one particle does not depend on the number of particles, and thus neither does this tree nor the process built on it. This proves that the laws of the individual particles converge and thus propagation of chaos. Under light symmetry assumptions the chaos hypothesis and its empirical measure consequence enable us to obtain a nonlinear martingale problem for the limit law, which characterizes it in the case of uniqueness.

We then study fluctuations on the two examples. The drift term of the first, a mean-field model, is closed by a simple algebraic factoring and linearization. The drift term of the second example does not close, because of local interaction. The simultaneous release intervenes strongly in the identification of the Gaussian martingale part. An auxiliary measure-valued process is introduced, and is the start of a hierarchy. In the mean-field model this hierarchy is also closed by algebraic factoring, but in the second model complex fine estimates on the interaction graph are needed.

Notation and preliminary results

For integers p and q, $(p)_q$ denotes the number $p(p-1)\dots(p-q+1)$ of ordered subsets of size q in a set of size p and $\binom{p}{q}$ the number $(p)_q/q!$ of subsets.
D denotes the Skorokhod space, and $\mathcal{P}$, $\mathcal{M}_+$ and $\mathcal{M}$ the spaces of probability, bounded positive and bounded signed measures.
$\langle\ ,\ \rangle$ denotes either martingale Doob-Meyer brackets or function-measure duality brackets $\langle f, m\rangle = \langle f(x), m(dx)\rangle = \int f(x)m(dx)$. We denote $m(\{x\})$ by $m(x)$.
$|\ |$ is the total variation norm $|m| = \sup\{\langle\phi, m\rangle : \|\phi\|_\infty \leq 1\}$ or the cardinal of a set or sequence, $|\ |_T$ the total variation norm on $D([0,T])$.
For a function ϕ, we set $\phi^+(x) = \phi(x+1) - \phi(x)$, $\phi^-(x) = \phi(x-1) - \phi(x)$.

Lemma 0.1. *Let X be a pure-jump process in $\mathbb{R}^d$ with jump measure J, and ϕ a function on $\mathbb{R}^d$. $\phi(X_t) - \phi(X_0) - \int_0^t \langle \phi(X_s+h) - \phi(X_s), J(X_s, dh)\rangle\, ds$ is a local martingale with Doob-Meyer bracket $\int_0^t \langle (\phi(X_s+h) - \phi(X_s))^2, J(X_s, dh)\rangle\, ds$.*

1. Various network models and their difficulties

We describe two networks of increasing complexity, then a general class of networks. The variables of interest will not be Markovian, and will be exchangeable in the first network but only identically distributed in the second.

We will show why classical techniques to show propagation of chaos do not work on most networks due to lack of symmetry. This has lead many authors to consider exchangeable caricatures of the networks. We refer to work in Graham and Méléard [9, 10, 11, 12].

1.1 A generalized star-shaped loss network

There are n links each of a capacity of C channels. Calls arrive on a given subset of K links at rate ν_n. The call is accepted only if each link has spare capacity, else it is lost. Calls end at rate 1, releasing all K channels simultaneously. The global call attempt rate seen by a link is constant equal to ν, and $\nu_n = \nu/\binom{n-1}{K-1} = O(1/n^{K-1})$. The birth of calls is a K-body mean-field interaction, while their death introduces strong interaction.

These symmetric networks model many situations of simultaneous service, for instance telecommunication or computer networks, locking of items in data-bases, parallel computing, and job processing in factories, and also chemical reversible processes and polymer formation. For $K = 2$ we can imagine that the network ensures connections through a central hub, hence the term "star-shaped". These network models were introduced by Whitt [24], and studied among others by Ziedins and Kelly [25], Hunt [13, 14], Kelly [16, 17].

R^n denotes the set of routes $\{r = i_1 \dots i_K : 1 \le i_1 < \dots < i_K \le n\}$ and $R^n_{j_1 \dots j_p}$ denotes the set $\{i_1 \dots i_K \in R^n : j_1, \dots, j_p \in \{i_1, \dots, i_K\}\}$ of routes involving the links $j_1, \dots, j_p$. On routes $r = i_1 \dots i_K$ we have independent Poisson processes $N^n_r = N^n_{i_1 \dots i_K}$ of call arrivals, of rate $\nu_n = \nu/\binom{n-1}{K-1}$.

For $1 \le i \le n$, X^n_i is the process of the number of occupied channels on link i. $(X^n_i)_{1 \le i \le n}$ is exchangeable (if the initial values are), but **not** Markovian because of the simultaneous releases. For distinct $1 \le i_1, \dots, i_K \le n$, $Y^n_{i_1 \dots i_K}$ is the process of calls involving the set $\{i_1, \dots, i_K\}$. All these processes belong to $D(\mathbb{R}+, \{0, 1, \dots, C\})$. The process $(Y^n_r)_{r \in R^n}$ is Markovian and for $1 \le i \le n$

$$X^n_i = \sum_{r \in R^n_i} Y^n_r = \sum_{\substack{i_2 < \dots < i_K \\ i \notin \{i_2, \dots, i_K\}}} Y^n_{i i_2 \dots i_K}. \tag{1.1}$$

We define the k-body empirical measures on path space

$$\mu^{k,n} = \frac{1}{(n)_k} \sum_{|i_1, \dots, i_k| = k} \delta_{X^n_{i_1}, \dots, X^n_{i_k}}, \quad \mu^{k,n}_i = \frac{1}{(n-1)_k} \sum_{\substack{|i_1, \dots, i_k| = k \\ i \notin \{i_1, \dots, i_k\}}} \delta_{X^n_{i_1}, \dots, X^n_{i_k}} \tag{1.2}$$

and $\mu^n = \mu^{1,n} = \frac{1}{n} \sum_{i=1}^n \delta_{X^n_i}$. Using Lemma 0.1, for any function ϕ

$$\begin{aligned}
&\phi(X^n_i(t)) - \phi(X^n_i(0)) - \int_0^t \phi^+(X^n_i(s)) \mathbb{1}_{X^n_i(s) < C} \\
&\quad \times \left(\frac{\nu}{\binom{n-1}{K-1}} \sum_{\substack{i_2 < \dots < i_K \\ i \notin \{i_2, \dots, i_K\}}} \mathbb{1}_{X^n_{i_2}(s) < C, \dots, X^n_{i_K}(s) < C} \right) + \phi^-(X^n_i(s)) X^n_i(s) \, ds \\
&= \phi(X^n_i(t)) - \phi(X^n_i(0)) - \int_0^t \phi^+(X^n_i(s)) \nu \mathbb{1}_{X^n_i(s) < C} \langle \mathbb{1}^{\otimes K-1}_{\cdot < C}, \mu^{K-1,n}_{i,s} \rangle \\
&\qquad + \phi^-(X^n_i(s)) X^n_i(s) \, ds
\end{aligned} \tag{1.3}$$

is a martingale. The flow $(\mu^n_t)_{t \ge 0}$ satisfies that for any function ϕ

$$\begin{aligned}
&\langle \phi, \mu^n_t \rangle - \langle \phi, \mu^n_0 \rangle \\
&- \int_0^t \langle \phi^+(x_1) \nu \mathbb{1}_{x_1 < C, \dots, x_K < C}, \mu^{K,n}_s(dx_1 \dots dx_K) \rangle + \langle \phi^-(x) x, \mu^n_s(dx) \rangle \, ds
\end{aligned} \tag{1.4}$$

is a martingale with Doob-Meyer bracket

$$\int_0^t \frac{1}{n}\langle(\phi^+(x_1)+\cdots+\phi^+(x_K))^2 \frac{\nu}{K} \mathbb{1}_{x_1<C,\ldots,x_K<C}, \mu_s^{K,n}(dx_1\ldots dx_K)\rangle + \frac{1}{n^2} \sum_{i_1<\ldots<i_K} (\phi^-(X_{i_1}^n)+\cdots+\phi^-(X_{i_K}^n))^2 Y_{i_1\ldots i_K}^n(s)\,ds. \quad (1.5)$$

Simultaneous release is seen in the Doob-Meyer bracket (1.5). Its second term is not immediately seen to vanish as in simple mean-field models. We guess the limit nonlinear martingale problem. Existence and uniqueness is proved by a contraction argument: see Graham [6, 7], Shiga and Tanaka [20].

Definition 1.1. *The probability measure* $\tilde{P}$ *on* $D(\mathbb{R}+, \{0,1,\ldots,C\})$ *solves the nonlinear martingale problem (1.6) if for any function* ϕ

$$\phi(X_t) - \phi(X_0) - \int_0^t \phi^+(X_s)\nu \mathbb{1}_{X_s<C}\left(1-\tilde{P}_s^C\right)^{K-1} + \phi^-(X_s)X_s\,ds \quad (1.6)$$

is a $\tilde{P}$*-martingale.* $\tilde{P}_s^C = \tilde{P}_s(C) = \tilde{P}(X_s = C)$ *is the nonlinearity.*

Mean-field interaction and factorization

$$\begin{aligned}\mu^{k,n} - (\mu^n)^{\otimes k} &= \frac{1}{(n)_k} \sum_{|i_1,\ldots,i_k|=k} \delta_{X_{i_1}^n,\ldots,X_{i_k}^n} - \frac{1}{n^k} \sum_{i_1,\ldots,i_k} \delta_{X_{i_1}^n,\ldots,X_{i_k}^n} \\ &= \left(\frac{1}{(n)_k} - \frac{1}{n^k}\right) \sum_{|i_1,\ldots,i_k|=k} \delta_{X_{i_1}^n,\ldots,X_{i_k}^n} - \frac{1}{n^k} \sum_{|i_1,\ldots,i_k|<k} \delta_{X_{i_1}^n,\ldots,X_{i_k}^n} \quad (1.7)\end{aligned}$$

and in variation norm, uniformly on the randomness,

$$|\mu^{k,n} - (\mu^n)^{\otimes k}| \le 2\left(1 - \frac{(n)_k}{n^k}\right) = O\left(\frac{1}{n}\right). \quad (1.8)$$

Hence (1.4) is a closed expression and (1.3) is expressed in terms of μ^n, up to a small remainder term. Sznitman [21, 22] exploited this general feature of mean-field interaction. He proved for exchangeable systems on a Polish space:

Theorem 1.2. $(\mathcal{L}(X_1^n,\ldots,X_k^n))_{n\ge k}$ *converges weakly to* $Q^{\otimes k}$ *for any fixed* k *if and only if* $(\mu^n)_{n\ge 0}$ *converges in law (and in probability) to* Q.

He then proved tightness of the empirical measures, deducing it from tightness of $(X_1^n)_{n\ge 0}$, and that the laws in the support of the accumulation points solve a nonlinear martingale problem, using the fact that the martingale problem for the empirical measure is asymptotically closed. Uniqueness for this martingale problem allowed to conclude to convergence. This could be applied to the star-shaped network but we shall get better results using interaction graphs.

1.2 A fully-connected loss network with alternate routing

We consider a fully-connected telecommunication network of n cities or nodes, denoted by lower-case letters $1 \leq a, b, \ldots \leq n$. There are $n(n-1)/2 = \binom{n}{2}$ links of capacity C between nodes denoted by $ab, \ldots$ with $ab = ba$. Calls arrive on ab at rate ν and occupy one circuit if the link is not saturated. If it is, a third node c is chosen uniformly among the $n-2$ others; the call is routed through links ac and bc if both are not full, else it is lost. Calls end at rate 1, and alternately routed calls release both channels simultaneously.

Marbukh [19] assumed a priori the chaos hypothesis. He obtained a limit nonlinear ODE for the process $(\mu_t^n(0), \ldots, \mu_t^n(C))_{t\geq 0}$ which presents multiple equilibria for certain values of ν and C. This hints to metastable long-time behavior of the finite system; see Dobrushin [4], Gibbens et al. [5], Kelly [17], Marbukh [19].

Exchangeable caricatures have been studied in Anantharam [1], Gibbens et al. [5], Hunt [13], and Kelly [17], leading to a variant of the star-shaped network. The chaos hypothesis was first proved in Graham and Méléard [9]. A functional law of large numbers is in Crametz and Hunt [2]. Kushner [18] studied the network in heavy traffic and increasing size.

The process of the number of occupied channels between a and b will be denoted by X_{ab}^n, and the number of calls in progress between a and b that were rerouted through c by $Y_{ab}^{c,n}$. This forms a Markov process. We denote by $Y_{ab}^n = X_{ab}^n - \sum_{c\neq a,b}(Y_{ac}^{b,n} + Y_{bc}^{a,n})$ the number of direct calls in progress.

There is a graph structure to the network, and the natural symmetry assumption is that the laws are invariant under permutation of the n nodes. The $\binom{n}{2}$ processes $(X_{ab}^n)_{1\leq a<b\leq n}$ are identically distributed but **not** exchangeable: couples that share a node do not have the same distribution as couples that do not. The symmetry group of μ^n is of order $\binom{n}{2}!$ while the one of $\mathcal{L}((X_{ab}^n)_{1\leq a<b\leq n})$ is only of order $n!$, and convergence of the μ^n does not imply propagation of chaos.

We define the "local around node a" and the "global" empirical measures

$$\mu_a^n = \frac{1}{n-1}\sum_{b:b\neq a}\delta_{X_{ab}^n}, \quad \mu^n = \frac{1}{\binom{n}{2}}\sum_{a<b}\delta_{X_{ab}^n} = \frac{1}{n}\sum_a \mu_a^n \tag{1.9}$$

and the "local" empirical measures over triangles and over triangles with side ab

$$\mu^{(3)n} = \frac{1}{(n)_3}\sum_{|a,b,c|=3}\delta_{X_{ab}^n, X_{ac}^n, X_{bc}^n}, \quad \mu_{ab}^{(2)n} = \frac{1}{n-2}\sum_{c:c\neq a,b}\delta_{X_{ac}^n, X_{bc}^n}. \tag{1.10}$$

μ^n averages over all the network while μ_a^n averages from the point of view of node a. The μ_a^n give much more information than μ^n, for instance about nodes isolated by saturated links within a network in a mean satisfactory state.

For any function ϕ on $\{0, 1, \ldots, C\}$

$$\phi(X^n_{ab}(t)) - \phi(X^n_{ab}(0)) - \int_0^t \phi^+(X^n_{ab}(s))\mathbb{1}_{X^n_{ab}(s)<C}$$
$$\times\Big(\nu + \frac{\nu}{n-2}\sum_{c:c\neq a,b}\big(\mathbb{1}_{X^n_{ac}(s)=C,X^n_{bc}(s)<C} + \mathbb{1}_{X^n_{bc}(s)=C,X^n_{ac}(s)<C}\ bigr)\Big)$$
$$+\phi^-(X^n_{ab}(s))X^n_{ab}(s)\,ds$$
$$= \phi(X^n_{ab}(t)) - \phi(X^n_{ab}(0)) - \int_0^t \phi^+(X^n_{ab}(s))\nu\mathbb{1}_{X^n_{ab}(s)<C}$$
$$\times(1+\langle\mathbb{1}_{x=C,y<C}+\mathbb{1}_{x<C,y=C}, \mu^{(2)n}_{ab,s}(dx,dy)\rangle)+\phi^-(X^n_{ab}(s))X^n_{ab}(s)\,ds \quad (1.11)$$

is a martingale. The empirical process $(\mu^n_t)_{t\geq 0}$ satisfies that for all ϕ

$$\langle\phi,\mu^n_t\rangle - \langle\phi,\mu^n_0\rangle - \int_0^t \langle\phi^+(x)\nu\mathbb{1}_{x<C} + \phi^-(x)x, \mu^n_s(dx)\rangle$$
$$+2\langle\phi^+(x)\nu\mathbb{1}_{x<C}\mathbb{1}_{y<C}\mathbb{1}_{z=C}, \mu^{(3)n}_s(dx,dy,dz)\rangle\,ds \quad (1.12)$$

is a martingale with Doob-Meyer bracket

$$\int_0^t \frac{1}{\binom{n}{2}}\langle\phi^+(x)^2\nu\mathbb{1}_{x<C} + \phi^-(x)^2x, \mu^n_s(dx)\rangle$$
$$+\frac{1}{\binom{n}{2}}\langle(\phi^+(x)+\phi^+(y))^2\nu\mathbb{1}_{x<C}\mathbb{1}_{y<C}\mathbb{1}_{z=C}, \mu^{(3)n}_s(dx,dy,dz)\rangle$$
$$+\frac{1}{\binom{n}{2}^2}\sum_{|a,b,c|=3}\phi^-(X^n_{ab}(s))\phi^-(X^n_{ac}(s))Y^{a,n}_{bc}(s)\,ds. \quad (1.13)$$

Definition 1.3. *The probability measure* $\tilde{\mathrm{P}}$ *on* $\mathrm{D}(\mathbb{R}+,\{0,1,\ldots,C\})$ *solves the nonlinear martingale problem (1.14) if for any function* ϕ

$$\phi(X_t)-\phi(X_0)-\int_0^t \phi^+(X_s)\nu\mathbb{1}_{X_s<C}(1+2\tilde{\mathrm{P}}^C_s(1-\tilde{\mathrm{P}}^C_s))+\phi^-(X_s)X_s\,ds \quad (1.14)$$

is a $\tilde{\mathrm{P}}$*-martingale.* $\tilde{\mathrm{P}}^C_s = \tilde{\mathrm{P}}_s(C) = \tilde{\mathrm{P}}(X_s = C)$ *is the nonlinearity.*

Lack of symmetry and local interaction Neither $\mu^{(3)n}$ nor $\mu^{(2)n}_{ab}$ factors as in (1.7) and (1.8) in terms of μ^n or μ^n_a, and (1.12) is far from closed and (1.11) cannot be expressed in terms of μ^n. This is the mark of strong local interaction: processes may interact directly only if they correspond to links forming a triangle.

The martingale problem for $\mu^{(3)N}$ involves an empirical measure $\mu^{(5)N}$ over quintuples (ab,ac,ad,bc,bd), and the one for $\mu^{(5)N}$ involves empirical measures over septuples of the form (ab,ac,ad,ae,bc,bd,be) and (ab,ac,ad,bc,bd,be,ce). This gives a complex hierarchy. We really have to prove the chaos hypothesis.

The following combinatorial result is from Graham and Méléard [9]. We do not use it since the local empirical measures interact in a very complex way.

Theorem 1.4. *$\mathcal{L}\big((X^n_{ab})_{1\le a<b\le k}\big)$ converges weakly to $\tilde{\mathrm{P}}^{\otimes k(k-1)/2}$ if and only if for any a, μ^n_a converges in law to $\tilde{\mathrm{P}}$.*

1.3 A general class of networks

Many generalizations of the previous example are considered in practice. There could be a maximal number k of retries over routes of length $l_1, \ldots, l_k$, and rejection and routing could depend on some optimization criterion on these routes, such as trunk reservation or choice of least-busy route.

We thus describe a large class of networks for which we shall prove chaoticity. It generalizes the class in Graham and Méléard [11]. We consider a system of interacting processes $(X^n_i)_{i\in I^n}$ for $n \ge 0$, with $I^n \subset I^{n+1}$. The processes are in $D(\mathbb{R}+, \mathbb{R}^d)$ and represent the amounts of different kinds of resources that are in use. We make no symmetry nor Markov assumption.

Tasks must be performed by the network and depend on the state of the resources. A given task will only concern a route r, which is an ordered subset of I^n. We denote by R^n the set of all the routes pertaining to the tasks and by A^n_r the process of arrivals of all tasks requiring route r. For $r \in R^n$ the process Y^n_r in $D(\mathbb{R}+, \mathbb{R}^d)$ represents the status of tasks that have required and obtained route r, and is used to govern simultaneous release of resources and similar phenomena. We assume $(X^n_i, Y^n_r)_{i\in I^n, r\in R^n}$ is a Markov process with a "local" evolution due to task arrival and completion:

Assumption on task arrival.

The evolution of $(X^n_i, Y^n_r)_{i\in I^n, r\in R^n}$ due to the arrival of a task with route r will depend on and influence only the states of X^n_i for $i \in r$ and Y^n_u for $u \subset r$. For $|r| \ge 2$ we have independent Poisson processes N^n_r of rates λ^n_r, and the jump times of A^n_r are included in those of N^n_r.

Assumption on task completion.

The evolution of X^n_i due to task completion depends only on X^n_i and on the Y^n_r such that $i \in r$. We normalize Y^n_r for $|r| \ge 2$ so that nothing happens when $Y^n_r = 0$. Holding times are arbitrary.

Interaction thus only happens due to routes r with $|r| \ge 2$, because of task arrival or when $Y^n_r \ne 0$ of task completion. We set for distinct i and j in I^n

$$\lambda^n_{ij} = \sum_{r:i,j\in r} \lambda^n_r, \qquad \lambda^n = \max_{i\ne j} \lambda^n_{ij}. \tag{1.16}$$

λ^n is the maximal rate at which any two particles may interact due the arrival of a task involving both. It must clearly go to zero for particles to become independent, which corresponds to a "dilution" of the interaction. For i in I^n

$$\lambda^n_i = \sum_{j:j\ne i} \lambda^n_{ij} = \sum_{r:i\in r, |r|\ge 2} \lambda^n_r, \quad \Lambda^n = \max_{i\in I^n} \lambda^n_i, \quad K^n = \max_{r:\lambda^n_r>0} |r|. \tag{1.17}$$

Λ^n is the maximal rate at which any particle may interact with others because of task arrival, and K^n the maximal number of particles involved.

Main result.

If λ^n converges to 0 and Λ^n and K^n are uniformly bounded, Theorem 2.1 states that the chaos hypothesis holds in variation norm for independent initial conditions and $Y_r^n(0) = 0$ for $|r| \geq 2$. Theorem 2.3 deduces a convergence result for multibody empirical measures. Theorem 2.4 shows that the law of one process is close in variation norm to the law of a Boltzmann process constructed on a Boltzmann tree. In many models the global effect of the system on one particle converges, and propagation of chaos to the Boltzmann process follows. We characterize under weak symmetry assumptions the limit law as the solution to a nonlinear martingale problem using the convergence of empirical measures.

Precise estimates are in Graham and Méléard [11]. A result for general chaotic initial conditions is in Theorem 1.4 in Graham and Méléard [9]. Holding times are arbitrary. The two examples enter this class:

The generalized star-shaped loss network.

$I_n = \{1, \ldots, n\}$ and R^n is composed of the subsets of K distinct indices in increasing order. $A_r^n = N_r^n$ and $\lambda^n = \lambda_{ij}^n = \binom{n-2}{K-2}\nu / \binom{n-1}{K-1} = (K-1)\nu/(n-1)$, $\Lambda^n = \nu$, $K^n = K$.

The fully-connected loss network with alternate routing.

$I^n = \{ab : 1 \leq a < b \leq n\}$, and a route will either be a triangle formed by the link that attempts a call and the two links that may be used for alternative routing, or a single link for call duration of direct calls. $R^n = \{(ab), (ab, ac, bc) : 1 \leq a < b \leq n, c \neq a, b\}$, $Y_{(ab)}^n = Y_{ab}^n$ and $Y_{(ab,ac,bc)}^n = Y_{ab}^{c,n}$. We take independent Poisson processes $N_{ab}^{c,n}$ of rate $\nu/(n-2)$ for $1 \leq a < b \leq n, c \neq a, b$, at each jump of which a call arrives on ab and in case of saturation attempts alternate routing through c. $A_{(ab,ac,bc)}^n = N_{(ab,ac,bc)}^n = N_{ab}^{c,n}$ and $\lambda^n = \lambda_{ij}^n = \lambda_r^n = \nu/(n-2)$, $\Lambda^n = 2\nu$, $K^n = 3$.

2. Asymptotic behavior of the networks

We now investigate the behavior of the network as it size increases, obtain chaotic results in variation norm on $[0, T]$ and asymptotics of empirical measures, and apply these results on the two examples. We are concerned with the behavior of a given subsystem or subnetwork as the global system increases in size.

2.1 The chaos hypothesis and the empirical measures

Theorem 2.1. *Assume* $(X_i^n(0), Y_i^n(0))_{i \in I^n}$ *independent,* $Y_r^n(0) = 0$ *for* $|r| \geq 2$, $\Lambda^n \leq \Lambda$, $K^n \leq K$. *Then there is an explicit bound* $B(\Lambda, T, K, k)$

such that for any distinct $i_1, \ldots, i_k$ *in some* I^n, $|\mathcal{L}(X^n_{i_1}, \ldots, X^n_{i_k}) - \mathcal{L}(X^n_{i_1}) \otimes \ldots \otimes \mathcal{L}(X^n_{i_k})|_T \leq \lambda^n B(\Lambda, T, K, k)$. *When* $\lim_{n\to\infty} \lambda^n = 0$, $\mathcal{L}(X^n_{i_1}, \ldots, X^n_{i_k}) - \mathcal{L}(X^n_{i_1}) \otimes \ldots \otimes \mathcal{L}(X^n_{i_k})$ *converges to zero for* $|\ |_T$ *and weakly for the Skorokhod topology on* $\mathbb{D}(\mathbb{R}+, (\mathbb{R}^d)^k)$ *and the chaos hypothesis holds.*

Proof. This is a simple adaptation of Graham and Méléard [11]. A sample-path construction in direct time is straightforward: when a task arrives on r at time t we perform the required Markovian evolution of $(X^n_i(t-))_{i\in r}$ and $Y^n_r(t-)$. The evolution of a process will be directly affected by other processes with which it performs tasks, and recursively these processes will be affected by other processes which thus influence indirectly our first process. Since for $|r| \geq 2$, $Y^n_r(0) = 0$ and the jump times of A^n_r are included in those of N^n_r, direct interaction between two distinct indices i and j can only happen after one of the processes N^n_r for $i, j \in r$ has jumped.

The interaction graphs

We now build the processes on $[0,T]$ in reverse time so that it can be done with the least amount of superfluous knowledge. This will define the past history of a particle, which contains all the particles which may have influenced it.

We describe this using interaction graphs, which are random marked subsets of $[0,T] \times I^n$. We imagine time as being vertical and directed upwards, and the links as being on a horizontal level. We work our way backward from time T and build a graph rooted on a given subset of I^n. Every time we encounter a jump of a Poisson process N^n_r which may influence the particles we consider, which means that the route r contains an index already in the graph, we include in the graph all the indices in r, and mark the graph by r itself and the variables needed for future evolution. This branching is deterministic given the Poisson processes and engenders $|r|$ branches; see Figure reffigure-carl1.

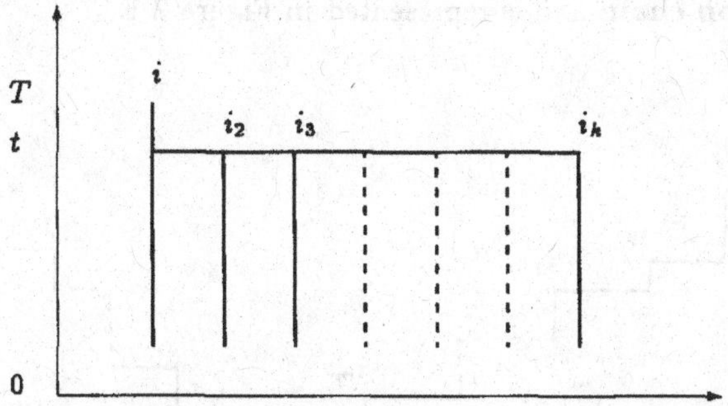

Fig. 2.1. Branching when N^n_r jumps at t, for $r = (i, i_2, i_3, \ldots, i_k)$.

Once an index is selected, the whole vertical line from the point of its first selection down to zero belongs to the graph, and we proceed recursively from the selected indices to define the graph. We do not have a tree, since a particle may influence another one several times. We can thus build graphs down from time T, rooted on any given set of indices $\{i_1, \ldots, i_k\}$, denoted by $G^n_{i_1 \ldots i_k}$.

At time zero, we consider the set of indices and the set of routes appearing in the graph and initial values $X^n_i(0)$ and $Y^n_r(0)$ for these sets. We then construct sample paths of the processes from 0 to T using as little superfluous knowledge as possible. External information for the evolution of $X_{i_1}, \ldots, X_{i_k}$ is brought into the graph only as needed. If the subgraphs rooted on two indices become disjoint in the limit, it will be possible using a coupling argument to show the particles become independent.

The coupling and priority rules

We detail the proof for two distinct indices i and j. We use two independent copies of all the variables needed, which we denote by superscripts i and j. We build two independent graphs $G^{i,n}_i$ and $G^{j,n}_j$ and independent processes $X^{i,n}_i$ and $X^{j,n}_j$ by using the above construction with the two independent copies. Then we build a single interaction graph G^n_{ij} in such a way that its subgraph G^n_i is as close as possible to $G^{i,n}_i$, and G^n_j to $G^{j,n}_j$.

This involves setting priority rules in order to build N^n_r from $N^{i,n}_r$ and $N^{j,n}_r$. If $i \in r$ and $j \notin r$ then we set $N^n_r = N^{i,n}_r$, if $j \in r$ and $i \notin r$ we set $N^n_r = N^{j,n}_r$, and if $i, j \in r$ we toss a fair coin. Because of the special properties of the Poisson process we are able to set the other priorities in reverse time and still obtain independent Poisson N^n_r of rate λ^n_r. If starting back from T, the graph $G^{i,n}_i$ intersects r before $G^{j,n}_j$ does, we set $N^n_r = N^{i,n}_r$, and if not $N^n_r = N^{j,n}_r$ (independence prevents a tie). We likewise couple all the variables we need.

There will be a problem only if one of a pair of Poisson processes for which we have set a priority as above jumps, since we then have to either disregard the jump or use it on both subtrees in G^n_{ij}. This event C^n_{ij} is called an **interaction chain** and is represented in Figure 2.1.

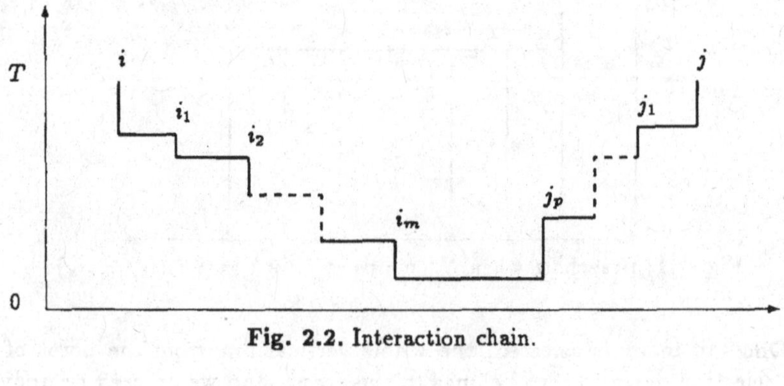

Fig. 2.2. Interaction chain.

The steps going down from i must be in $G_i^{i,n}$, those going down from j must be in $G_j^{j,n}$, except for the bottom step which must belong to one or the other. This represents the spread of influence in the network: in the past, two indices i_m and j_p had a direct interaction by competing for the same resource, and after that in direct time, i_m interacted directly with i_{m-1} and so on to i before T, and similarly j_p interacted directly with j_{p-1} and so on to j.

On the complement of C_{ij}^n, $(X_i^n, X_j^n) = (X_i^{i,n}, X_j^{j,n})$, and since $\mathcal{L}(X_i^{i,n}) = \mathcal{L}(X_i^n)$ and $\mathcal{L}(X_j^{j,n}) = \mathcal{L}(X_j^n)$, $|\mathcal{L}(X_i^n, X_j^n) - \mathcal{L}(X_i^n) \otimes \mathcal{L}(X_j^n)| \leq 2\mathrm{P}(C_{ij}^n)$.

Estimates and convergence

A recursive estimate on the probability of occurrence of such interaction chains gives the bound in Theorem 2.1. The iteration is on the length $q \geq 1$ of the interaction chains, where $q = m + p + 1$ in Figure 2. If Q_T denotes a bound on the probability of occurrence of a chain interaction and $Q_T(q)$ a bound for chains of length q, then $Q_T \leq \sum_{q \geq 1} Q_T(q)$.

$q = 1$ is the case of direct interaction in which a $N_r^{i,n}$ or a $N_r^{j,n}$ for $i, j \in r$ jumps between 0 and T, an event of probability bounded by $1 - \exp(-2\lambda^n T)$.

For $q \geq 2$, an interaction chain happens only if $G_i^{i,n}$ or $G_j^{j,n}$ branches on a link which is then (in reverse time) joined by a chain of length $q - 1$ to the root of the other graph. The top branching occurs at rate bounded by $2\Lambda^n$ and generates at most $K^n - 1$ new branches. Hence

$$Q_T(q) \leq (K^n - 1)\, 2\Lambda^n \int_0^T Q_{T-t}(q-1) e^{-2\Lambda^n t}\, dt$$

and we get the estimate using convolution and the gamma functions. □

The uniform convergence in Theorem 2.1 is very strong. We immediately get results on empirical measures without any symmetry assumption.

Theorem 2.2. *Under the assumptions of Theorem 2.1, let J_k^n be a set of disjoint k-tuples of distinct indices in I^n. Then uniformly for $\|\phi\|_\infty \leq 1$*

$$\mathrm{E}\left(\left\langle \phi, \frac{1}{|J_k^n|} \sum_{j_1 \ldots j_k \in J_k^n} \delta_{X_{j_1}^n, \ldots, X_{j_k}^n} - \mathcal{L}(X_{j_1}^n) \otimes \ldots \otimes \mathcal{L}(X_{j_k}^n) \right\rangle^2\right) = O\left(\lambda^n \vee \frac{1}{|J_k^n|}\right).$$

Proof. We develop the square. Each of the $|J_k^n|$ diagonal terms is bounded by $1/|J_k^n|^2$. We must study more precisely the contribution of the product terms

$$\frac{1}{|J_k^n|^2} \sum_{i_1 \ldots i_q \neq j_1 \ldots j_q \in J_k^n} \mathrm{E}\big((\phi(X_{i_1}^n, \ldots, X_{i_q}^n) - \langle \phi, \mathcal{L}(X_{i_1}^n) \otimes \ldots \otimes \mathcal{L}(X_{i_k}^n)\rangle)$$
$$\times (\phi(X_{j_1}^n, \ldots, X_{j_k}^n) - \langle \phi, \mathcal{L}(X_{j_1}^n) \otimes \ldots \otimes \mathcal{L}(X_{j_k}^n)\rangle)\big)$$

which is $O(\lambda^n)$ because of the uniform bound in Theorem 2.1. □

Theorem 2.3. *Under the assumptions of Theorem 2.2, let* $\lim_{n\to\infty} \lambda^n = 0$ *and* $\lim_{n\to\infty} |J_k^n| = \infty$. *Then in probability for the weak topology for the Skorokhod topology on* $D(\mathbb{R}+, (\mathbb{R}^d)^k)$,

$$\frac{1}{|J_k^n|} \sum_{j_1\ldots j_k \in J_k^n} \delta_{X_{j_1}^n,\ldots,X_{j_k}^n} - \frac{1}{|J_k^n|} \sum_{j_1\ldots j_k \in J_k^n} \mathcal{L}(X_{j_1}^n) \otimes \ldots \otimes \mathcal{L}(X_{j_k}^n) = o(1).$$

We have $O(\sqrt{\lambda^n} \vee |J_k^n|^{-1/2})$ *estimates in appropriate metrics.*

Proof. Since the Skorokhod space is Polish, there exists a sequence of continuous functions $(\phi_i)_{i\geq 1}$ bounded by 1 such that m^n converges weakly to m on $\mathcal{P}(D)$ if and only for any $i \geq 1$, $\langle \phi_i, m^n \rangle$ converges to $\langle \phi_i, m \rangle$. We can then metrize weak convergence by $d(m, m')^2 = \frac{1}{i^2} \sum_{i\geq 1} \langle \phi_i, m - m' \rangle^2$ and we see using Theorem 2.2 that we have actually convergence in L^2 with speed of convergence. □

2.2 The limit Boltzmann tree and Boltzmann processes

As n goes to infinity, it is natural to compare a particle, constructed pathwise on the interaction graph, to another constructed on a Boltzmann tree on which all the branches are distinct and carry independent jump variables. Heuristically, there are so many particles that each interaction brings in new ones from an "ocean" of independent ones.

For given n and index i, we build a multitype interaction tree T_i^n inspired from G_i^n. Each branch of the tree shall have a distinct label, and it is practical to use finite sequences in I^n giving the successive filiation from the root. The last index in the sequence represents the index in the graph. We need independent Poisson processes and random jump amplitude variables thus indexed.

We start from a root labeled i and work backward in time from T. At the first encounter with a jump of a $N_r^{i,n}$ for a r containing i the branch dies and branches are born labeled ij for all j in r including i. Recursively, from a branch $ii_1 \ldots i_k$, at the first encounter with a jump of a $N_r^{ii_1\ldots i_k,n}$ for a r containing i_k, the branches born are labeled $ii_1 \ldots i_k j$ for j in r. Once zero is reached, we use the initial values indexed by the last indices of the labeling, and jump amplitude processes labeled by the branches, and construct processes $\bar{X}_{ii_1\ldots i_k}^n$ on each branch as in Section 2.1.

We then build a process $\tilde{X}_i^n$ alive on $[0, T]$ by collecting all the processes $\bar{X}_i^n$, $\bar{X}_{ii}^n$, ..., $\bar{X}_{i\ldots i}^n$, ... indexed by finite sequences consisting only of i. This Boltzmann process is thus without "self-interactions".

Theorem 2.4. *We assume* $Y_r^n(0) = 0$ *for all routes* r *with* $|r| \geq 2$, $\Lambda^n \leq \Lambda$, *and* $K^n \leq K$. *Then there is an explicit bound* $B(\Lambda, T, K)$ *such that for any* i *in some* I^n, $|\mathcal{L}(X_i^n) - \mathcal{L}(\tilde{X}_i^n)|_T \leq \lambda^n B(\Lambda, T, K)$. *When* $\lim_{n\to\infty} \lambda^n = 0$, $\mathcal{L}(X_i^n) - \mathcal{L}(\tilde{X}_i^n)$ *converges to 0 for* $|\ |_T$ *and weakly for the Skorokhod topology on* $D(\mathbb{R}+, \mathbb{R}^d)$.

Proof. The proof relies on building a coupled interaction graph from a tree, and the same for the processes. This is possible without modifying the process except if an event called an **interaction loop** happens, meaning that at a certain depth of the tree two branches are born which are then joined by an interaction chain. The probability of such an event is estimated recursively on the depth in the tree in which it happens, using Q_T^n, convolutions, and the gamma functions. □

2.3 Propagation of chaos and slight symmetry

It is hopeless to get a limit law if we do not make assumptions on the relationship between particle systems for different n. In many models the trees constructed for different n are equivalent. When the $(X_i^n(0))_{i\in I^n}$ are independent and their laws and the jump amplitude laws do not depend on n or converge, it is the same for the laws of the the Boltzmann processes $\tilde{X}_i^n$. Then Theorems 2.1 and 2.4 give propagation of chaos in variation norm with rate of convergence.

When in the martingale problem for X_i^n the process interacts with other processes with certain indices, the branching for the tree is multitype for these indices. It is then enough that the interaction converges. This is the case in our examples, and in most network models in which the offered load for a link remains constant. In general the Y_r^n are not relevant and are used only once.

There is often some symmetry in the particle system, for instance over the n nodes in the network with alternate routing. The symmetry and markovian generators give martingale problems involving certain empirical measures, which may give a limit nonlinear martingale problem using convergence in variation and Theorem 2.3. It is simple to thus show that the limit of the processes solving respectively the martingale problems (1.3) and (1.11) solve respectively the nonlinear martingale problems (1.6) and (1.14). General chaotic initial assumptions are in Theorem 1.4 in Graham and Méléard [9].

Theorem 2.5. *Assume all $X_i^n(0) = 0$. Then $|\mathcal{L}(X_{i_1}^n, \ldots, X_{i_k}^n) - \tilde{\mathrm{P}}^{\otimes k}|_T = O(1/n)$ uniformly for distinct $i_1, \ldots, i_k$, where $\tilde{\mathrm{P}}$ is the unique solution to the nonlinear martingale problem (1.6) starting at δ_0.*

Theorem 2.6. *Assume $(X_{ab}^n(0))_{1\le a<b\le n}$ i.i.d. of law $\tilde{\mathrm{P}}_0$ and all $Y_{ab}^{c,n}(0) = 0$. Then $|\mathcal{L}(X_{i_1}^n, \ldots, X_{i_k}^n) - \tilde{\mathrm{P}}^{\otimes k}|_T = O(1/n)$ uniformly for distinct $i_1, \ldots, i_k$ in some $I^n = \{ab : 1 \le a < b \le n\}$, where $\tilde{\mathrm{P}}$ is the unique solution to the nonlinear martingale problem (1.14) starting at $\tilde{\mathrm{P}}_0$.*

The flow of time-marginals $(\tilde{\mathrm{P}}_t)_{t\ge 0}$ of the limits follow an evolution equation obtained by taking expectations in (1.6) and (1.14). $(\tilde{\mathrm{P}}_t(0), \ldots, \tilde{\mathrm{P}}_t(C))_{t\ge 0}$ follows a nonlinear ODE on the simplex, of which the fixed points give a hint on the limiting stationary and blocking probabilities.

3. Speed of convergence and fluctuations

We are interested in finding the speed of convergence for the law of large numbers of the empirical measures in the two examples. Because of simultaneous jumps we cannot use a Girsanov transform and work on path space as in Shiga and Tanaka [20], nor directly use techniques in Dawson and Zheng [3] and Uchiyama [23] which require a "closed form" for the fluctuations.

The flow of empirical fluctuations $(\eta_t^n)_{t\geq 0}$ is in $D(\mathbb{R}+, \mathcal{M}(\{0,1,\ldots,C\}))$ and is isomorphic to the $\mathbb{R}^{C+1}$-valued process $(\eta_t^n(0),\ldots,\eta_t^n(C))_{t\geq 0}$. The evolution equations can be easily rewritten as ODEs on $\mathbb{R}^{C+1}$. We proceed in three steps:

1. We write a martingale problem satisfied by $(\eta_t^n)_{t\geq 0}$. It contains terms due to local interaction and simultaneous release.
2. We prove bounds which give tightness, using finite-dimensional criteria.
3. We study the limit of the martingale problem for $(\eta_t^n)_{t\geq 0}$. The mean-field terms will be simple but not the local interaction and simultaneous release ones.

We prove convergence to an Ornstein-Uhlenbeck process for the star-shaped network, and characterize the Gaussian martingale part for the much more difficult network with alternate routing.

3.1 The star-shaped network

For simplicity we assume an asymptotically empty initial network. Theorems 2.1, 2.3 and 2.5 show that $\lim_{n\to\infty}\mu^n = \tilde{\mathrm{P}}$. We define the empirical fluctuations

$$\eta^n = \sqrt{n}(\mu^n - \tilde{\mathrm{P}}) = \frac{1}{\sqrt{n}}\sum_{i=1}^{n}\left(\delta_{X_i^n} - \tilde{\mathrm{P}}\right), \quad \eta^{K,n} = \sqrt{n}(\mu^{K,n} - \tilde{\mathrm{P}}^{\otimes K}). \quad (3.1)$$

3.1.1 The martingale problem.

Proposition 3.1. *For all functions ϕ on $\{0,1,\ldots,C\}$,*

$$\langle\phi,\eta_t^n\rangle - \langle\phi,\eta_0^n\rangle$$
$$- \int_0^t \langle\phi^+(x_1)\nu\mathbb{1}_{x_1<C,\ldots,x_K<C}, \eta_s^{K,n}(dx_1\ldots dx_K)\rangle + \langle\phi^-(x)x, \eta_s^n(dx)\rangle\, ds \quad (3.2)$$

is a martingale with Doob-Meyer bracket

$$\int_0^t \langle(\phi^+(x_1)+\cdots+\phi^+(x_K))^2\frac{\nu}{K}\mathbb{1}_{x_1<C,\ldots,x_K<C}, \mu_s^{K,n}(dx_1\ldots dx_K)\rangle$$
$$+\frac{1}{n}\sum_{i_1<\ldots<i_K}(\phi^-(X_{i_1}^n)+\cdots+\phi^-(X_{i_K}^n))^2 Y_{i_1\ldots i_K}^n(s)\, ds. \quad (3.3)$$

See Section 1.1. (3.2) is closed using the factoring result (1.8), which implies $|\eta^{K,n} - \sqrt{n}((\mu^n)^{\otimes K} - \tilde{\mathrm{P}}^{\otimes K})| = O(1/\sqrt{n})$, and a simple linearization:

$$\sqrt{n}((\mu^n)^{\otimes K} - \tilde{\mathrm{P}}^{\otimes K}) = \sum_{i=1}^{K} (\mu^n)^{\otimes K-i} \otimes \sqrt{n}(\mu^n - \tilde{\mathrm{P}}) \otimes \tilde{\mathrm{P}}^{\otimes i-1}. \tag{3.4}$$

Propagation of chaos closes (3.4) in terms of $\eta^n = \sqrt{n}(\mu^n - \tilde{\mathrm{P}})$, thus closing (3.2), and gives the limit of the first term of (3.3). The second term in (3.3) is specific to the strong interaction due to simultaneous release.

3.1.2 Tightness estimates. We could obtain tightness estimates by the Gronwall Lemma since (3.2) is almost closed. We use the estimates in Theorem 2.2 for simplicity:

Proposition 3.2. *$\mathrm{E}(\langle\phi, \eta^n\rangle^2)$ is bounded for ϕ in $\mathcal{L}^\infty(\mathrm{D}([0,T], \{0,1,\ldots,C\}))$ and $\mathrm{E}(\langle\psi, \eta^{K,n}\rangle^2)$ is bounded for ψ in $\mathcal{L}^\infty(\mathrm{D}([0,T], \{0,1,\ldots,C\}^K))$ uniformly in $||\phi||_\infty$ and $||\psi||_\infty$.*

This bound and $\sum_{r\in R_i^n} Y_r^n = X_i^n \le C$ give tightness using Proposition 3.2.3 in Joffe and Métivier [15]. Continuity of limit points is obvious.

Theorem 3.3. *The $(\eta_t^n)_{t\ge 0}$ are tight in $\mathrm{D}(\mathbb{R}+, \mathbb{R}^{C+1})$ and any accumulation point is continuous.*

3.1.3 Convergence. Convergence would follow from uniqueness of the accumulation points. It is natural to show that any accumulation point must satisfy a martingale problem limit of (3.2), (3.3) with unique solution.

A process governing simultaneous release.
$Y_{ij}^n = \sum_{r\in R_{ij}^n} Y_r^n$ denotes the number of calls in progress that involve both i and j. We introduce the process on $\mathcal{M}_+(\{0,1,\ldots,C\}^2)$

$$\theta^n = \frac{1}{n}\sum_{i\ne j} Y_{ij}^n \delta_{X_i^n, X_j^n}. \tag{3.5}$$

θ_0^n converges to zero if $X_i^n(0)$ does, and θ^n is uniformly bounded since

$$|\theta_t^n| = \frac{1}{n}\sum_{i=1}^{n}\sum_{j:j\ne i} Y_{ij}^n, \quad \sum_{j:j\ne i} Y_{ij}^n = (K-1)X_i \le (K-1)C. \tag{3.6}$$

Theorem 3.4. *We assume that $|\mathcal{L}(X_1^n,\ldots,X_n^n) - \delta_0| = o(1)$ and η_0^n converges to η_0. Let $\tilde{\mathrm{P}}$ be the solution to the nonlinear martingale problem (1.6) starting at 0. Then $(\eta_t^n)_{t\ge 0}$ converges in law to the unique Gaussian Ornstein-Uhlenbeck process such that for each function ϕ*

$$\langle\phi, \eta_t\rangle - \langle\phi, \eta_0\rangle - \int_0^t \nu(1-\tilde{\mathrm{P}}_s^C)^{K-1}\langle\phi^+(x)\mathbb{1}_{x<C}, \eta_s(dx)\rangle + \langle\phi^-(x)x, \eta_s(dx)\rangle$$
$$+\nu(K-1)(1-\tilde{\mathrm{P}}_s^C)^{K-2}\langle\phi^+(x)\mathbb{1}_{x<C}, \tilde{\mathrm{P}}_s(dx)\rangle\eta_s(0,\ldots,C-1)\,ds \tag{3.7}$$

is a Gaussian continuous martingale with deterministic Doob-Meyer bracket

$$\int_0^t \nu(1-\tilde{P}_s^C)^{K-1}\langle \phi^+(x)^2 1\!\!1_{x<C}, \tilde{P}_s(dx)\rangle + \langle \phi^-(x)^2 x, \tilde{P}_s(dx)\rangle$$
$$+\nu(K-1)(1-\tilde{P}_s^C)^{K-2}\langle \phi^+(x)1\!\!1_{x<C}, \tilde{P}_s(dx)\rangle^2 + \langle \phi^-\otimes\phi^-, \theta_s\rangle\, ds, \quad (3.8)$$

where θ is the deterministic process on $\mathcal{M}_+(\{0,\ldots,C\}^2)$ solving the affine evolution problem with unique solution that for any functions α and β on $\{0,\ldots,C\}$

$$\langle \alpha\otimes\beta, \theta_t\rangle = \int_0^t \langle \nu(1-\tilde{P}_s)^{K-1}(\alpha^+(x)\beta(y)1\!\!1_{x<C} + \alpha(x)\beta^+(y)1\!\!1_{y<C})$$
$$+\alpha^-(x)\beta(y)(x-1) + \alpha(x)\beta^-(y)(y-1) - \alpha(x)\beta(y), \theta_s(dx,dy)\rangle$$
$$+\nu(K-1)(1-\tilde{P}_s)^{K-2}\langle \alpha(x+1)1\!\!1_{x<C}, \tilde{P}_s(dx)\rangle\langle \beta(x+1)1\!\!1_{x<C}, \tilde{P}_s(dx)\rangle\, ds (3.9)$$

and describes the limit behavior of the simultaneous releases.

Proof. Classical martingale characterization techniques show that for any accumulation point of $(\eta_t^n)_{t\geq 0}$, the limit of (3.2) should be a martingale with Doob-Meyer bracket given by the limit of (3.3), if these limits exist. (3.2) characterizes the drift term and (3.3) the martingale part of the limit.

We have seen in Section 3.1.1 that (3.2) is almost closed. (1.8), (3.4), propagation of chaos, and symmetry give (3.7). This is general for mean-field interaction. The first term in (3.3) converges by propagation of chaos to

$$\int_0^t \langle (\phi^+(x_1)+\cdots+\phi^+(x_K))^2 \frac{\nu}{K} 1\!\!1_{x_1<C,\ldots,x_K<C}, \tilde{P}_s^{\otimes K}(dx_1\ldots dx_K)\rangle\, ds \quad (3.10)$$

in which we develop the square and use symmetry to get the ϕ^+ terms in (3.8). For mean-field interaction, this would be all, but we must consider the second term in (3.3) coming from the strong interaction due to simultaneous release

$$\frac{1}{n}\sum_{i_1<\ldots<i_K} (\phi^-(X_{i_1}^n)+\cdots+\phi^-(X_{i_K}^n))^2 Y_{i_1\ldots i_K}^n(s).$$

We develop the square. The diagonal terms give

$$\frac{1}{n}\sum_{i_1<\ldots<i_K}\sum_{i\in\{i_1,\ldots,i_K\}} \phi^-(X_i^n)^2 Y_{i_1\ldots i_K}^n = \frac{1}{n}\sum_{i=1}^n \phi^-(X_i^n)^2 \sum_{r\in R_i^n} Y_r^n$$
$$= \frac{1}{n}\sum_{i=1}^n \phi^-(X_i^n)^2 X_i^n = \langle \phi^-(x)^2 x, \mu^n(dx)\rangle \quad (3.11)$$

and converge to $\langle \phi^-(x)^2 x, \tilde{P}(dx)\rangle$. The product terms give

$$\frac{1}{n}\sum_{i_1<\ldots<i_K}\sum_{\substack{i\neq j\\ i,j\in\{i_1,\ldots,i_K\}}}\phi^-(X_i^n)\phi^-(X_j^n)Y_{i_1\ldots i_K}^n=\frac{1}{n}\sum_{i\neq j}\phi^-(X_i^n)\phi^-(X_j^n)\sum_{r\in R_{ij}^n}Y_r^n$$
$$=\frac{1}{n}\sum_{i\neq j}\phi^-(X_i^n)\phi^-(X_j^n)Y_{ij}^n=\langle\phi^-\otimes\phi^-,\theta^n\rangle. \quad (3.12)$$

We write a martingale problem for $\langle\alpha\otimes\beta,\theta_t\rangle$ and show the martingale part vanishes and the drift term converges to (3.9).

$$\begin{aligned}
&Y_{ij}^n(t)\alpha(X_i^n(t))\beta(X_j^n(t))-Y_{ij}^n(0)\alpha(X_i^n(0))\beta(X_j^n(0))\\
&-\int_0^t\frac{\nu}{\binom{n-1}{K-1}}\Bigg(Y_{ij}^n(s)\alpha^+(X_i^n(s))\beta(X_j^n(s))1\!\!1_{X_i^n(s)<C}\sum_{\substack{i_2<\ldots<i_K\\ i,j\notin\{i_2,\ldots,i_K\}}}1\!\!1_{X_{i_2}^n(s)<C,\ldots,X_{i_K}^n(s)<C}\\
&\qquad+Y_{ij}^n(s)\alpha(X_i^n(s))\beta^+(X_j^n(s))1\!\!1_{X_j^n(s)<C}\sum_{\substack{i_2<\ldots<i_K\\ i,j\notin\{i_2,\ldots,i_K\}}}1\!\!1_{X_{i_2}^n(s)<C,\ldots,X_{i_K}^n(s)<C}\\
&\qquad+\big((Y_{ij}^n(s)+1)\alpha(X_i^n(s)+1)\beta(X_j^n(s)+1)-Y_{ij}^n(s)\alpha(X_i^n(s))\beta(X_j^n(s))\big)\\
&\qquad\qquad\times1\!\!1_{X_i^n(s)<C,X_j^n(s)<C}\sum_{\substack{i_3<\ldots<i_K\\ i,j\notin\{i_3,\ldots,i_K\}}}1\!\!1_{X_{i_3}^n(s)<C,\ldots,X_{i_K}^n(s)<C}\Bigg)\\
&+Y_{ij}^n(s)\alpha^-(X_i^n(s))\beta(X_j^n(s))(X_i^n(s)-Y_{ij}^n(s))\\
&+Y_{ij}^n(s)\alpha(X_i^n(s))\beta^-(X_j^n(s))(X_j^n(s)-Y_{ij}^n(s))\\
&+\big((Y_{ij}^n(s)-1)\alpha(X_i^n(s)-1)\beta(X_j^n(s)-1)-Y_{ij}^n(s)\alpha(X_i^n(s))\beta(X_j^n(s))\big)Y_{ij}^n(s)\,ds
\end{aligned}\quad(3.13)$$

is a martingale M_{ij}^n. $M^n=\frac{1}{n}\sum_{i\neq j}M_{ij}^n$ is the martingale for $\langle\alpha\otimes\beta,\theta_t^n\rangle$.

Lemma 3.5. $\mathrm{E}(\sup_{0\le t\le T}M^n(t)^2)=O(1/n)$, *uniformly in* T, $||\alpha||_\infty$ *and* $||\beta||_\infty$.

Proof. This is implied by $\mathrm{E}([M^n]_T)=O(1/n)$, where $[M^n]_T$ is the sum of the squares of the jumps of M and thus has jumps of size $O(1/n^2)$. There are nC circuits in the network and thus there are $O(n)$ calls present at the start, and their contribution to $\mathrm{E}([M^n]_T)$ is $O(1/n)$. The ends of the calls arrived afterward contribute at most as much as their arrival, so we now only need to show that the contribution from call arrivals is $O(n)$, which is the case since there are $\binom{n}{K}$ routes each with an arrival rate $\nu/\binom{n-1}{K-1}$. □

Continuation of the proof of Theorem 3.4 . θ^n jumps as $[M^n]$ except that its jumps are of size $O(1/n)$. Thus $(\theta_t^n)_{t\geq0}$ is tight and its accumulation points are absolutely continuous and have no martingale part. We only need to close the hierarchy implicit in (3.13) to obtain the deterministic affine evolution problem satisfied by the limit process $(\theta_t)_{t\geq0}$. This martingale problem involves the higher-order measure-valued processes

$$\begin{cases} \theta^{K,n} & = \dfrac{K-1}{(n)_K} \displaystyle\sum_{|i,j,i_3,\ldots,i_K|=K} Y^n_{ij}\delta_{X^n_i,X^n_j,X^n_{i_3},\ldots,X^n_{i_K}} \\ \theta^{K+1,n} & = \dfrac{1}{(n)_K} \displaystyle\sum_{|i,j,i_2,\ldots,i_K|=K+1} Y^n_{ij}\delta_{X^n_i,X^n_j,X^n_{i_2},\ldots,X^n_{i_K}} \end{cases} \tag{3.14}$$

the first of which vanishes since its total mass is seen to be $O(1/n)$: there are not enough routes containing both i and j for them to be seen at the limit. Instead, $\theta^{K+1,n}$ necessitates a delicate factorization.

Lemma 3.6. $\mathrm{E}(\langle\alpha_0\otimes\cdots\otimes\alpha_K, \theta^{K+1,n}_t - \theta^n_t\otimes\tilde{\mathrm{P}}^{\otimes K-1}_t\rangle^2) = O(1/n)$ *uniformly on* $[0,T]$, $\|\alpha_0\|_\infty, \ldots, \|\alpha_K\|_\infty$.

Proof. $\mathrm{E}(\langle\alpha_0\otimes\cdots\otimes\alpha_K, \theta^n_t\otimes\tilde{\mathrm{P}}^{\otimes K-1}_t - \theta^n_t\otimes\mu^{K-1,n}\rangle^2) = O(1/n)$ using Theorem 2.2, where $\mu^{K-1,n}$ is defined in (1.2).

$$\begin{aligned} \theta^{K+1,n} - \theta^n_t \otimes \frac{n}{n-K+1}\mu^{K-1,n} &= \frac{1}{(n)_K}\Bigg(\sum_{|i,j,i_2,\ldots,i_K|=K+1} Y^n_{ij}\delta_{X^n_i,X^n_j,X^n_{i_2},\ldots,X^n_{i_K}} \\ &\qquad - \sum_{i\neq j}\ \sum_{|i_2,\ldots,i_K|=K-1} Y^n_{ij}\delta_{X^n_i,X^n_j,X^n_{i_2},\ldots,X^n_{i_K}} \Bigg) \end{aligned} \tag{3.15}$$

is bounded in variation norm by

$$\frac{1}{(n)_K}\sum_{i\neq j} Y^n_{ij} |\{i_2,\ldots,i_K\} : |i_2,\ldots,i_K| = K-1,\ |i,j,i_2,\ldots,i_K| \leq K| \tag{3.16}$$

which is uniformly $O(1/n)$ using (3.6). □

Continuation of the proof of Theorem 3.4. We now consider (3.13) and find the limit deterministic evolution equation (3.9) which has an unique solution. The affine part comes from the 1 in $Y^n_{ij}+1$ in the term involving jumps of Y^n_{ij} and propagation of chaos and we use

Lemma 3.7. *Uniformly for t in $[0,T]$, for $i\neq j$,* $\mathrm{P}(Y^n_{ij}(t) > 0) = O(1/n)$, $\mathrm{E}(Y^n_{ij}(t)) = O(1/n)$ *and* $\mathrm{E}(|(Y^n_{ij}(t))^2 - Y^n_{ij}(t)|) = o(1/n)$, *and if $k\neq l$ and* $\{i,j\}\neq\{k,l\}$ *then* $\mathrm{P}(Y^n_{ij}(t)Y^n_{kl}(t) > 0) = O(1/n^2)$ *and* $\mathrm{E}(Y^n_{ij}(t)Y^n_{kl}(t)) = O(1/n^2)$.

Proof. $\mathrm{P}(Y^n_{ij}(t) > 0) \leq \mathrm{E}(Y^n_{ij}(t)) = O(1/n)$ because of (3.6). By developing $(X^n_i)^2$ we see that for $|i,j,k| = 3$, $\mathrm{E}(Y^n_{ij}(t)Y^n_{ik}(t)) = O(1/n^2)$ and thus $\mathrm{P}(Y^n_{ij}(t)Y^n_{ik}(t) > 0) = O(1/n^2)$. By developing $X^n_iX^n_j$ for distinct i and j, we see that for $|i,j,k,l| = 4$, $\mathrm{E}(Y^n_{ij}(t)Y^n_{kl}(t)) = O(1/n^2)$ and thus $\mathrm{P}(Y^n_{ij}(t)Y^n_{kl}(t) \neq 0) = O(1/n^2)$. Since $\mathrm{P}(Y^n_{ij}(0) > 0) \leq \mathrm{E}(X^n_i(0)) = o(1)$ and Y^n_{ij} increases at rate $O(1/n)$, $\mathrm{P}(Y^n_{ij}(t) \geq 2) = o(1)$. □

Theorem 3.4 is now proved since a continuous martingale with known deterministic Doob-Meyer bracket is Gaussian with determined law. □

3.2 The network with alternate routing

Every step of the proof is much more difficult and we attain only partial results using fine estimates obtained with interaction graphs and precise couplings. All the algebraic factoring properties above have to be proved the hard way using these estimates. $X^n_{ab}(0)$ is i.i.d. and $Y^{c,n}_{ab}(0) = 0$. We define

$$\eta^n = \sqrt{\tbinom{n}{2}}(\mu^n - \tilde{\mathrm{P}}), \quad \eta^{(3)n} = \sqrt{\tbinom{n}{2}}(\mu^{(3)n} - \tilde{\mathrm{P}}^{\otimes 3}) \tag{3.17}$$

and could likewise define a local empirical fluctuation at node a.

3.2.1 The martingale problem.

Proposition 3.8. *For all functions* ϕ *on* $\{0,1,\ldots,C\}$

$$\begin{aligned}\langle \phi, \eta^n_t \rangle - \langle \phi, \eta^n_0 \rangle - \int_0^t & \langle \phi^+(x)\nu 1\!\!1_{x<C} + \phi^-(x)x, \eta^n_s(dx)\rangle \\ & + 2\langle \phi^+(x)\nu 1\!\!1_{x<C} 1\!\!1_{y<C} 1\!\!1_{z=C}, \eta^{(3)n}_s(dx,dy,dz)\rangle\, ds \end{aligned} \tag{3.18}$$

is a martingale with Doob-Meyer bracket

$$\begin{aligned}\int_0^t & \langle \phi^+(x)^2\nu 1\!\!1_{x<C} + \phi^-(x)^2 x, \mu^n_s(dx)\rangle \\ & + \langle (\phi^+(x)+\phi^+(y))^2 \nu 1\!\!1_{x<C} 1\!\!1_{y<C} 1\!\!1_{z=C}, \mu^{(3)n}_s(dx,dy,dz)\rangle \\ & + \frac{1}{\binom{n}{2}} \sum_{|a,b,c|=3} \phi^-(X^n_{ab}(s))\phi^-(X^n_{ac}(s)) Y^{a,n}_{bc}(s)\, ds. \end{aligned} \tag{3.19}$$

See Section 1.2. It is to be stressed that (3.18) **cannot** be closed easily, since $\mu^{(3)n}$ differs widely from $(\mu^n)^{\otimes 3}$ and the factoring (1.8) and subsequent linearization (3.4) cannot be applied. Propagation of chaos gives the asymptotics of the first terms in (3.19) but not of the last.

3.2.2 Tightness estimates. We cannot apply the Gronwall Lemma since (3.18) is not closed. We have to use specific fine estimates on interaction graphs, the least of which states that if a, b, c, d are distinct then $\mathcal{L}(X^n_{ab}, X^n_{cd}) - \mathcal{L}(X^n_{ab}) \otimes \mathcal{L}(X^n_{cd}) = O(1/n^2)$ since there cannot be a direct interaction. This is needed merely to bound η^n_t.

3.2.3 Convergence. We get a partial result, and the factoring corresponding to Lemmas 3.5 and 3.6 is no longer algebraic but must be proved by complex interaction graph estimates. (3.5) is replaced by the local measure

$$\theta^n = \frac{1}{(n)_2} \sum_{|a,b,c|=3} Y^{a,n}_{bc} \delta_{X^n_{ab}, X^n_{ac}}. \tag{3.20}$$

Theorem 3.9. *The accumulation points for the fluctuation processes* $(\eta_t^n)_{t\geq 0}$ *are continuous semimartingales. The martingale part is Gaussian, and its law is characterized by its deterministic Doob-Meyer bracket*

$$\int_0^t \langle \phi^+(x)^2 \nu \mathbb{1}_{x<C}(1+2\tilde{\mathrm{P}}_s^C(1-\tilde{\mathrm{P}}_s^C)) + \phi^-(x)^2 x), \tilde{\mathrm{P}}_s(dx)\rangle$$
$$+2\nu\tilde{\mathrm{P}}_s^C\left(\int \phi^+(x)\mathbb{1}_{x<C}\,\tilde{\mathrm{P}}_s(dx)\right)^2 + 2\langle\phi^-\otimes\phi^-,\theta_t\rangle\, ds \quad (3.21)$$

where $(\theta_t)_{t\geq 0}$ *is the deterministic process on* $\mathcal{M}_+(\{0,\ldots,C\}^2)$ *solving the affine evolution problem with unique solution*

$$\langle\alpha\otimes\beta,\theta_t^n\rangle = \int_0^t \langle\left(\alpha^+(x)\beta(y)\mathbb{1}_{x<C} + \alpha(x)\beta^+(y)\mathbb{1}_{y<C}\right)\left(\nu + 2\nu\tilde{\mathrm{P}}_s^C(1-\tilde{\mathrm{P}}_s^C)\right)$$
$$+\alpha^-(x)\beta(y)(x-1) + \alpha(x)\beta^-(y)(y-1) - \alpha(x)\beta(y)), \theta_s(dx,dy)\rangle$$
$$+\nu\tilde{\mathrm{P}}_s^C\langle\alpha(x+1)\mathbb{1}_{x<C}, \tilde{\mathrm{P}}_s(dx)\rangle\langle\beta(x+1)\mathbb{1}_{x<C}, \tilde{\mathrm{P}}_s(dx)\rangle\, ds \quad (3.22)$$

for any functions α *and* β *on* $\{0,\ldots,C\}$. *A similar result holds for the local empirical measures.*

References

1. Anantharam, V., "A mean-field limit for a lattice caricature of dynamic routing in circuit switched networks", *Ann. Appl. Prob.* 1, 481-503 (1991).
2. Crametz, J.P. Hunt, P.J., "A limit result respecting graph structure for a fully connected network with alternative routing", *Ann. Appl. Prob.* 1, 436-444 (1991).
3. Dawson, D.A., Zheng, X., "Law of large numbers and central limit theorem for unbounded jump mean-field models", *Adv. Appl. Math.* 12, 293-326, (1991).
4. Dobrushin, R.L., "Queuing networks - without explicit solutions and without computer simulation", Keynote lecture, Conference on applied probability in engineering, computer and communication sciences INRIA-ORSA-TIMS-SMAI 1993, Paris, (1993).
5. Gibbens, R.J., Hunt, P.J., Kelly, F.P., "Bistability in communication networks", in *Disorder in Physical Systems*, G.R. Grimmett and D.J.A. Welsh (eds.), Oxford Univ. Press, 113-128, (1990).
6. Graham, C., "McKean-Vlasov Ito-Skorohod equations, and nonlinear diffusions with discrete jump sets", *Stoch. Processes Appl.* 40, 69-82, (1992).
7. Graham, C., "Nonlinear diffusion with jumps", *Annales Institut Henri Poincaré, Prob.-Stat.*, 28 3, 393-402, (1992).
8. Graham, C., *Systèmes de particules stochastiques, problèmes non-linéaires et théorèmes limites pour des modèles physiques, chimiques et informatiques*, Habilitation à diriger les recherches de l'Université Paris 6, 7 novembre 1994, (1994).
9. Graham, C., Méléard, S., "Propagation of chaos for a fully connected loss network with alternate routing", *Stoch. Processes Appl.* 44, 159-180, (1993).

10. Graham, C., Méléard, S., "Fluctuations for a fully connected loss network with alternate routing", *Stoch. Processes Appl.* 53, 97-115, (1994).
11. Graham, C., Méléard, S., "Chaos hypothesis for a system interacting through shared resources", *Prob. Th. and Rel. Fields* 100, 157-173, (1994).
12. Graham, C., Méléard, S., "Dynamic asymptotic results for a generalized star-shaped loss network", *Ann. Appl. Prob.* 5, (1995).
13. Hunt, P.J., *Limit theorems for stochastic loss networks*, PhD dissertation, Christ's College, University of Cambridge (1990).
14. Hunt, P.J., "Losunts networks under diverse routing, I: The symmetric star network", Research report no. 92-13, Statistical Laboratory, University of Cambridge (1992).
15. Joffe, A., Métivier, M., "Weak convergence of sequences of semimartingales with applications to multitype branching processes", *Adv. Appl. Prob.* 18, 20-65, (1986).
16. Kelly, F.P., "Blocking probabilities in large circuit-switched networks", *Adv. Appl. Prob.* 18, 473-505, (1986).
17. Kelly, F.P., "Loss networks", *Ann. Appl. Prob.* 1, 319-378, (1991).
18. Kushner, H., "Approximation of large trunk line systems under heavy traffic", *Advances Appl. Prob.* 26, 1063-1094, (1994).
19. Marbukh, V.V., "Investigation of a fully connected channel-switching network with many nodes and alternative routes", *Simulation of Behaviour and Intelligence*, 1601-1608, (1984).
20. Shiga, T., Tanaka, H., "Central limit theorem for a system of Markovian particules with mean-field interactions", *Z. WahrschR. Verw. Geb.* 69, 439-459, (1985).
21. Sznitman, A.S., "Équations de type de Boltzmann, spatialement homogènes", *Z. Wahrsch. verw. Gebeite* 66, 559-592, (1984).
22. Sznitman, A.S., "Propagation of chaos", in *École d'été Saint-Flour 1989*, Lecture Notes in Mathematics 1464, 165-251, New-York: Springer-Verlag, (1991).
23. Uchiyama, K., "Fluctuations in a Markovian system of pairwise interacting particles", *Prob. Th. Rel. Fields* 79, 289-302, (1988).
24. Whitt, W., "Blocking when service is required from several facilities simultaneously", *AT&T Tech. J.* 64, 1807-1856, (1985).
25. Ziedins, I.B., Kelly, F.P., "Limit theorems for loss networks with diverse routing", *Adv. Appl. Prob.* 21, 804-830, (1989).

Probabilistic Numerical Methods for Partial Differential Equations: Elements of Analysis

Denis Talay

INRIA, 2004 Route des Lucioles, BP 93, F-06902 Sophia-Antipolis Cedex, France

The objective of these notes is to present recent results on the convergence rate of Monte–Carlo methods for linear Partial Differential Equations and integrodifferential Equations, and for stochastic particles methods for some nonlinear evolution problems (McKean–Vlasov equations, Burgers equation, convection-reaction-diffusion equations). The given bounds for the numerical errors are non asymptotic: one wants to estimate the global errors of the methods for different possible values taken by their parameters (discretization step, number of particles or of simulations, etc).

Only a selection of existing results is presented. Most of the proofs are only sketched but the methodologies are described carefully. Deeper information should be available in Talay and Tubaro [49]. A companion review paper of these notes, with an emphasis on applications in Random Mechanics, is Talay [48].

PART I - Monte Carlo Methods for Parabolic PDE's

1. Notation

We fix a filtered probability space $(\Omega, \mathcal{F}, (\mathcal{F}_t), \mathbf{P})$, and a r-dimensional Brownian motion (W_t) on this space.

Usually a time interval $[0, T]$ will be fixed.

The notation $(X_t(x))$ stands for a process (X_t) such that $X_0 = x$ *a.s.*

Given a smooth function φ and a multiindex α of the form

$$\alpha = (\alpha_1, \ldots, \alpha_k) \ , \ \alpha_i \in \{1, \ldots, d\}$$

the notation $\partial^x_\alpha \varphi(t, x, y)$ means that the multiindex α concerns the differentiation with respect to the coordinates of x, the variables t and y being fixed.

When $\gamma = (\gamma^i_j)$ is a matrix, $\hat{\gamma}$ denotes the determinant of γ, and γ_j denotes the $j - th$ column of γ.

When V is a vector, ∂V denotes the matrix $(\partial_i V_j)^i_j$.

Finally, the same notation $K(\cdot)$, q, Q, μ, etc is used for different functions and positive real numbers, having the common property of being independent of T and of the approximation parameters (discretization step, number of simulations or number of particles, etc).

2. The Euler and Milshtein schemes for SDE's

Let (X_t) be the process taking values in $\mathbb{R}^d$ solution to

$$X_t = X_0 + \int_0^t b(X_s)ds + \int_0^t \sigma(X_s)dW_s\ , \tag{2.1}$$

where (W_t) is a r-dimensional Brownian motion.

Our objective is to approximate the unknown process (X_t) by an approximate process whose trajectories can easily be simulated on a computer. Typically we must simulate a large number of independent trajectories of this process. Therefore, the cost of the simulation of one trajectory must be so low as possible.

An efficient procedure consists in choosing a discretisation step $\frac{T}{n}$ of the time interval $[0,T]$ and in simulating the Euler scheme defined by

$$\begin{cases} X_0^n & = X_0\ , \\ X_{(p+1)T/n}^n & = X_{pT/n}^n + b(X_{pT/n}^n)\frac{T}{n} \\ & \quad +\sigma(X_{pT/n}^n)(W_{(p+1)T/n} - W_{pT/n}). \end{cases} \tag{2.2}$$

To simulate one trajectory of $(X_t^n, 0 \le t \le T)$, one simply has to simulate the family

$$(W_{T/n}, W_{2T/n} - W_{T/n}, \ldots, W_T - W_{T-T/n})$$

of independent Gaussian random variables. For $\frac{kT}{n} \le t < \frac{(k+1)T}{n}$, X_t^n is defined by

$$X_t^n = X_{kT/n}^n + b(X_{kT/n}^n)\left(t - \frac{kT}{n}\right) + \sigma(X_{kT/n}^n)(W_t - W_{kT/n}). \tag{2.3}$$

The convergence rate of this scheme has been studied according to various convergence criterions. In the sequel we will present estimates on the discretization error according to several different criterions, all of them being related to probabilistic numerical procedures for Partial Differential Equations. The proofs of most of these estimates use an elementary result concerning the convergence in $L^p(\Omega)$.

Proposition 2.1. *Suppose that the functions $b(\cdot)$ and $\sigma(\cdot)$ are globally Lipschitz.*

Let $p \ge 1$ be an integer such that $\mathbf{E}|X_0|^{2p} < \infty$.

Then there exists an increasing function $K(\cdot)$ such that, for any integer $n \ge 1$,

$$\mathbf{E}\left[\sup_{t\in[0,T]} |X_t - X_t^n|^{2p}\right] \le \frac{K(T)}{n^p}. \tag{2.4}$$

The function $K(\cdot)$ depends on the Lipschitz constants of the functions $b(\cdot)$ and $\sigma(\cdot)$, on the dimension d, on p and on $\mathbf{E}|X_0|^{2p}$.

Sketch of the proof. Let L_b be the Lipschitz constant of $b(\cdot)$:

$$|b(x)| \leq L_b|x| + |b(0)| ;$$

a similar inequality holds for $\sigma(\cdot)$. Thus, from (2.3) and Itô's formula, an induction on k shows: there exists an increasing function $K(\cdot)$ such that for any $n \in \mathbb{N}^*$,

$$\mathbf{E}\left[\sup_{t\in[0,T]} |X_t^n|^{2p}\right] \leq K(T)(1+\mathbf{E}|X_0|^{2p})\exp(K(T)). \tag{2.5}$$

Here, the function K depends on L_b, L_σ, p and the dimension d.

Consider, for $t \in [kT/n, (k+1)T/n]$, the process

$$\begin{aligned}\varepsilon_t \;:=\; & X_{kT/n} - X_{kT/n}^n + \int_{kT/n}^t (b(X_s) - b(X_{kT/n}^n))ds \\ & + \int_{kT/n}^t (\sigma(X_s) - \sigma(X_{kT/n}^n))dW_s. \end{aligned} \tag{2.6}$$

Apply the Itô formula to $|\varepsilon_t|^{2p}$ between $t = \frac{kT}{n}$ and $t = \frac{(k+1)T}{n}$; standard computations and (2.5) then show that, for a new increasing function $K(\cdot)$,

$$\mathbf{E}|\varepsilon_{(k+1)T/n}|^{2p} \leq \left(1+\frac{K(T)}{n}\right)\mathbf{E}|\varepsilon_{kT/n}|^{2p} + \frac{K(T)}{n^{p+1}}.$$

Noting that $\varepsilon_0 = 0$, an induction on k provides the estimate

$$\sup_{0\leq k\leq n} \mathbf{E}|\varepsilon_{kT/n}|^{2p} \leq \frac{C_1\exp(C_2T)}{n^p}.$$

To conclude, it remains to use (2.6) again. □

Applying the Borel-Cantelli lemma, one readily deduces the

Proposition 2.2. *Suppose that the functions $b(\cdot)$ and $\sigma(\cdot)$ are globally Lipschitz. Suppose that $\mathbf{E}|X_0|^{2p} < \infty$ for any integer p. Then*

$$\forall 0 \leq \alpha < \frac{1}{2}\,,\; n^\alpha \sup_{t\in[0,T]} |X_t - X_t^n| \stackrel{n\to\infty}{\longrightarrow} 0 \;a.s. \tag{2.7}$$

The details of the easy proofs of the two preceding propositions can be found in Faure's thesis [13] or in Kanagawa [23] e.g.

Concerning the path by path convergence of the Euler scheme, one has an even better information, which we briefly now present; we refer to Roynette [43] for a complete exposition.

Let $g(\cdot)$ be a function from $[0,T]$ to $\mathbb{R}$ and let p be a strictly positive integer. Set

$$\omega_p(g,t) := \sup_{|h|\leq t} \left(\int_{I_h} |g(x+h) - g(x)|^p dx \right)^{1/p}$$

with $I_h := \{x \in [0,T]; x+h \in [0,T]\}$. For $0 < \alpha < 1$ and $1 \leq q \leq +\infty$, set

$$\|g\|_{\alpha,p,q} := \| g \|_{L^p(\mathbb{R})} + \left(\int_0^T \left(\frac{\omega_p(g,t)}{t^\alpha} \right)^q \frac{dt}{t} \right)^{1/q}.$$

The Besov space $\mathcal{B}^\alpha_{p,q}$ is the Banach space of the functions $g(\cdot)$ such that $\|g\|_{\alpha,p,q} < \infty$, endowed with the norm $\| \cdot \|_{\alpha,p,q}$. The Besov space $\mathcal{B}^\alpha_{\infty,\infty}$ is the usual space of Hölder functions of order α.

Theorem 2.3 (Roynette [43]). *Suppose that the functions $b(\cdot)$ and $\sigma(\cdot)$ are globally Lipschitz. Suppose that $\mathbf{E}|X_0|^{2p} < \infty$ for any integer p. Let $\tilde{X}^n := X^{2^n}$. For any integer $p > 1$, for any n large enough, there exists a constant $C_T(p)$ uniform with respect to n such that, for any $\gamma < \frac{1}{2}$,*

$$\| X_\cdot - \tilde{X}^n_\cdot \|_{1/2,p,\infty} \leq C_T(p) 2^{-n\gamma} \ a.s. \tag{2.8}$$

Sketch of the proof. Consider the process $\varepsilon(\cdot)$ defined in (2.6). Define

$$b^n_s := b(X_s) - b(\tilde{X}^n_s) \ , \ \sigma^n_s := \sigma(X_s) - \sigma(\tilde{X}^n_s).$$

Thus,

$$\varepsilon_t := \int_0^t b^n_s ds + \int_0^t \sigma^n_s dW_s.$$

From the estimates (2.4) and (2.7), one can easily show that, for any $\gamma < \frac{1}{2}$, for any n large enough,

$$\sup_{0\leq s\leq T} (|b^n_s| + |\sigma^n_s|) \leq C 2^{-n\gamma} \ a.s.,$$

and for any integer $p \geq 1$,

$$\sup_{0\leq s\leq T} \mathbf{E}|\sigma^n_s|^{2p} \leq C_T(p) 2^{-np}.$$

The technical proposition 1 in Roynette [43] then implies: for any integer $p > 1$, there exists a (new) constant $C_T(p)$ uniform w.r.t. n such that, for any $\gamma < \frac{1}{2}$,

$$\| \varepsilon_\cdot \|_{1/2,p,\infty} \leq C_T(p) 2^{-n\gamma} \ a.s.$$

□

The asymptotic distribution of the normalized Euler scheme error

$$U^n := \sqrt{n}(X_\cdot - X^n_\cdot)$$

is analysed in Kurtz and Protter [25] (see also their contribution to this volume): (X, U^n) converges in law to the process (X, U) where U is the solution to

$$U_t := \int_0^t \partial b(X_s)U_s ds + \sum_{j=1}^{r}\int_0^d \partial\sigma_j(X_s)U_s dW_s^j + \frac{1}{\sqrt{2}}\sum_{i,j=1}^{r} \partial\sigma_i(X_s)\sigma_j(X_s)dB_s^{ij},$$

where $(B^{ij}, 1 \leq i,j \leq r)$ is a r^2-dimensional standard Brownian motion independent of X.

In the sequel, when $d = r = 1$, we also use the Milshtein scheme

$$\begin{cases} X_0^n & = X_0 , \\ X^n_{\frac{(p+1)T}{n}} & = X^n_{pT/n} + b(X^n_{pT/n})\frac{T}{n} \\ & \quad + \sigma(X^n_{pT/n})(W_{(p+1)T/n} - W_{pT/n}) \\ & \quad + \frac{1}{2}\sigma\left(X^n_{pT/n}\right)\sigma'\left(X^n_{pT/n}\right)\left((W_{(p+1)T/n} - W_{pT/n})^2 - \frac{T}{n}\right). \end{cases} \tag{2.9}$$

Our reason for considering that scheme here comes from the

Proposition 2.4. *Suppose that the functions $b(\cdot)$ and $\sigma(\cdot)$ are twice continuously differentiable with bounded derivatives.*

Let $p \geq 1$ be an integer such that $\mathbf{E}|X_0|^{4p} < \infty$. Then there exists an increasing function $K(\cdot)$ such that

$$\sup_{t\in[0,T]} \mathbf{E}|X_t - X_t^n|^{2p} \leq \frac{K(T)}{n^{2p}}. \tag{2.10}$$

If $\mathbf{E}|X_0|^{4p} < \infty$ holds for any integer p, then for any $0 \leq \alpha < 1$,

$$n^\alpha \sup_{t\in[0,T]} \mathbf{E}|X_t - X_t^n|^{2p} \leq \frac{K(T)}{n^{2p}}. \tag{2.11}$$

The proof follows the same guidelines as the proof of Propositions 2.1 and 2.2. See Faure[13]. Note that the Milshtein scheme has better convergence rates than the Euler scheme for the convergence in $L^p(\Omega)$ and the almost sure convergence. A similar remark is true for the convergence in Besov spaces, see Roynette [43].

Here we consider the Milshtein scheme only when $d = r = 1$. In the multidimensional case, generally it requires double stochastic integrals which are not simple to simulate: see Talay [48] for a discussion and Gaines and Lyons [16] for a method of resolution.

3. Monte Carlo methods for parabolic PDE's

3.1 Principle of the method

Define the $d \times d$ matrix valued function $(a_j^i(\cdot))$ by

$$a(\cdot) := \sigma(\cdot)\sigma^*(\cdot).$$

Define the second-order differential operator $\mathcal{L}$ by

$$\mathcal{L} := \sum_{i=1}^{d} b_i(\cdot)\partial_i + \frac{1}{2}\sum_{i,j=1}^{d} a_j^i(\cdot)\partial_{ij}. \tag{3.1}$$

Consider the problem

$$\begin{cases} \frac{\partial u}{\partial t}(t,x) + \mathcal{L}u(t,x) & = & 0 \ in \ [0,T) \times \mathbb{R}^d , \\ u(T,x) & = & f(x) , \ x \in \mathbb{R}^d . \end{cases} \tag{3.2}$$

In the two different sets of hypotheses that we will consider for $b(\cdot)$, $\sigma(\cdot)$ and $f(\cdot)$, the following holds: the problem (3.2) has a unique solution which belongs to the set $C^{1,2}([0,T) \times \mathbb{R}^d)$ and is continuous on $[0,T] \times \mathbb{R}^d$. This unique solution is given by

$$u(t,x) = \mathbf{E}_x f(X_{T-t}) = P_{T-t}f(x) , \ \forall (t,x) \in [0,T] \times \mathbb{R}^d$$

where P_θ denotes the transition operator of the Markov process (X_t).

Let $\{X^{(i)}, i \in \mathbb{N}\}$ be a sequence of independent trajectories of the process X. If the Strong Law of Large Numbers applies for the sequence $f(X_t^{(i)}(x))$, then

$$u(t,x) = \lim_{N\to\infty} \frac{1}{N}\sum_{i=1}^{N} f(X_{T-t}^{(i)}(x)) , a.s.$$

In practice one must approximate the $X_t^{(i)}(x)$'s. We consider the simplest approximation method: the Euler scheme. As we will see, this simple method has very interesting properties in the present context, even from the point of view of the convergence rate and of the numerical efficiency. The "Monte Carlo+Euler" approximation of $u(t,x)$ is

$$u^{n,N}(t,x) := \frac{1}{N}\sum_{i=1}^{N} f(X_{T-t}^{n,(i)}(x)) , \tag{3.3}$$

where $\{X^{n,(i)} , 1 \le i \le N\}$ denotes a set of independent trajectories of the process X^n.

Why are we interested in this method? Is it competitive with the usual deterministic algorithms of resolution of (3.2)? Of course there is no general answer to such a question. The answer depends on the dimension d and on the functions $b(\cdot)$, $\sigma(\cdot)$. Roughly speaking, the Monte Carlo method seems unuseful when a finite difference method, a finite element method, a finite volume method or a suitable deterministic algorithm is numerically stable and does not require too a long computation time.

Nevertheless we can give examples of situations where a Monte Carlo method is efficient.

First, the computational cost of the deterministic algorithms growths exponentially with the dimension d of the state space: these algorithms use grids whose number of points growths exponentially with d. Thus, when d is large ($d \geq 4$, say), the numerical resolution of (3.2) may even be impossible without a Monte Carlo procedure, whose computational cost growths only linearly with the dimension of the process X^n to simulate.

A Monte Carlo algorithm may also be interesting when one wants to compute $u(t,\cdot)$ at only a few points. This situation occurs in financial problems (evaluation of an option price in terms of the spot prices of the stocks) or in Physics (computation of the probability that a random process reaches given thresholds). One can also think to use a Monte Carlo method to compute $u(t,\cdot)$ on artificial boundaries in view of a decomposition of domains procedure: one divides the whole space in a set of subdomains; then the objective is to solve the problem (3.2) in each subdomain with Dirichlet boundary conditions by deterministic methods; these Dirichlet boundary conditions, i.e. the values of $u(t,\cdot)$ along the boundaries, can be approximated by a Monte Carlo algorithm. This combination of numerical methods may have several advantages. The resolution in the subdomains can be distributed to different processors. The convergence rate results for the Monte Carlo+Euler method suppose much weaker assumptions than the strong ellipticity condition of the operator $\mathcal{L}$; moreover if $f(\cdot)$ is a smooth function, no assumption on $\mathcal{L}$ is required; therefore, if the matrix $a(\cdot)$ degenerates locally, the domain of the numerical integration by a deterministic method can be reduced to the nondegeneracy region by an approximation of $u(t,\cdot)$ along its boundaries, which may considerably improve the efficiency of the deterministic method.

3.2 Introduction to the error analysis

Our objective is to give estimates for

$$|u(T,x) - u^{n,N}(T,x)|.$$

A natural decomposition of this error is as follows:

$$\begin{aligned} |u(T,x) - u^{n,N}(T,x)| &\leq |u(T,x) - \mathbf{E}f(X_T^n(x))| \\ &\quad + |\mathbf{E}f(X_T^n(x)) - u^{n,N}(T,x)| \\ &=: \alpha^n + \beta^{n,N}. \end{aligned}$$

The analysis of $\beta^{n,N}$ is related to usual considerations on the Strong Law of Large Numbers: Central Limit Theorems, Berry-Esseen inequalities, etc. The difficulty here is to obtain estimates uniform w.r.t. to n. This can be solved by the convergence in $L^p(\Omega)$ of X^n to X which holds under the hypotheses we make below.

Consequently we concentrate our attention to α^n.

When $f(\cdot)$ is a Lipschitz function, one can bound α^n from above by using the estimate (2.4) for $p = 2$. This gives the estimate:

$$|\alpha^n| \le \frac{C}{\sqrt{n}}.$$

We now show that one can be much more clever.

For the rest of the section we suppose

(H) The functions b and σ are C^∞ functions, whose derivatives of any order are bounded (but b and σ are not supposed bounded themselves).

Define $\Psi(t,\cdot)$ by

$$\begin{aligned}\Psi(t,\cdot) &= \frac{1}{2}\sum_{i,j=1}^{d} b^i(\cdot)b^j(\cdot)\partial_{ij}u(t,\cdot) + \frac{1}{2}\sum_{i,j,k=1}^{d} b^i(\cdot)a^j_k(\cdot)\partial_{ijk}u(t,\cdot)\\ &+ \frac{1}{8}\sum_{i,j,k,l=1}^{d} a^i_j(\cdot)a^k_l(\cdot)\partial_{ijkl}u(t,\cdot) + \frac{1}{2}\frac{\partial^2}{\partial t^2}u(t,\cdot)\\ &+ \sum_{i=1}^{d} b^i(\cdot)\frac{\partial}{\partial t}\partial_i u(t,\cdot) + \frac{1}{2}\sum_{i,j=1}^{d} a^i_j(\cdot)\frac{\partial}{\partial t}\partial_{ij}u(t,\cdot). \qquad (3.4)\end{aligned}$$

Lemma 3.1. *It holds that*

$$\mathbf{E}f(X^n_T(x)) - \mathbf{E}f(X_T(x)) = \frac{T^2}{n^2}\sum_{k=0}^{n-2}\mathbf{E}\Psi\left(\frac{kT}{n}, X^n_{kT/n}(x)\right) + \sum_{k=0}^{n-1} R^n_k(x)\,, \qquad (3.5)$$

where

$$R^n_{n-1}(x) := \mathbf{E}f(X^n_T(x)) - \mathbf{E}(P_{T/n}f)(X^n_{T-T/n}(x))\,,$$

and for $k < n-1$, $R^n_k(x)$ can be explicited under a sum of terms, each of them being of the form

$$\begin{aligned}\mathbf{E}\Big[\varphi^{\natural}_\alpha(X^n_{kT/n}(x)) \int_{kT/n}^{(k+1)T/n}\int_{kT/n}^{s_1}\int_{kT/n}^{s_2} &(\varphi^{\sharp}_\alpha(X^n_{s_3}(x))\partial_\alpha u(s_3, X^n_{s_3}(x))\\ &+\varphi^{\flat}_\alpha(X_{s_3}(x))\partial_\alpha u(s_3, X_{s_3}(x)))ds_3ds_2ds_1\Big]\,, \qquad (3.6)\end{aligned}$$

where $|\alpha| \le 6$, and the $\varphi^{\natural}_\alpha$'s, $\varphi^{\sharp}_\alpha$'s, $\varphi^{\flat}_\alpha$'s are products of functions which are partial derivatives up to the order 3 of the a^{ij}'s and b^i's.

Proof. For $z \in \mathbb{R}^d$ define the differential operator $\mathcal{L}_z$ by

$$\mathcal{L}_z g(\cdot) := \sum_{i=1}^{d} b^i(z)\partial_i g(\cdot) + \frac{1}{2}\sum_{i,j=1}^{d} a^i_j(z)\partial_{ij}.$$

As $u(t,\cdot) = P_{T-t}f(\cdot) = \mathbf{E}f(X_{T-t}(\cdot))$, one has

$$\mathbf{E}_x f(X^n_T) - \mathbf{E}_x f(X_T) = \mathbf{E}_x u(T, X^n_T) - u(0,x) = \sum_{k\le n-1}\delta^n_k$$

with

$$\delta_k^n := \mathbf{E}_x \left[u\left(\frac{(k+1)T}{n}, X^n_{(k+1)T/n} \right) - u\left(\frac{kT}{n}, X^n_{kT/n} \right) \right]. \tag{3.7}$$

The Itô formula implies

$$\delta_k^n = \mathbf{E}_x \int_{\frac{kT}{n}}^{\frac{(k+1)T}{n}} \left(\partial_t u(t, X_t^n) + \mathcal{L}_z u(t, X_t^n) \Big|_{z = X^n_{kT/n}} \right) dt ,$$

from which, using (3.2), one gets

$$\delta_k^n = \mathbf{E}_x \int_{\frac{kT}{n}}^{\frac{(k+1)T}{n}} \left(-\mathcal{L} u(t, X_t^n) + \mathcal{L}_z u(t, X_t^n) \Big|_{z = X^n_{kT/n}} \right) dt.$$

Denote

$$I_k^n(t) := \mathcal{L}_z u(t, X_t^n) \Big|_{z = X^n_{kT/n}} - \mathcal{L}_z u\left(\frac{kT}{n}, X^n_{kT/n} \right) \Big|_{z = X^n_{kT/n}}$$

and

$$\begin{aligned} J_k^n(t) &:= \mathcal{L}_z u\left(\frac{kT}{n}, X^n_{kT/n} \right) \Big|_{z = X^n_{kT/n}} - \mathcal{L} u(t, X_t^n) \\ &= \mathcal{L} u\left(\frac{kT}{n}, X^n_{kT/n} \right) - \mathcal{L} u(t, X_t^n). \end{aligned}$$

We have:

$$\delta_k^n = \mathbf{E}_x \int_{\frac{kT}{n}}^{\frac{(k+1)T}{n}} (I_k^n(t) + J_k^n(t)) dt.$$

We now consider $I_k^n(t)$ and $J_k^n(t)$ as smooth functions of the process (X_t^n) and recursively apply the Itô formula, using the fact that the function u solves (3.2), so that $\mathcal{L}u$ solves a similar PDE. □

The expansion (3.5) can be rewritten as follows:

$$\begin{aligned} &\mathbf{E}_x f(X_T^n) - \mathbf{E}_x f(X_T) \\ &= \frac{T}{n} \int_0^T \mathbf{E}_x \Psi(s, X_s) ds \\ &+ \frac{T^2}{n^2} \sum_{k=0}^{n-2} \mathbf{E}_x \Psi\left(\frac{kT}{n}, X_{kT/n} \right) - \frac{T}{n} \int_0^T \mathbf{E}_x \Psi(s, X_s) ds \\ &+ \frac{T^2}{n^2} \sum_{k=0}^{n-2} \mathbf{E}_x \left(\Psi\left(\frac{kT}{n}, X^n_{kT/n} \right) - \Psi\left(\frac{kT}{n}, X_{kT/n} \right) \right) \\ &+ \sum_{k=0}^{n-2} r_k^n(x) + \mathbf{E}_x f(X_T^n) - \mathbf{E}_x (P_{T/n} f)(X^n_{T-T/n}) \end{aligned}$$

$$= \frac{T}{n}\int_0^T \mathbf{E}_x\Psi(s,X_s)ds$$
$$+A^n + B^n + \sum_{k=0}^{n-2} r_k^n(x) + C^n. \qquad (3.8)$$

From this expansion, it is reasonable to expect that the error

$$\mathbf{E}_x f(X_T^n) - \mathbf{E}_x f(X_T)$$

is *equal* to

$$\frac{T}{n}\int_0^T \mathbf{E}_x\Psi(s,X_s)ds$$

plus a remainder of order n^{-2}, because

$$\sum_{k=0}^{n-2}\left(\mathbf{E}_x\Psi\left(\frac{kT}{n},X_{kT/n}^n\right) - \mathbf{E}_x\Psi\left(\frac{kT}{n},X_{kT/n}\right)\right)$$

should be uniformly bounded w.r.t. n since each term of the sum should be of order $\frac{1}{n}$.

More precisely, for $1 \le k \le n-2$, one applies the expansion (3.8), substituting the function

$$f_{n,k}(\cdot) := \Psi\left(\frac{kT}{n},\cdot\right)$$

to $f(\cdot)$. Set $u_{n,k}(t,x) := P_{kT/n-t}f_{n,k}(\cdot)$ and denote by $\Psi_{n,k}(t,\cdot)$ the function defined in (3.4) with $u_{n,k}$ instead of u and kT/n instead of T; thus, for some functions $g_\lambda(\cdot) \in C_b^\infty(\mathbb{R}^d)$ one has that, for $t \le \frac{kT}{n}$,

$$\Psi_{n,k}(t,\cdot) = \sum_\lambda g_\lambda(\cdot)\partial_\lambda\left[P_{kT/n-t}\Psi\left(\frac{kT}{n},\cdot\right)\right].$$

There holds

$$\mathbf{E}_x\Psi\left(\frac{kT}{n},X_{kT/n}^n\right) - \mathbf{E}_x\Psi\left(\frac{kT}{n},X_{kT/n}\right) = \frac{T^2}{n^2}\sum_{j=0}^{k-2}\mathbf{E}_x\Psi_{n,k}\left(\frac{jT}{n},X_{jT/n}^n\right)$$
$$+\sum_{j=0}^{k-1} r_j^{n,k}(x), \qquad (3.9)$$

where the $r_j^{n,k}(x)$'s are sums of terms of type (3.6) with $u_{n,k}$ instead of u.

It is now clear that one key problem is as follows. Let γ and λ be multiindices, let $g_\gamma(\cdot)$ and $g_\lambda(\cdot)$ be smooth functions with polynomial growth. Set

$$\varphi(\theta,\cdot) := g_\gamma(\cdot)\partial_\gamma P_{T-\theta}f(\cdot).$$

We want to prove that quantities of the type

$$\left|\mathbf{E}_x\left[g_\lambda(X_t^n)\partial_\lambda P_{\theta-t}\varphi(\theta,\cdot)(z)\Big|_{z=X_t^n}\right]\right| \tag{3.10}$$

can be bounded uniformly w.r.t. $n \in \mathbb{N}^*$, $\theta \in [0, T - \frac{T}{n}]$, $t \in [0, \theta - \frac{T}{n}]$.

We distinguish two different situations. When $f(\cdot)$ is a smooth function, we make no assumption on the operator $\mathcal{L}$. When $f(\cdot)$ is only measurable and bounded, we suppose that $\mathcal{L}$ satisfies an assumption of the Hörmander type.

3.3 Smooth functions $f(\cdot)$

Let $\mathcal{H}_T$ be the class of functions $\phi : [0,T] \times \mathbb{R}^d \to \mathbb{R}$ with the following properties: ϕ is of class C^∞ and for any multiindex α there exist a positive integer s and an increasing function $K(\cdot)$ such that

$$\forall \theta \in [0,T] \ , \ \forall x \in \mathbb{R}^d \ , \ |\partial_\alpha \phi(\theta, x)| \leq K(T)(1 + |x|^s). \tag{3.11}$$

A function ϕ of $\mathcal{H}_T$ is said *homogeneous* if it does not depend on the time variable: $\phi(\theta, x) = \phi(x)$.

In this subsection we suppose

(H1) The function $f(\cdot)$ is a *homogeneous* function of $\mathcal{H}_T$.

It is well known that the condition (H) implies that there exists a smooth version of the stochastic flow $x \longrightarrow X_t(x)$. For the sake of simplicity we denote this smooth version $X_t(\cdot)$. Besides, for any integer $k > 0$ the family of the processes equal to the partial derivatives of the flow up to the order k solves a system of stochastic differential equations with Lipschitz coefficients: see, e.g., Kunita [24] and Protter [37]. Thus, for any $0 \leq t \leq T$,

$$\partial_i u(t,x) = \partial_i \mathbf{E} f(X_{T-t}(x)) = \mathbf{E} \sum_{j=1}^{d} \partial_j f(X_{T-t}(x)) \partial_i X_{T-t}(x). \tag{3.12}$$

From (H) and (H1) one easily deduces that, for some increasing function $K(\cdot)$ and some integer m,

$$|\partial_i u(t,x)| \leq K(T)(1 + |x|^m).$$

Differentiations of (3.12) provide a probabilistic interpretation of $\partial_\alpha u(t,x)$ for any multiindex α. It is easy to prove by induction that, for any multiindex α, there exist an increasing function $K_\alpha(\cdot)$ and an integer m_α such that

$$|\partial_\alpha u(t,x)| \leq K_\alpha(T)(1 + |x|^{m_\alpha}). \tag{3.13}$$

For $\phi \in \mathcal{H}_T$ and θ fixed in $[0,T]$ the function $u(\theta; t, x)$ defined by

$$u(\theta; t, x) := \mathbf{E}\phi(\theta, X_{T-t}(x)) = \mathbf{E}_x \phi(\theta, X_{T-t})$$

belongs to $\mathcal{H}_T$ and satisfies

$$\begin{cases} \frac{\partial u}{\partial t} + \mathcal{L}u &= 0\,, 0 \le t < T, \\ u(\theta; T, x) &= \phi(\theta, x). \end{cases} \tag{3.14}$$

Similarly to (3.13) one has: for any multiindex α, there exists an increasing function $K_\alpha(\cdot)$ and an integer m_α such that

$$\forall \theta \in [0, T], \mid \partial_\alpha u(\theta; t, x) \mid \le K_\alpha(T)(1 + \mid x \mid^{k_\alpha}).$$

This result can be used to prove:

Lemma 3.2. *Suppose (H) and (H1).*

Let γ et λ be multiindices, let $g(\cdot)$ and $g_\gamma(\cdot)$ be smooth functions with polynomial growth. Set

$$\varphi(\theta, \cdot) := g_\gamma(\cdot)\partial_\gamma P_{T-\theta} f(\cdot).$$

There exist an increasing function $K(\cdot)$ and an integer m such that

$$\left| \mathbf{E}_x \left[g(X_t^n)\partial_\lambda P_{\theta-t}\varphi(\theta, \cdot)(z)\Big|_{z=X_t^n} \right] \right| \le K(T)(1 + |x|^m). \tag{3.15}$$

Coming back to (3.8) and (3.9) one deduces the

Theorem 3.3 (Talay and Tubaro [50]). *Suppose (H) and (H1). The Euler scheme error satisfies*

$$u(T, x) - \mathbf{E}_x f(X_T^n) = -\frac{T}{n} \int_0^T \mathbf{E}_x \Psi(s, X_s) ds + \frac{Q_T^n(x)}{n^2} \tag{3.16}$$

and there exist an increasing function $K(\cdot)$ and an integer m such that

$$|Q_T^n(x)| \le K(T)(1 + |x|^m). \tag{3.17}$$

Here $\Psi(\cdot,\cdot)$ is defined by (3.4).

Observe that in the preceding statement the differential operator $\mathcal{L}$ may be degenerate.

3.4 Non smooth functions $f(\cdot)$

Theorem 3.3 supposes that $f(\cdot)$ is a smooth function. From an applied point of view this is a stringent condition: often one wants to compute quantities of the type

$$\mathbf{P}\left[|X_T(x)| > y\right]$$

for a given threshold $y > 0$. Our objective now is to show that an expansion of the type (3.16) still holds even when $f(\cdot)$ is only supposed measurable and bounded. In the proof that we give, the boundedness could be relaxed: as in the preceding section we could suppose that $f(\cdot)$ belongs to the set $\mathcal{H}_T$. To realize this programme a nondegeneracy condition is supposed. As we now see, this condition is less restrictive than the uniform strong ellipticity of the operator $\mathcal{L}$.

We need some basic elements of the Malliavin calculus. For a complete exposition of this theory we refer to Nualart [34] (we use the notation of this book) and Ikeda-Watanabe [22]; the applications to the existence of a density for the law of a diffusion process can also be found in Pardoux [36].

For $h(\cdot) \in L^2(\mathbb{R}_+, \mathbb{R}^r)$, $W(h)$ denotes the quantity $\int_0^T < h(t), dW_t >$. $\mathcal{S}$ is the space of "simple " functionals of the Wiener process W, i.e. the subspace of $L^2(\Omega, \mathcal{F}, \mathrm{P})$ of random variables F which can be written under the form

$$F = f(W(h_1), \ldots, W(h_n))$$

for some n, some polynomial function $f(\cdot)$, some $h_i(\cdot) \in L^2(\mathbb{R}_+, \mathbb{R}^r)$.

For $F \in \mathcal{S}$, $(D_t F)$ denotes the $\mathbb{R}^r$-dimensional process defined by

$$D_t F = \sum_{i=1}^{n} \frac{\partial f}{\partial x_i}(W(h_1), \ldots, W(h_n)) h_i(t).$$

The operator D is closable as an operator from $L^p(\Omega)$ to $L^p(\Omega; L^2(0,T))$, for any $p \geq 1$. Its domain is denoted by $\mathbf{D}^{1,p}$. Define the norm

$$\|F\|_{1,p} := \left[\mathbf{E}|F|^p + \|DF\|^p_{L^p(\Omega; L^2(0;T))}\right]^{1/p},$$

The j-th component of $D_t F$ is denoted by $D_t^j F$. The k-th order derivative is the the random vector on $[0,T]^k \times \Omega$ whose coordinates are

$$D^{j_1,\ldots,j_k}_{t_1,\ldots,t_k} F := D^{j_k}_{t_k} \ldots D^{j_1}_{t_1} F ,$$

and $\mathbf{D}^{N,p}$ denotes the completion of $\mathcal{S}$ with respect to the norm

$$\|F\|_{N,p} := \left[\mathbf{E}|F|^p + \sum_{k=1}^{N} \mathbf{E}\|D^k F\|^p_{L^2((0;T)^k)}\right]^{1/p} .$$

$\mathbf{D}^\infty$ denotes the space $\bigcap_{p\geq 1} \bigcap_{j \geq 1} \mathbf{D}^{j,p}$.

For $F := (F^1, \ldots, F^m) \in (\mathbf{D}^\infty)^m$, γ_F denotes the Malliavin covariance matrix associated to F, i.e. the $m \times m$-matrix defined by

$$(\gamma_F)^i_j := < DF^i, DF^j >_{L^2(0,T)} .$$

Definition 3.4. *We say that the random vector F satisfies the nondegeneracy assumption if the matrix γ_F is a.s. invertible, and the inverse matrix $\Gamma_F := \gamma_F^{-1}$ satisfies*

$$|\det(\Gamma_F)| \in \bigcap_{p \geq 1} L^p(\Omega).$$

Remark 3.5. The above condition can also be written as follows:

$$\frac{1}{\det(\gamma_F)} \in \bigcap_{p\geq 1} L^p(\Omega).$$

The main ingredient of our analysis is the following integration by parts formula (cf. the section V-9 in Ikeda-Wanabe [22]):

Proposition 3.6. *Let $F \in (\mathbf{D}^\infty)^m$ satisfy the nondegeneracy condition 3.4, let g be a smooth function with polynomial growth, and let G in $\mathbf{D}^\infty$. Let $\{H_\beta\}$ be the family of random variables depending on multiindices β of length strictly larger than 1 and with coordinates $\beta_j \in \{1,\ldots,m\}$, recursively defined in the following way:*

$$\begin{aligned} H_i(F,G) &= H_{(i)}(F,G) \\ &:= -\sum_{j=1}^{m}\Big\{G < D\Gamma_F^{ij}, DF^j >_{L^2(0,T)} \\ &\qquad + \Gamma_F^{ij} < DG, DF^j >_{L^2(0,T)} \qquad (3.18) \\ &\qquad + \Gamma_F^{ij}\cdot G\cdot \hat{L}F^j\Big\}, \\ H_\beta(F,G) &= H_{(\beta_1,\ldots,\beta_k)}(F,G) \\ &:= H_{\beta_k}(F, H_{(\beta_1,\ldots,\beta_{k-1})}(F,G)), \qquad (3.19) \end{aligned}$$

where $\hat{L}$ is the so called Ornstein-Uhlenbeck operator whose domain includes $\mathbf{D}^\infty$. Then, for any multiindex α,

$$\mathbf{E}[(\partial_\alpha g)(F)G] = \mathbf{E}[g(F)H_\alpha(F,G)]. \qquad (3.20)$$

One has the following estimate:

Proposition 3.7. *For any $p > 1$ and any multiindex β, there exist a constant $C(p,\beta) > 0$ and integers $k(p,\beta)$, $m(p,\beta)$, $m'(p,\beta)$, $N(p,\beta)$, $N'(p,\beta)$, such that, for any measurable set $A \subset \Omega$ and any F, G as above, one has*

$$\begin{aligned} \mathbf{E}[|H_\beta(F,G)|^p\ \mathbb{1}_A]^{\frac{1}{p}} \ \leq\ & C(p,\beta)\ \|\Gamma_F\ \mathbb{1}_A\|_{k(p,\beta)}\ \|G\|_{N(p,\beta),m(p,\beta)} \\ & \|F\|_{N'(p,\beta),m'(p,\beta)}. \qquad (3.21) \end{aligned}$$

We now state another classical result, which concerns the solutions of stochastic differential equations considered as functionals of the driving Wiener process. $[A, A']$ denotes the Lie brackett of two vector fields A and A'.

Definition 3.8. *Denote by $A_0, A_1, \ldots, A_r$ the vector fields defined by*

$$\begin{aligned} A_0(x) &:= \sum_{i=1}^{d} b^i(x)\partial_i \ , \\ A_j(x) &:= \sum_{i=1}^{d} \sigma_j^i(x)\partial_i \ , \quad j=1,\dots,r. \end{aligned}$$

For a multiindex $\alpha=(\alpha_1,\dots,\alpha_k)\in\{0,1,\dots r\}^k$, define the vector fields A_i^α $(1\le i\le r)$ by induction: $A_i^\emptyset := A_i$ and for $0\le j\le r$, $A_i^{(\alpha,j)} := [A^j, A_i^\alpha]$.

Finally set

$$V_L(x,\eta) := \sum_{i=1}^{r} \sum_{|\alpha|\le L-1} < A_i^\alpha(x),\eta >^2 .$$

Set

$$V_L(x) = 1 \wedge \inf_{\|\eta\|=1} V_L(x,\eta). \tag{3.22}$$

Under the hypothesis (H), $X_t(x)\in \mathbf{D}^\infty$ for any $x\in\mathbb{R}^d$. Let $\gamma_t(x)$ denote the Malliavin covariance matrix of $X_t(x)$ and let $\Gamma_t(x)$ denote its inverse.

We replicate Corollary 3.25 in Kusuoka and Stroock [26] in a weakened form.

Proposition 3.9. *Suppose (H) and*

(UH) $C_L := \inf_{x\in\mathbb{R}^d} V_L(x) > 0$ for some integer L.

Let L be an integer such that (UH) holds. Then

$$\|\Gamma_t(x)\| \in \bigcap_{p\ge 1} L^p(\Omega) \ , \forall x\in\mathbb{R}^d \ ,$$

and for any $p\ge 1$, for some constant μ and some increasing function $K(\cdot)$,

$$\|\Gamma_t(x)\|_p \le K(T)\frac{1+|x|^\mu}{t^{dL}} \ , \ \forall x\in\mathbb{R}^d \ , \ \forall 0<t\le T. \tag{3.23}$$

Thus, for any $t>0$ and any $x\in\mathbb{R}^d$ the law of $X_t(x)$ has a smooth density $p_t(x,\cdot)$. Besides, for any integers m,k and any multiindices α and β such that $2m+|\alpha|+|\beta|\le k$, there exist an integer $M(k,L)$, a non decreasing function $K(\cdot)$ and real numbers C,q,Q depending on L,T,m,k,α,β and on the bounds associated to the coefficients of the stochastic differential equation and their derivatives up to the order $M(k,L)$, such that the following inequality holds[1]*:*

$$|\partial_t^m \partial_x^\alpha \partial_y^\beta p_t(x,y)| \le \frac{K(T)(1+|x|^Q)}{t^q(1+|y-x|^2)^k} \exp\left(-C\frac{(|x-y|\wedge 1)^2}{t(1+|x|)^2}\right) \ , \ \forall 0<t\le T. \tag{3.24}$$

[1] The constant γ_0 of the statement of Kusuoka and Stroock is equal to 1 under (H).

Equipped with this result we can prove the

Theorem 3.10 (Bally and Talay [1]). *Let $f(\cdot)$ be a measurable and bounded function. Under the hypotheses (UH) and (H), the Euler scheme error satisfies*

$$\mathbf{E}f(X_T(x)) - \mathbf{E}f(X_T^n(x)) = -\frac{C_f(T,x)}{n} + \frac{Q_n(f,T,x)}{n^2}. \tag{3.25}$$

The terms $C_f(T,x) := \int_0^T \mathbf{E}\Psi(s, X_s(x))ds$ and $Q_n(f,T,x)$ have the following property: there exists an integer m, a non decreasing function $K(\cdot)$ depending on the coordinates of a and b and on their derivatives up to the order m, and positive real numbers q, Q such that

$$|C_f(T,x)| + sup_n |Q_n(f,T,x)| \leq K(T)\|f\|_\infty \frac{1+\|x\|^Q}{T^q}. \tag{3.26}$$

Sketch of the proof. As for Theorem 3.3, the main part of the proof consists in bounding terms of the type (3.10) from above. In the present context there is a serious difficulty: when $f(\cdot)$ is not smooth, the spatial derivatives of $u(t,\cdot)$ explode when t goes to T. Indeed,

$$u(t,x) = \int_{R^d} p_{T-t}(x,y) f(y) dy$$

and the estimate (3.24) shows that for any $|\gamma| \geq 1$,

$$|\partial_\gamma^x p_{T-t}(x,y)| \leq \frac{K(T-t)}{(T-t)^q}(1+\|x\|^Q)\frac{1}{(1+\|y-x\|^2)^{|\gamma|}},$$

from which

$$|\partial_\alpha^x u(t,x)| \leq K(T)\frac{\|f\|_\infty}{(T-t)^q}(1+\|x\|^Q). \tag{3.27}$$

It can be shown that there is no hope to improve the explosion rate in power of $T-t$.

But a miracle occurs: in (3.10) the derivatives of the function

$$P_{\theta-t}\varphi(\theta,\cdot)$$

are integrated w.r.t. the law of $X_t^n(x)$. Let us give an intuition of what happens. Consider the case $\theta = t$ and replace $X_t^n(x)$ by $X_t(x)$. Then, in view of (3.10) the problem becomes to bound from above an expression of the type

$$|\mathbf{E}_x[g_\gamma(X_t)\partial_\gamma u(t,X_t)]| = |\mathbf{E}_x[g_\gamma(X_t)\partial_\gamma(P_{T-t}f)(X_t)]|$$

uniformly w.r.t. $t \in [0,T)$. When t is "small" i.e $t \leq \frac{T}{2}$ the transition operator P_{T-t} has smoothened enough the initial condition $f(\cdot)$: the inequality (3.27) implies

$$|\partial_\alpha^x u(t,x)| \leq K(T)\frac{\|f\|_\infty}{T^q}(1+\|x\|^Q).$$

When t is "large" i.e $t \geq \frac{T}{2}$ the estimate (3.27) cannot be used. Instead, we observe that the matrix $\Gamma_t(x)$ has L^p-norms which satisfy (see (3.23))

$$\|\Gamma_t(x)\| \leq K(T)\frac{1+|x|^\mu}{T^{dL}}.$$

Thus, one can apply the integration by part formula (3.20) with

$$g(\cdot) = (P_{T-t}f)(\cdot) = u(t,\cdot)$$

and

$$F = X_t(x).$$

Using (3.21) one deduces that for $T \geq t \geq \frac{T}{2}$,

$$|\mathbf{E}_x\left[g_\gamma(X_t)\partial_\gamma u(t,X_t)\right]| \leq K(T)\|f\|_\infty \frac{1+|x|^\mu}{T^{dL}}.$$

This would be perfect if we would not have to consider $F = X_t^n(x)$ rather than $F = X_t(x)$: we must take care that $X_t^n(x)$ does not satisfy the nondegeneracy condition (3.4). We now explain the reason.

On one hand, one can easily prove the following: for any $p > 1$ and $j \geq 1$, there exist an integer Q and a non decreasing function $K(\cdot)$ such that

$$\sup_{n\geq 1}\|X_t^n(x)\|_{j,p} < K(t)(1+\|x\|^Q) \tag{3.28}$$

and

$$\sup_{n\geq 1}\|X_t(x) - X_t^n(x)\|_{j,p} < \frac{K(t)}{\sqrt{n}}(1+\|x\|^Q). \tag{3.29}$$

On the other hand this result is far from satisfactory in view of the condition (3.4): indeed, if (Z^n) is a sequence of random variables, the convergence to a random variable Z in $L^p(\Omega)$ does not imply that $\frac{1}{Z^n}$ is in $L^p(\Omega)$.

At this step of the proof a localization argument seems necessary. Let γ_t^n denote the Malliavin covariance matrix of X_t^n and let Γ_t^n denotes its inverse (where it is defined). We recall that we are considering the case $T \geq t \geq \frac{T}{2}$. Let Ω_0 be the set of events where $|\hat{\gamma}_t^n - \hat{\gamma}_t|$ is larger than $\frac{\hat{\gamma}_t}{4}$. Using (3.23) and (3.29) one proves that $\mathbf{P}(\Omega_0)$ is small. On the complementary set of Ω_0, $|\hat{\gamma}_t^n - \hat{\gamma}_t|$ is small, which (roughly speaking) means that the Malliavin covariance matrix of $X_t^n(x)$ behaves like that of $X_t(x)$ (see (3.23)), which allows integrations by parts of the type (3.20) with a good control of the L^p-norms of the variables H_α. □

3.5 Extensions

In the preceding proof, we have integrated by parts in order to make appear $f(\cdot)$ instead of derivatives of $u(t,\cdot)$. One can refine the method to get an expansion for

$$p_T(x,y) - \tilde{p}_T^n(x,y)$$

where $p_T(x,y)$ denotes the density of $X_T(x)$ and $\tilde{p}_t^n(x,\cdot)$ denotes the density of the law of a suitable small perturbation of $X_t^n(x)$ (the law of $X_t^n(x)$ may have no density, see our remark above on $\Gamma_t^n(x)$). To treat this problem, it is natural to fix y, choose $f_\delta(\xi) = \rho_\delta(y-\xi)$ where the $\rho_\delta(\cdot)$'s are such that the sequence of measures $(\rho_\delta(\xi)d\xi)$ converges weakly to the Dirac measure at 0, and make δ tend to 0. Theorem 3.10 is not sufficient since, when δ tends to 0, $(\| f_\delta \|_\infty)$ tends to infinity. Nevertheless, if F_δ is the distribution function of the measure $f_\delta(\xi)d\xi$, the sequence $(\| F_\delta \|_\infty)$ is constant: this gives the idea of proving inequalities of the type (3.26) with $\| F \|_\infty$ instead of $\| f \|_\infty$ when $f(\cdot)$ has a compact support, $F(\cdot)$ being the distribution function of the measure $f(\xi)d\xi$. A supplementary difficulty is to prove that, instead of $(1+\|x\|^Q)$ appears a function which satisfies an exponential upper bound and that the function $C_{f_\delta}(T,x)$ itself satisfies an exponential upper bound: such estimates permit to conclude that, when the differential operator $\mathcal{L}$ in (3.1) is strongly uniformly elliptic, that the density $p_T^n(x,y)$ of $X_t^n(x)$ (which does exist in this case) satisfies:

$$\forall (x,y) \in \mathbb{R}^d \times \mathbb{R}^d,\ p_T(x,y) - p_T^n(x,y) = -\frac{1}{n}\pi_T(x,y) + \frac{1}{n^2}R_T^n(x,y) \quad (3.30)$$

and there exists a strictly positive constant c, an integer q and an increasing function $K(\cdot)$ such that

$$|\pi_T(x,y)| + |R_T^n(x,y)| \le \frac{K(T)}{T^q}\exp\left(-c\frac{\|x-y\|^2}{T}\right) . \quad (3.31)$$

For a complete exposition and a precise result, see Bally and Talay [2].

Observe also that the expansion of the error (3.25) justifies the Romberg extrapolation procedure. Indeed, for some function $e(\cdot)$ one has

$$\mathbf{E}f(X_T) - \mathbf{E}f\left(X_T^n\right) = \frac{T}{n}e(T) + \mathcal{O}\left(\frac{1}{n^2}\right) ,$$

and

$$\mathbf{E}f(X_T) - \mathbf{E}f\left(X_T^{2n}\right) = \frac{T}{2n}e(T) + \mathcal{O}\left(\frac{1}{n^2}\right) .$$

Consider the new approximate value

$$Z_T^n := 2\mathbf{E}f(X_T^{2n}) - \mathbf{E}f(X_T^n) , \quad (3.32)$$

then

$$\mathbf{E}f(X_T) - Z_T^n = \mathcal{O}(n^{-2}).$$

Thus, a precision of order n^{-2} is achieved by a linear combination of the results produced by the Euler scheme with 2 different step sizes. For numerical examples and comments, see Talay and Tubaro [50].

In the context of the present subsection, the Milshtein scheme (2.9) (for $d = r = 1$) has the same convergence rate as the Euler scheme (contrarily to the approximation in $L^p(\Omega)$). The expansion of the Milshtein scheme error makes appear a different function $\Psi(\cdot)$.

The results given above only concern SDE's driven by a Wiener process. One can extend both the convergence rate analysis and the simulation technique to SDE's driven by Lévy processes, which corresponds to the analysis of Monte-Carlo methods for integro-differential equations of the type

$$\frac{\partial u}{\partial t}(t,x) = \tag{3.33}$$

$$Lu(t,x) + \int_{\mathbb{R}^d} \{u(t,x+z) - u(t,x) - < z, \nabla u(t,x) > \mathbb{1}_{[\|z\|\leq 1]}\} M(x,dz)$$

where L is the elliptic operator (3.1) and the measure $M(x,\cdot)$ is defined as follows: let ν be a measure on $\mathbb{R}^d - \{0\}$ such that

$$\int_{\mathbb{R}^d} (\| x \|^2 \wedge 1)\nu(dx) < \infty$$

and let $g(\cdot)$ be a $d \times r$-matrix valued function defined in $\mathbb{R}^d$; then, for any Borel set $B \subset \mathbb{R}^d$ whose closure does not contain 0, set

$$M(x,B) := \nu\{z \,;\, < g(x), z > \in B\}.$$

Consider a Lévy process (Z_t) and (X_t) solution to

$$X_t = X_0 + \int_0^t g(X_{s-})dZ_s. \tag{3.34}$$

For $K > 0$, $m > 0$ and $p \in \mathbb{N} - \{0\}$, set

$$\begin{aligned} \rho_p(m) \;&:=\; 1 + \|\beta\|^2 + \|\sigma\|^2 + \int_{-m}^{m} \|z\|^2\nu(dz) + \|\beta\|^p + \|\sigma\|^p \\ &\quad + \left(\int_{-m}^{m} \|z\|^2\nu(dz)\right)^{p/2} + \int_{-m}^{m} \|z\|^p\nu(dz) \end{aligned} \tag{3.35}$$

where ν is the Lévy measure of (Z_t), and

$$\eta_{K,p}(m) := \exp\left(K\rho_p(m)\right). \tag{3.36}$$

For $m > 0$ we define

$$h(m) := \nu(\{x; \|x\| \geq m\}). \tag{3.37}$$

Theorem 3.11 (Protter and Talay [38]). *Suppose:*

(H1) the function $f(\cdot)$ is of class C^4; $f(\cdot)$ and all derivatives up to order 4 are bounded;

(H2) the function $g(\cdot)$ is of class C^4; $g(\cdot)$ and all derivatives up to order 4 are bounded;

(H3) $X_0 \in L^4(\Omega)$.

Then there exists a strictly increasing function $K(\cdot)$ depending only on d, r and the L^∞-norm of the partial derivatives of $f(\cdot)$ and $g(\cdot)$ up to order 4 such that, for any discretization step of type $\frac{T}{n}$, for any integer m,

$$|\mathbf{E}g(X_T) - \mathbf{E}g(\bar{X}_T^n)| \leq 4\|g\|_{L^\infty(\mathbb{R}^d)}(1 - \exp(-h(m)T)) + \frac{\eta_{K(T),8}(m)}{n}. \quad (3.38)$$

Thus, the convergence rate is governed by the rate of increase to infinity of the functions $h(\cdot)$ and $\eta_{K(T),8}(\cdot)$.

Stronger hypotheses permit to get much more precise results:

Theorem 3.12. *Suppose:*

(H1') the function $f(\cdot)$ is of class C^4; all derivatives up to order 4 of $f(\cdot)$ are bounded;

(H2') the function $g(\cdot)$ is of class C^4 and moreover $|\partial_I g(x)| = \mathcal{O}(\|x\|^{M'})$ for $|I| = 4$ and some $M' \geq 2$;

(H3') $\int_{\|x\|\geq 1} \|x\|^\gamma \nu(dx) < \infty$ for $2 \leq \gamma \leq M'^* := \max(2M', 8)$ and $X_0 \in L^{M'^*}(\Omega)$.

Then there exists an increasing function $K(\cdot)$ such that, for all $n \in \mathbb{N} - \{0\}$,

$$|\mathbf{E}g(X_T) - \mathbf{E}g(\bar{X}_T^n)| \leq \frac{\eta_{K(T),M'^*}(\infty)}{n}. \quad (3.39)$$

Suppose now:

(H1") the function $f(\cdot)$ is of class C^8; all derivatives up to order 8 of $f(\cdot)$ are bounded;

(H2") the function $g(\cdot)$ is of class C^8 and moreover $|\partial_I g(x)| = \mathcal{O}(\|x\|^{M''})$ for $|I| = 8$ and some $M'' \geq 2$;

(H3") $\int_{\|x\|\geq 1} \|x\|^\gamma \nu(dx) < \infty$ for $2 \leq \gamma \leq M''^* := 2\max(2M'', 16)$ and $X_0 \in L^{M''^*}(\Omega)$.

Then there exists a function $C(\cdot)$ and an increasing function $K(\cdot)$ such that, for any discretization step of type $\frac{T}{n}$, one has

$$\mathbf{E}g(X_T) - \mathbf{E}g(\bar{X}_T^n) = \frac{C(T)}{n} + R_T^n \quad (3.40)$$

and $\sup_n n^2|R_T^n| \leq \eta_{K(T),M''^*}(\infty)$.

3.6 Newton's variance reduction technique

In Newton [33] are presented variance reduction techniques for the Monte Carlo computation of quantities of the type

$$\mathbf{E}\Phi(X_.)$$

where $\Phi(\cdot)$ is a real valued functional defined on $C([0,T];\mathbb{R}^d)$. In the preceding subsection we have considered a much less general situation:

$$\Phi(\omega_.) = f(\omega_T). \tag{3.41}$$

Newton proposes a general methodology to reduce the variance of the Monte Carlo procedure. His rather complex approach is based upon Haussmann's integral representation theorem applied to $\Phi(X_.)$. The analysis is considerably simplified in the context (3.41) to which we limit ourselves here. In this context, the principle of Newton's method is as follows. Write

$$f(X_T(x)) = \mathbf{E}f(X_T(x)) + \int_0^T (\partial u)(t, X_t(x))\sigma(X_t(x))dW_t \text{ , a.s.}$$

and set

$$Z := f(X_T(x)) - \int_0^T (\partial u)(t, X_t(x))\sigma(X_t(x))dW_t.$$

Of course Z is an unbiased estimator of $\mathbf{E}f(X_T(x))$ and the variance of the error is 0. Now suppose that one knows an approximation $\bar{v}$ of ∂u. Then it is natural to consider

$$\bar{Z} := f(X_T(x)) - \int_0^T \bar{v}(t, X_t(x))\sigma(X_t(x))dW_t.$$

$\bar{Z}$ is an unbiased estimator of $\mathbf{E}f(X_T(x))$; the error of the variance is

$$\begin{aligned}
&\mathbf{E}|\bar{Z} - \mathbf{E}f(X_T(x))|^2 \\
&= \mathbf{E}\left| f(X_T(x)) - \int_0^T \bar{v}(t, X_t(x))\sigma(X_t(x))dW_t - \mathbf{E}f(X_T(x))\right|^2 \\
&= \mathbf{E}\int_0^T |((\partial u)(t, X_t(x)) - \bar{v}(t, X_t(x)))\sigma(X_t(x))|^2\, dt.
\end{aligned}$$

Thus, the variance may be small is $\bar{v}(\cdot,\cdot)$ is a good approximation of $u(\cdot,\cdot)$ in the sense that the right hand side of the preceding inequality is small ($\bar{v}(\cdot,\cdot)$ can be seen as an approximation of $\partial u(\cdot,\cdot)$ in a suitable Hilbert space). In such a case, one approximates

$$\int_0^T (\partial u)(t, X_t(x))\sigma(X_t(x))dW_t$$

by the sum

$$\sum_{p=1}^{n} \bar{v}(pT/n, X^n_{pT/n}(x))\sigma(X^n_{pT/n}(x))(W_{(p+1)T/n} - W_{pT/n}).$$

Such a variance reduction technique is called a "control variate" technique. Newton also proposes a methodology to construct "importance sampling" methods. See [33].

3.7 Lépingle's reflected Euler scheme

Elliptic and parabolic PDE's with a Dirichlet condition at the boundary lead to probabilistic interpretations in terms of diffusion processes stopped at the boundary. If the boundary condition is of the Neumann type then the probabilistic interpretations involve reflected diffusion proceses. See Bensoussan and Lions [3] or Freidlin [14] e.g.

We do not discuss here the approximation of stopped diffusions. Only a few convincing results are available, see Milshtein [32].

For reflected diffusions on the boundary of the half-space, Lépingle [27] has constructed and analysed a version of the Euler scheme which mimics the reflection and is numerically efficient in the sense that the random variables involved in the scheme are easy to simulate.

We first define a diffusion process obliquely reflected at the boundary of the half-space.

For $d > 1$ consider the domain $D := \mathbb{R}^{d-1} \times \mathbb{R}^*_+$. Suppose that $X_0 \in \bar{D}$ *a.s.*. Fix a vector

$$\gamma := (\gamma_1, \ldots, \gamma_{d-1}, 1)$$

in $\mathbb{R}^d$.

Suppose that the functions $b(\cdot)$ and $\sigma(\cdot)$ are globally Lipschitz. Then there exists a unique adapted continuous process X with values in $\bar{D}$ and a unique adapted continuous nondecreasing process L such that for any $t \in [0, T]$,

$$\begin{aligned} X_t &= X_0 + \int_0^t b(X_s)ds + \int_0^t \sigma(X_s)dW_s + \gamma L_t \ , \\ L_t &= \int_0^t \mathbb{1}_{\{X_s^d = 0\}} dL_s. \end{aligned}$$

The process L is given by

$$L_t = \sup_{0 \le s \le t} (X^d_s - L_s)^- .$$

The reflected Euler scheme is as follows:

$$\left\{ \begin{aligned} X^n_0 &= X_0 \ , \\ X^n_{(p+1)T/n} &= X^n_{pT/n} + b(X^n_{pT/n})\tfrac{T}{n} \\ &\quad + \sigma(X^n_{pT/n})(W_{(p+1)T/n} - W_{pT/n}) \\ &\quad + \gamma \max(0, A^n_{T/n}(p) - X^{n,d}_{pT/n}) \end{aligned} \right. \qquad (3.42)$$

where

$$A_\theta^n(p) := \sup_{pT/n \le s \le \theta + pT/n} \{-b^d(X_{pT/n}^n)(s - pT/n) - \sum_{j=1}^r \sigma_j^d(X_{pT/n}^n)(W_s^j - W_{pT/n}^j)\}.$$

The simulation of the reflected Euler scheme requires the simulation of the pair

$$\left(W_{(p+1)T/n} - W_{pT/n}, A_{T/n}^n(p)\right)$$

at each step. This can be efficiently done, as proven in Lépingle [27]:

Proposition 3.13. *Let $\alpha = (\alpha_1, \ldots, \alpha_r)$ be a vector of $\mathbb{R}^r$ and let c be a real number. Set*

$$S_t := \sup_{s \le t}(< \alpha, W_s > + cs)\ .$$

Let $U = (U_1, \ldots, U_r)$ be a Gaussian vector of zero mean with covariance matrix $t\ Id$, and let V be an exponential random variable with parameter $(2t)^{-1}$ independent of U. Set

$$Y := \frac{1}{2}(< \alpha, U > + ct + (|\alpha|^2 V + (< \alpha, U > + ct)^2)^{1/2}).$$

Then the vectors (W_t, S_t) and (U, Y) have the same law.

Now define a continuous-time version of the preceding scheme, coinciding with $X_{pT/n}^n$ at each time pT/n: for $\frac{kT}{n} \le t < \frac{(k+1)T}{n}$,

$$\begin{cases} X_0^n = X_0 \ , \\ X_t^n = X_{kT/n}^n + b(X_{kT/n}^n)\left(t - \frac{kT}{n}\right) + \sigma(X_{kT/n}^n)(W_t - W_{kT/n}) \\ \qquad\qquad + \gamma \sup_{pT/n \le s \le t}(A_{s-pT/n}^n(p) - X_s^n). \end{cases} \tag{3.43}$$

One has the following convergence result, similar to (2.4):

Theorem 3.14 (Lépingle [27]). *Suppose $b(\cdot)$ and $\sigma(\cdot)$ are Lipschitz functions and that $\mathbf{E}|X_0|^2 < \infty$. Then, for some constants C_1 and C_2 uniform w.r.t. n,*

$$\mathbf{E}\left[\sup_{t \in [0,T]} |X_t - X_t^n|^2\right] \le \frac{C_1 \exp(C_2 T)}{n}. \tag{3.44}$$

In [28] Lépingle extends his analysis to the case of hypercubes with normal reflections.

An estimate for the approximation of $\mathbf{E}f(X_T)$ would be useful: this work is in progress. A result of this nature has just appeared in a manuscript by Costantini, Pacchiarotti and Sartoretto [12].

An original numerical procedure is proposed by Liu [29]. This procedure is based upon a penalization technique.

For Monte Carlo methods coupled with the simulation of obliquely reflecting Brownian motions, Calzolari, Costantini and Marchetti [8] give confidence intervals.

Other approximation problems are investigated by Slominski [44] and [45] for much wider classes of semimartingales and much larger types of reflections. As expected the approximating processes are less easy to simulate than Lépingle's scheme and the convergence rates are lower. Other references can be found in [27].

3.8 The stationary case

In this subsection we assume

(H3) the functions b, σ are of class C^∞ with bounded derivatives of any order; the function σ is bounded;

(H4) the operator L is uniformly elliptic: there exists a strictly positive constant α such that

$$\forall x, \ \xi \in \mathbb{R}^d \ , \ \sum_{i,j} a_j^i(\xi) x^i x^j \geq \alpha |x|^2 \ ;$$

(H5) there exists a strictly positive constant β and a compact set K such that:

$$\forall x \in \mathbb{R}^d - K \ , \ x \cdot b(x) \leq -\beta |x|^2 .$$

It is well known that (H3)-(H5) are (even too strong) sufficient conditions for the ergodicity of (X_t): see for instance Hasminskii [21]. Thus, (X_t) has a unique invariant probability measure μ. The hypothesis (H4) implies the existence of a smooth density $p(\cdot)$ for μ. This density solves the stationary parabolic PDE

$$L^* p(\cdot) = 0. \tag{3.45}$$

Our objective is to approximate

$$\int_{\mathbb{R}^d} f(y) p(y) dy$$

for a given function $f(\cdot)$ in $L^1(\mu)$.

Theorem 3.15 (Talay and Tubaro [50]). *Assume (H3)-(H5).*

The Euler scheme defines an ergodic Markov chain.

Let $f(\cdot)$ be a real function of class $C^\infty(\mathbb{R}^d)$. Assume that $f(\cdot)$ and any of its partial derivatives have a polynomial growth at infinity.

Let Ψ be defined as in (3.4). Set

$$\lambda := \int_0^{+\infty} \int_{\mathbb{R}^d} \Psi(t, y) \mu(dy) dt.$$

Then the Euler scheme with step size $\frac{1}{n}$ satisfies: for any deterministic initial condition $\xi = \overline{X}_0^h$,

$$\int f(y)\mu(dy) - a.s. \lim_{N\to+\infty} \frac{1}{N}\sum_{p=1}^{N} f(X_{p/n}^n(\xi)) = -\frac{\lambda}{n} + \mathcal{O}\left(\frac{1}{n^2}\right). \quad (3.46)$$

Sketch of the proof. The ergodicity of the Euler scheme can be proven by using a sufficient criterion due to Tweedie [51]: first, one can check that there exits a compact set which is reached in finite time by the chain $(X_{p/n}^n)$ with a strictly positive probability; second, it is easy to check that for any n large enough, there exists $\epsilon > 0$ such that for all deterministic starting point $X_0^n = x$ outside this compact set,

$$\mathbf{E}|X_{1/n}^n|^2 \leq |x|^2 - \epsilon.$$

Next, we oberve that the measure μ has finite moments of any order. Similarly,

$$\begin{aligned} &\forall p \in \mathbb{N}\ ,\ \exists C_p > 0\ ,\ \exists \gamma_p > 0\ ,\ \exists n_0 > 0\ ,\ \ \forall n \geq n_0\ , \\ &\mathbf{E}|X_t^n(x)|^p \leq C_p(1 + |x|^p e^{-\gamma_p t})\ ,\ \forall t > 0\ ,\ \forall x \in \mathbb{R}. \end{aligned} \quad (3.47)$$

Note that (3.47) imply that $\mathbf{E}_x f(X_t^n)$ is well defined.

Equipped with these preliminary results, our main ingredient to prove (3.46) is the following. Set

$$u(t,x) := \mathbf{E}_x f(X_t) - \int f(y)d\mu(y).$$

Then, for any multiindex α there exist an integer s_α, there exist strictly positive constants Γ_α and γ_α such that

$$|\partial_\alpha u(t,x)| \leq \Gamma_\alpha(1 + |x|^{s_\alpha})e^{-\gamma_\alpha t}\ ,\ \forall t > 0\ ,\ \forall x \in \mathbb{R}^d. \quad (3.48)$$

The proof of this estimate is technical (see Talay [47]). One step is to show that for any multiindex I, if M_I is defined by

$$|I| = \textit{integer part of } (M_I - d/2)\ ,$$

and if

$$\pi_s(x) := \frac{1}{(1+|x|^2)^s}\ ,$$

there holds, for $s \in \mathbb{N}$ large enough:

$$\begin{aligned} &\exists C_I > 0\ ,\ \exists \lambda_I > 0\ ,\ \forall |\alpha| \leq M_I\ ,\ \forall t > 0\ , \\ &\int |\partial_\alpha u(t,x)|^2 \pi_s(x)dx \leq C_I \exp(-\lambda_I t). \end{aligned} \quad (3.49)$$

An easy computation shows that the preceding inequality implies that

$$\exists C_I \ , \ \lambda_I \ : \ \forall|\alpha| \leq M_I \ , \ \forall t > 0 \ , \ \int |\partial_\alpha(u(t,x)\pi_s(x))|^2 dx \leq C_I \exp(-\lambda_I t).$$

We then can deduce (3.48) by using the Sobolev imbedding Theorem.

Next, one observes that

$$\frac{1}{N}\sum_{p=1}^{N} \mathbf{E}_x f(X^n_{p/n}) = \frac{1}{N}\sum_{p=1}^{N} u(p/n,x) + \frac{1}{Nn^2}\sum_{j=1}^{N}\sum_{p=0}^{N-j} \mathbf{E}_x \Psi(j/n, X^n_{p/n}) + \frac{1}{Nn^3}\sum_{p=1}^{N} \mathcal{R}^n_p$$

where $\mathcal{R}^n_p$ is a sum of terms, each term being a product of derivatives of $b(\cdot)$, $\sigma(\cdot)$ and $u(p/n,\cdot)$. Then one makes N tend to infinity. The exponential decay in (3.48) permits to control the sum of the remainders $\mathcal{R}^n_p$ and to prove that

$$\lim_{N\to\infty} \frac{1}{Nn}\sum_{j=1}^{N}\sum_{p=0}^{N-j} \mathbf{E}_x \Psi(j/n, X^n_{p/n}) = \int_0^{+\infty}\int_{\mathbb{R}^d} \Psi(t,y)\mu(dy)dt.$$

□

The Milshtein scheme (2.9) (for $d = r = 1$) has the same convergence rate as the Euler scheme. The expansion of the Milshtein scheme error makes appear a different function $\Psi(\cdot)$.

As in the non stationary case, the expansion of the error in terms of $\frac{1}{n}$ justifies a Romberg extrapolation which permits to accelate the convergence rate. See [50] for numerical experiments.

PART II - Stochastic Particle Methods

In this part, we analyse stochastic particle methods for nonlinear PDE's in a few special cases. Our objective is to establish the convergence rates which can be observed in numerical experiments for PDE's such that an explicit solution is known, especially the rate $N^{-1/2}$ where N is the number of particles of the algorithm.

Works in progress at Inria, based on the results presented below, have for objective the analysis of the random vortex methods for the incompressible 2D Navier-Stokes equation developed by Chorin, Hald, etc (Chorin [11], Chorin and Marsden [10], Goodman [17], Hald ([19] and [20]), Long [30], Puckett ([40], [39]) e.g.; see also the bibliography in [11] and in the different contributions of [18]).

From now on, we suppose

$$d = r = 1.$$

We also suppose

(H6) $b(\cdot)$ and $\sigma(\cdot)$ are bounded functions of class $C^\infty(\mathbb{R})$; any derivative of any order is assumed bounded;

(H7) $\sigma(x) \geq \sigma_0 > 0$, $\forall x \in \mathbb{R}$.

We continue to set $a(\cdot) := \sigma^2(\cdot)$.

4. Introduction to the stochastic particle methods

Let $V_0(\cdot)$ be the distribution function of a probability law. Consider the PDE in $(0,T] \times \mathbb{R}$

$$\begin{cases} \frac{\partial V}{\partial t}(t,x) &= \frac{1}{2}a(x)\frac{\partial^2}{\partial x^2}V(t,x) \\ & \quad + \left(\frac{1}{2}a'(x) - b(x)\right)\frac{\partial}{\partial x}V(t,x) \text{ in } (0,T] \times \mathbb{R}, \\ \lim_{t\to 0} V(t,x) &= V_0(x) \text{ at all continuity points of } V_0(\cdot). \end{cases} \tag{4.1}$$

It is well known (see for instance [15]) that under (H6)-(H7) the law of $X_t(x)$ has a smooth density $p_t(x,\cdot)$ for all $x \in \mathbb{R}$ and all $t > 0$; this density satifies

$$\begin{cases} \frac{\partial p}{\partial t}(t,x) = L^* p(t,x) \ , \ \forall t > 0 \ , \ \forall x \in \mathbb{R} \ , \\ \\ p_t(x,\xi)d\xi \xrightarrow{w} \delta_x . \end{cases} \tag{4.2}$$

Besides, there exists an increasing function $K(\cdot)$ and a constant $\lambda > 0$ such that, for all $(x,y) \in \mathbb{R}^2$,

$$p_t(x,y) \leq \frac{K(t)}{\sqrt{t}} \exp\left(-\frac{(x-y)^2}{2\lambda t}\right). \tag{4.3}$$

From (4.2) it is easy to see that $V(t,\cdot)$ is the distribution function of the law of X_t when the law of X_0 is $\mu_0(d\xi) := dV_0(\xi)$.

Let $H(\cdot)$ denote the Heaviside function ($H(z) = 0$ if $z < 0$, $H(z) = 1$ if $z \geq 1$). Set

$$\omega_0^i = \frac{1}{N}\ ,\ \text{for } i = 1,\ldots,N\ ;\quad V_0^N(x) = \sum_{i=1}^{N} \omega_0^i\, H(x - x_0^i)\ ,$$

where

$$\forall i > 1\ ,\ x_0^i := V_0^{-1}\left(\frac{i}{N}\right)\ ,\ x_0^1 := V_0^{-1}\left(\frac{1}{2N}\right). \tag{4.4}$$

Thus, $V_0^N(\cdot)$ is a piecewise constant approximation to $V_0(\cdot)$.

Now consider N independent copies of the process (W_t) and the corresponding N copies (X_t^i) of (X_t) $(1 \leq i \leq N)$, with $X_0^i = x_0^i$. Define

$$\mu_t^N := \frac{1}{N}\sum_{i=1}^{N} H(x - X_t^i). \tag{4.5}$$

The distribution function of the measure μ_t^N is

$$V^N(t,x) = \frac{1}{N}\sum_{i=1}^{N} H(x - X_t^i).$$

Proposition 4.1. *Suppose (H6)-(H7) and*

(H8) There exist strictly positive constants C_1, C_2 such that, for any x in $\mathbb{R}$, $|V_0(x)| \leq C_1 e^{-C_2 x^2}$.

Then there exists an increasing function $K(\cdot)$ such that, for all $N \in \mathbb{N}^*$,

$$\mathbf{E}\,\|\,V(t,\cdot) - V^N(t,\cdot)\,\|_{L^1(\mathbb{R})} \leq \frac{K(t)}{\sqrt{N}}. \tag{4.6}$$

Sketch of the proof. We have:

$$\begin{aligned} V(t,x) - V^N(t,x) &= V(t,x) - \mathbf{E}V^N(t,x) + \mathbf{E}V^N(t,x) - V^N(t,x) \\ &= \mathbf{P}_{\mu_0}(X_t \leq x) - \frac{1}{N}\sum_{i=1}^{N} \mathbf{P}(X_t^i \leq x) \\ &\quad + \frac{1}{N}\sum_{i=1}^{N} \{\mathbf{E}H(x - X_t^i) - H(x - X_t^i)\} \\ &=: A(x) + B(x). \end{aligned}$$

An integration by parts for Stieljes integrals leads to

$$\begin{aligned} A(x) &= \int \mathbf{P}_y(X_t \le x)dV_0(y) - \int \mathbf{P}_y(X_t \le x)dV_0^N(y) \\ &= \int (V_0(y) - V_0^N(y))\frac{d}{dy}\mathbf{P}_y(X_t \le x)dy. \end{aligned}$$

Therefore

$$\| A(\cdot) \|_{L^1(\mathbb{R})} \le \int\int \left|\frac{d}{dy}\mathbf{P}_y(X_t \le x)\right| dx |V_0(y) - V_0^N(y)|dy.$$

The function $y \to X_t(y)$ is a.s. increasing since its derivative is an exponential (see Kunita [24, Ch.2] e.g., for the diffeomorphism property of stochastic flows associated with stochastic differential equations). Thus, again denoting by $X_t(\cdot)$ the flow defined by (2.1),

$$P(X_t(y) \le x) = P(y \le X_t^{-1}(x)).$$

X_t^{-1} is the solution to a stochastic differential equation. The coefficients of this SDE are such that the above mentioned result of Friedman [15] applies: for some new function $K(\cdot)$,

$$\left|\frac{d}{dy}\mathbf{P}_x(X_t^{-1} \le y)\right| \le \frac{K(t)}{\sqrt{t}} \exp\left(-\frac{(x-y)^2}{2\lambda t}\right).$$

It comes:

$$\| A(\cdot) \|_{L^1(\mathbb{R})} \le K(t) \| V_0 - V_0^N \|_{L^1(\mathbb{R})} \le \frac{K(t)\sqrt{\log(N)}}{N}.$$

Now consider $B(x)$.

$$\mathbf{E} \| B(\cdot) \|_{L^1(\mathbb{R})} = \frac{1}{N}\int \mathbf{E}\left|\sum_{i=1}^N (H(x - X_t^i) - \mathbf{E}H(x - X_t^i))\right| dx.$$

The random variables $(H(x - X_t^i) - \mathbf{E}H(x - X_t^i))_{1 \le i \le N}$ have mean 0 and are independent. Thus, by the Cauchy-Schwarz inequality,

$$\begin{aligned} \mathbf{E} \| B(\cdot) \|_{L^1(\mathbb{R})} &\le \frac{1}{N}\int \sqrt{\sum_{i=1}^N \mathbf{E}(H(x - X_t^i) - \mathbf{E}H(x - X_t^i))^2 dx} \\ &= \frac{1}{N}\int \sqrt{\sum_{i=1}^N \mathbf{P}(X_t^i \le x)\mathbf{P}(X_t^i \ge x)dx} \\ &\le \frac{K(t)}{N}\int \sqrt{\sum_{i=1}^N \int_{\frac{x - x_0^i}{\sqrt{\lambda t}}}^{+\infty} \exp\left\{-\frac{y^2}{2}\right\} dy} \quad dx. \end{aligned}$$

For fixed x the function

$$\xi \stackrel{\psi_x}{\longmapsto} \frac{1}{\sqrt{2\pi}} \int_{\frac{x-V_0^{-1}(\xi)}{\sqrt{\lambda t}}}^{+\infty} \exp\left\{-\frac{y^2}{2}\right\} dy$$

is decreasing from $(0,1)$ to $(0,1)$; therefore, the definition of the x_0^i implies

$$\frac{1}{2N}\sum_{i=1}^{N} \int_{\frac{x-x_0^i}{\sqrt{\lambda t}}}^{+\infty} \exp\left\{-\frac{y^2}{2}\right\} dy \leq \int_0^1 \int_{\frac{x-V_0^{-1}(s)}{\sqrt{\lambda t}}}^{+\infty} \exp\left\{-\frac{y^2}{2}\right\} dy\, ds.$$

Easy computations wich use (H8) (see [4]) then lead to the following estimate:

$$\mathbf{E} \parallel B(\cdot) \parallel_{L^1(\mathbb{R})} \leq \frac{K(t)}{\sqrt{N}}.$$

□

In practice, one cannot use exact values of X_t^i. Thus, we consider N independent processes defined by the Euler or Milshtein scheme $(\bar{X}_t^i)$ with $X_0^i = x_0^i$ and the new approximate measure

$$\bar{\mu}_t^N := \frac{1}{N}\sum_{i=1}^{N} H(x - \bar{X}_t^i). \tag{4.7}$$

The distribution function of the measure $\bar{\mu}_t^N$ is

$$\bar{V}^N(t,x) := \frac{1}{N}\sum_{i=1}^{N} H(x - \bar{X}_t^i).$$

Proposition 4.2. *Suppose (H6)-(H7) and*

(H8) There exist strictly positive constants C_1, C_2 such that, for any x in $\mathbb{R}$, $|V_0(x)| \leq C_1 e^{-C_2 x^2}$.

Then there exists an increasing function $K(\cdot)$ such that, for all $N \in \mathbb{N}^*$,

$$\mathbf{E} \parallel V(T,\cdot) - \bar{V}^N(t,\cdot) \parallel_{L^1(\mathbb{R})} \leq K(T)\left(\frac{1}{\sqrt{N}} + \frac{1}{n^\alpha}\right) \tag{4.8}$$

with $\alpha = \frac{1}{2}$ for the Euler scheme and $\alpha = 1$ for the Milshtein scheme.

Proof. The conclusion readily follows from Section 2 of Part I, the preceding theorem and the inequality

$$\mathbf{E}\|H(x - X_T^i) - H(x - \bar{X}_T^i)\|_{L^1(\mathbb{R})} \leq \mathbf{E}|X_T^i - \bar{X}_T^i| \leq \frac{K(T)}{n^\alpha}.$$

□

The convergence rate $\frac{1}{\sqrt{N}}$ is optimal. Indeed, in the following example, the error estimate is equal to $\frac{1}{\sqrt{N}}$ plus a negligible term. Let X be a random variable taking the values 0 and 1 with probability $\frac{1}{2}$. Let μ^N be the empirical distribution of N independent copies of X and let V^N be the distribution function of μ^N. It is easy to see that

$$\|V - V^N\|_{L^1(\mathbb{R})} = \frac{1}{2^N}\sum_{k=0}^{N}\left|\frac{1}{2} - \frac{k}{N}\right|\frac{N!}{k!(N-k)!}.$$

For example, suppose that $N = 2n$. Then,

$$\|V - V^N\|_{L^1(\mathbb{R})} = 2^{-2n}\sum_{k=0}^{n}\frac{(2n)!}{k!(2n-k)!} - 2^{-2n}\sum_{k=0}^{n}\frac{k}{n}\frac{(2n)!}{k!(2n-k)!}.$$

Now, an easy induction shows that, for all $n > 0$,

$$\sum_{k=0}^{n} k\frac{(2n)!}{k!(2n-k)!} = n2^{2n-1}.$$

Besides,

$$2\sum_{k=0}^{n-1}\frac{(2n)!}{k!(2n-k)!} + \frac{(2n)!}{n!n!} = 2^{2n}.$$

Thus,

$$\|V - V^N\|_{L^1(\mathbb{R})} = \frac{1}{2^{2n+1}}\frac{(2n)!}{n!n!}.$$

Applying Stirling's formula, one deduces

$$\|V - V^N\|_{L^1(\mathbb{R})} = \frac{1}{\sqrt{N}}(1 + o(1)).$$

5. The Chorin-Puckett method for convection-reaction-diffusion equations

Let $f(\cdot)$ be a real function such that

(H9) f is a C^2 function on $[0,1]$ such that $f(0) = f(1) = 0$, $f(u) \geq 0$ for $u \in [0,1]$ (therefore, $\frac{f(u)}{u}$ is bounded in $(0,1]$ and continuous in 0).

Let $V_0(\cdot)$ be as in the preceding subsection. Consider the convection-reaction-diffusion PDE

$$\begin{cases} \dfrac{\partial u}{\partial t} = L\,u + f(u)\,, \\ u(0,\cdot) = u_0(\cdot) = 1 - V_0(\cdot). \end{cases} \tag{5.1}$$

In [4], Bernard, Talay and Tubaro have analysed a stochastic particle method introduced by Puckett [40]. They have proven Puckett's conjecture, based on numerical obervations, on the convegence rate of the method. The analysis is based on an original probabilistic interpretation of the solution.

Theorem 5.1. *Under (H7)-(H9), if u_0 is of class $C_b^\infty(\mathbb{R})$, we have the following representation:*

$$u(t,x) = \mathbf{E}\left[H(X_t - x)\exp\left(\int_0^t f' \circ u(s, X_s)\, ds\right)\right], \tag{5.2}$$

where (X_t) is the solution to

$$dX_t = \sigma(X_t)\, dB_t - \{b(X_t) - \sigma(X_t)\sigma'(X_t)\}\, dt. \tag{5.3}$$

Here, the law of X_0 has a density equal to $-u_0'$, and (B_t) is a standard Brownian motion.

Sketch of the proof. The function $v(t,x) := \frac{\partial u}{\partial x}(t,x)$ satifies the following equation:

$$\begin{cases} \dfrac{\partial v}{\partial t}(t,x) &= \dfrac{1}{2}\sigma^2(x)\dfrac{\partial^2 v}{\partial x^2}(t,x) + (b(x) + \sigma(x)\sigma'(x))\dfrac{\partial v}{\partial x}(t,x) \\ & \quad +(b'(x) + f' \circ u(t,x))\, v(t,x)\,, \\ v(0,x) &= u_0'(x). \end{cases}$$

By applying the Feynman-Kac formula, we obtain

$$v(t,x) = \mathbf{E}\left[u_0'(Y_t(x))\exp\left\{\int_0^t [b'(Y_s(x)) + f' \circ u(t-s, Y_s(x))]\, ds\right\}\right], \tag{5.4}$$

where (Y_t) is the solution to

$$dY_t = (b(Y_t) + \sigma(Y_t)\sigma'(Y_t))\, dt + \sigma(Y_t)\, dB_t. \tag{5.5}$$

One can easily check that $u(t,x) \to 1$ as $x \to -\infty$. Thus,

$$u(t,x) = -\mathbf{E}\int_x^{+\infty} u_0'(Y_t(y))\exp\left\{\int_0^t [b'(Y_s(y)) + f' \circ u(t-s, Y_s(y))]\, ds\right\} dy.$$

Let $\xi_{0,t}(\cdot)$ be the flow associated with the stochastic differential equation (5.5). Hence, we set $y = \xi_{0,t}^{-1}(z)$.

Using results of the second chapter of Kunita [24], we have, for $\theta < t$,

$$\xi_{\theta,t}^{-1}(z) = z - \int_\theta^t \sigma(\xi_{s,t}^{-1}(z))\, \hat{d}B_s - \int_\theta^t b(\xi_{s,t}^{-1}(z))\, ds,$$

where $\hat{d}B_\theta$ denotes the "backward" stochastic integral [2]. One infers that

$$\frac{\partial}{\partial z}\xi_{0,t}^{-1}(z)$$

[2] For a definition, cf. Kunita [24, end of Ch. I]

$$= \exp\left(\int_0^t \left\{ -b'(\xi_{\theta,t}^{-1}(z)) - \frac{1}{2}\sigma'^2(\xi_{\theta,t}^{-1}(z))\right\} d\theta - \int_0^t \sigma'(\xi_{\theta,t}^{-1}(z))\, \hat{d}B_\theta\right)$$

from which

$-u(t,x)$

$$= \mathbf{E}\left[\int_{\xi_{0,t}(x)}^{+\infty} u_0'(z) \exp\left\{\int_0^t \left(b'(\xi_{0,s}(\alpha)) + f'\circ u(t-s,\xi_{0,s}(\alpha))\right) ds\Big|_{\alpha=\xi_{0,t}^{-1}(z)}\right\}\right.$$

$$\left.\exp\left\{\int_0^t \left[-b'(\xi_{s,t}^{-1}(z)) - \frac{1}{2}\sigma'^2(\xi_{s,t}^{-1}(z))\right] ds - \int_0^t \sigma'(\xi_{s,t}^{-1}(z))\hat{d}B_s\right\} dz\right].$$

One now uses Kunita [24, Lemma 6.2, Ch. II]: for any continuous function $g(s,x)$ we have

$$\int_0^t g(s,\xi_{0,s}(\alpha))\Big|_{\alpha=\xi_{0,t}^{-1}(z)} ds = \int_0^t g(s,\xi_{s,t}^{-1}(z))\, ds.$$

Thus,

$$-u(t,x) = \mathbf{E}\left[\int_{-\infty}^{+\infty} H(-\xi_{0,t}(x)+z) \exp\left\{\int_0^t f'\circ u(t-s,\xi_{s,t}^{-1}(z)) ds\right\}\right.$$

$$\left. M_0^t(z) u_0'(z) dz\right]$$

where $(M_\theta^t(z))_{\theta\leq t}$ is the exponential (backward) $(\mathcal{F}_\theta^t)_{\theta\leq t}$-martingale defined by

$$M_\theta^t(z) = \exp\left\{-\frac{1}{2}\int_\theta^t \sigma'^2(\xi_{s,t}^{-1}(z))\, ds - \int_\theta^t \sigma'(\xi_{s,t}^{-1}(z))\hat{d}B_s\right\}.$$

The application $x \to \xi_{0,t}(x)$ is a.s. increasing (its derivative is an exponential), thus $H(-\xi_{0,t}(x)+z) = H(\xi_{0,t}^{-1}(z) - x)$.

Hence,

$-u(t,x)$

$$= \mathbf{E}\left[\int_{\mathbf{R}} H(\xi_{0,t}^{-1}(z) - x) \exp\left\{\int_0^t f'\circ u(s,\xi_{t-s,t}^{-1}(z)) ds\right\} M_0^t(z) u_0'(z) dz\right].$$

We observe that the law of the process $(\xi_{t-\theta,t}^{-1})_{0\leq\theta\leq t}$, on $(\Omega,\mathcal{F},\mathbf{P},\mathcal{F}_0^t)$, is identical to the law of the process $(X_\theta)_{0\leq\theta\leq t}$ solution to

$$dX_\theta = \sigma(X_\theta)\, dB_\theta - b(X_\theta)\, d\theta.$$

Hence, $\mathbf{E}_0$ denoting the expectation under the law P_0 for which the initial law of the process (X_θ) has a density equal to $-u_0'(z)$, and (M_t) denoting the exponential martingale defined by

$$M_t = \exp\left\{-\frac{1}{2}\int_0^t \sigma'^2(X_s)\,ds + \int_0^t \sigma'(X_s)dB_s\right\},$$

we have

$$u(t,x) = \mathbf{E}_0\left[H(X_t - x)\exp\left\{\int_0^t f'\circ u(t-s, X_s)\,ds\right\} M_t\right].$$

On $(\Omega, \mathcal{F}, \mathrm{P}_0, \mathcal{F}_0^T)$, one performs the Girsanov transformation defined by

$$\widetilde{\mathrm{P}}(A) := \mathbf{E}_0\left[1_A\, M_T\right] \quad , \quad A \in \mathcal{F}_0^T \ ;$$

then, for $t \le T$,

$$u(t,x) = \widetilde{\mathbf{E}}\left[H(X_t - x)\exp\left\{\int_0^t f'\circ u(s, X_s)\,ds\right\}\right].$$

Under $\widetilde{\mathrm{P}}$, (X_t) solves

$$dX_t = \sigma(X_t)\,d\tilde{B}_t - \{b(X_t) - \sigma(X_t)\,\sigma'(X_t)\}\,dt.$$

Here, $(\tilde{B}_\theta)$ defined by

$$\tilde{B}_\theta = B_\theta - \int_0^\theta \sigma'(X_s)\,ds,$$

is a Brownian motion under $\widetilde{\mathrm{P}}$. Obviously, the above representation of u is identical to (5.2). □

Define the initial weights and the initial approximation by

$$\omega_0^i = \frac{1}{N}, \ \text{ for } i = 1, \ldots, N \ ; \qquad \bar{u}_0(x) = \sum_{i=1}^N \omega_0^i\, H(x_0^i - x)\ ,$$

where

$$\forall i < N \ : \ x_0^i = u_0^{-1}\left(1 - \frac{i}{N}\right) \quad , \quad x_0^N = u_0^{-1}\left(\frac{1}{2N}\right). \tag{5.6}$$

Let X^n be defined by the Milshtein scheme (2.9). From now on, we write $\bar{X}$ instead of X^n. We set

$$\begin{aligned} \bar{X}^i_{(p+1)T/n} \ = \ & \bar{X}^i_{pT/n} - \left(b(\bar{X}^i_{pT/n}) - \sigma(\bar{X}^i_{pT/n})\sigma'(\bar{X}^i_{pT/n})\right)\frac{T}{n} \\ & + \sigma(\bar{X}^i_{pT/n})(B^i_{(p+1)T/n} - B^i_{pT/n}) \\ & + \frac{1}{2}\sigma(\bar{X}^i_{pT/n})\sigma'(\bar{X}^i_{pT/n})\left((B^i_{(p+1)T/n} - B^i_{pT/n})^2 - \frac{T}{n}\right). \end{aligned} \tag{5.7}$$

Let $\pi_k(i)$ denotes the label number of the particle located immediately at the right side of the particle of label i at the time kT/n. We define

$$\omega^i_{pT/n} =$$

$$\omega^i_{(p-1)T/n} \left(1 + \frac{T}{n}\frac{f\circ\bar{u}((p-1)T/n, \bar{X}^i_{(p-1)T/n}) - f\circ\bar{u}((p-1)T/n, \bar{X}^{\pi_{p-1}(i)}_{(p-1)T/n})}{\omega^i_{(p-1)T/n}}\right) \tag{5.8}$$

and

$$\bar{u}(pT/n, x) = \sum_{i=1}^{N} \omega^i_{pT/n} H(\bar{X}^i_{pT/n} - x) \tag{5.9}$$

for $p = 1, \ldots, n$.

Theorem 5.2 (Bernard, Talay and Tubaro [4]). *(i) Under (H7)-(H9), there exists an increasing function $K(\cdot)$ and an integer n_0 such that, for any $n > n_0$ and any $N \geq 1$,*

$$\|u(T,\cdot) - \bar{u}(T,\cdot)\|_{L^1(\mathbb{R}\times\Omega)} \leq K(T)\left(\frac{1}{\sqrt{N}} + \frac{1}{\sqrt{n}}\right).$$

(ii) When the functions $b(\cdot)$ and $\sigma(\cdot)$ are constant, then the rate of convergence is given by

$$\|u(T,\cdot) - \bar{u}(T,\cdot)\|_{L^1(\mathbb{R}\times\Omega)} \leq K(T)\left(\frac{1}{\sqrt{N}} + \frac{1}{\sqrt{n}}\right).$$

The same estimates hold for the standard deviation of $\|u(T,\cdot) - \bar{u}(T,\cdot)\|_{L^1(\mathbb{R})}$.

Sketch of the proof. The lengthy proof consists in observing that the algorithm is a discretization of the representation (5.2).

Indeed, the approximation of $-u_0'(z)\,dz$ by

$$\sum_{i=1}^{N} \omega^i_0 \delta_{x^i_0}$$

leads to

$$u(T,x) \simeq \sum_{i=1}^{N} \omega^i_0 \,\mathbf{E}\left[H(X_T(x^i_0) - x)\exp\left\{\int_0^T f'\circ u(s, X_s(x^i_0))\,ds\right\}\right].$$

Let $\{(B^i_\theta), i = 1, \ldots, N\}$ be N independent Brownian motions and let (X^i_θ) be the (independent) solutions to the following SDE's (in forward time):

$$\begin{cases} dX^i_\theta = \sigma(X^i_\theta)\,dB^i_\theta - \{b(X^i_\theta) - \sigma(X^i_\theta)\sigma'(X^i_\theta)\}\,d\theta, \\ X^i_0 = x^i_0. \end{cases}$$

One has

$$u(T,x) \simeq \sum_{i=1}^{N} \omega_0^i \, \mathbf{E}\left[H(X_T^i - x) \exp\left\{ \int_0^T f' \circ u(s, X_s^i)\, ds \right\} \right].$$

The particle algorithm replaces the expectation by a point estimation:

$$u(T,x) \simeq \sum_{i=1}^{N} \omega_0^i \, H(X_T^i - x) \exp\left\{ \int_0^T f' \circ u(s, X_s^i)\, ds \right\}.$$

Then, one approximates $\exp\left\{ \int_0^T f' \circ u(s, X_s^i)\, ds \right\}$. The integral is discretized with a step T/n and the Milshtein scheme is used to approximate the $X_{pT/n}^i$'s. Besides, the unknown function $u(pT/n, \cdot)$ is replaced by its approximation $\bar{u}(pT/n, \cdot)$. It is this substitution which introduces a dependency in the algorithm, because the computation of $\bar{u}(pT/n, \cdot)$ requires to sort the positions of the particles at each step of the algorithm (see the role of the functions $\pi_k(\cdot)$ in (5.8)). Without this substitution, the weights would be recursively defined by

$$\bar{\rho}_{(p+1)T/n}^i = \bar{\rho}_{pT/n}^i + \frac{T}{n} f \circ u(pT/n, \bar{X}_{pT/n}^i).$$

The following key estimate shows that the true weights are not far from being independent, which explains that the global error of the algorithm is of order $N^{-1/2}$ as if the weights were independent. Set $\alpha_p^i := \mathbf{E}|\omega_p^i - \rho_p^i|^2$, and $\alpha_p := \sup_i \alpha_p^i$. One can show (the proof is very technical) :

$$\forall p = 1, \ldots, n \; , \; \alpha_p \leq \frac{C}{nN^2} + \frac{C}{N^3}. \tag{5.10}$$

Then, one must carefully estimate the error produced by each one of the successive approximations that have just been described. In particular, the difficulty is to avoid the summation over p of the "statistical error" involved in the algorithm which identifies $\bar{u}(pT/n, \cdot)$ and $\mathbf{E}\bar{u}(pT/n, \cdot)$, because such a summation would lead to an estimate on the global error of order $\frac{n}{\sqrt{N}}$. In fact, a more clever analysis shows that the algorithm propagates the error

$$u(pT/n, \cdot) - \mathbf{E}\bar{u}(pT/n, \cdot)$$

in a rather complex way whereas the "statistical error" can be taken into account only at time T; this latter error can be controlled owing to the estimate (5.10). □

6. One-dimensional Mc-Kean Vlasov equations

Consider two Lipschitz kernels $b(x,y)$, $s(x,y)$ from $\mathbb{R}^2$ to $\mathbb{R}$, a probability distribution function V_0 and the nonlinear problem

$$\begin{cases} \frac{\partial V}{\partial t}(t,x) = \frac{1}{2}\frac{\partial}{\partial x}\left[\left(\int_{\mathbb{R}} s(x,y)\frac{\partial V}{\partial x}(t,y)dy\right)^2 \frac{\partial V}{\partial x}(t,x)\right] \\ \qquad\qquad - \left[\int_{\mathbb{R}} b(\cdot,y)\frac{\partial V}{\partial x}(t,y)dy\right]\frac{\partial V}{\partial x}(t,x) , \\ V(0,x) = V_0(x) , \end{cases} \tag{6.1}$$

Later on, we will see that Burgers equation can be interpreted as a special case of this family of problems.

Our objective is to develop an algorithm of simulation of a discrete time particle system $\{Y^i_{kT/n}, i = 1, \ldots, N\}$ such that the empirical distribution

$$\bar{V}_{kT/n}(x) := \frac{1}{N}\sum_{i=1}^{N} H(x - Y^i_{kT/n})$$

approximates the solution $V(t,x)$ of (6.1). Contrarily to the stochastic particle method of the previous subsection, the weights are constant but, in counterpart, the positions of the particles are given by dependent stochastic processes.

Consider the system of weakly interacting particles described by

$$\begin{cases} dX^{i,N}_t = \int_{\mathbb{R}} b(X^{i,N}_t, y)\mu^N_t(dy)\,dt + \int_{\mathbb{R}} s(X^{i,N}_t, y)\mu^N_t(dy)dW^i_t , \\ X^{i,N}_0 = X^i_0 \ , i = 1, \ldots, N , \end{cases} \tag{6.2}$$

where $(W^1_t), \ldots, (W^N_t)$ are independent one-dimensional Brownian motions and μ^N_t is the random empirical measure

$$\mu^N_t = \frac{1}{N}\sum_{i=1}^{N} \delta_{X^{i,N}_t}.$$

The functions b and s are the "interaction kernels". When the initial distribution of the particles is symmetric and when the kernels are Lipschitz, one has the propagation of chaos property: the sequence of random probability measures (μ^N) on the space of trajectories defined by

$$\mu^N = \frac{1}{N}\sum_{i=1}^{N} \delta_{X^{i,N}_\cdot}$$

converges in law as N goes to infinity to a deterministic probability measure μ. Besides, if for each t we denote by μ_t the one-dimensional distribution of

μ (μ_t is the limit in law of μ_t^N), then there exists a unique strong solution (X_t) to the nonlinear stochastic differential equation

$$\begin{cases} X_t = X_0 + \int_0^t \int_{\mathbb{R}} b(X_\theta, y)\mu_\theta(dy)dt + \int_0^t \int_{\mathbb{R}} s(X_\theta, y)\mu_\theta(dy)dW_\theta \,, \\ \mu_t \text{ is the law of the random variable } X_t, \text{ for all } t \geq 0 \end{cases} \tag{6.3}$$

(see S. Méléard's contribution to this volume or Sznitman [46] e.g.). One consequence of the propagation of chaos is that the law of one particle, for example the law of $(X_t^{1,N})$, tends to the law of the process (X_t) when N goes to infinity.

Defining the differential operator $L(\mu)$ by

$$L(\mu)f(x) = \frac{1}{2}\left(\int_{\mathbb{R}} s(x,y)d\mu(y)\right)^2 f''(x) + \left(\int_{\mathbb{R}} b(x,y)d\mu(y)\right) f'(x) \,,$$

Itô's formula shows that μ_t is the solution to the McKean-Vlasov equation

$$\begin{cases} \frac{d}{dt} < \mu_t, f > = < \mu_t, L(\mu_t)f > \,, \ \forall f \in C_K^\infty(\mathbb{R}) \,, \\ \mu_{t=0} = \mu_0. \end{cases} \tag{6.4}$$

Consequently, the distribution function $V(t,x)$ of μ_t solves (6.1) where $V_0(\cdot)$ is the distribution function of μ_0.

We suppose that the following assumptions hold:

(H10) There exists a strictly positive constant s_* such that

$$s(x,y) \geq s_* > 0 \,, \ \forall (x,y).$$

(H11) The kernels $b(\cdot,\cdot)$ and $s(\cdot,\cdot)$ are uniformly bounded functions of $\mathbb{R}^2$; $b(\cdot,\cdot)$ is globally Lipschitz and $s(\cdot,\cdot)$ has uniformly bounded first partial derivatives.

(H12) The initial law μ_0 has a continuous density $u_0(\cdot)$ satisfying: there exist constants $M > 0$, $\eta \geq 0$ and $\alpha > 0$ such that

$$u_0(x) \leq \eta \exp(-\alpha \frac{x^2}{2}) \text{ for } |x| > M.$$

The initial distribution function $V(0,\cdot) = V_0(\cdot)$ is approximated as in the preceding subsection. We set

$$y_0^i := V_0^{-1}(i/N).$$

Consider the system (6.2) with the initial condition $X_0^{i,N} = y_0^i$, and denote its solution by $(X_t^i \,,\ 1 \leq i \leq N)$. There holds

$$\begin{cases} dX_t^i = \frac{1}{N}\sum_{j=1}^{N} b\left(X_t^i, X_t^j\right)\, dt + \frac{1}{N}\sum_{j=1}^{N} s\left(X_t^i, X_t^j\right)\, dw_t^i\ ,\ t\in[0,T]\ , \\ X_0^i = y_0^i\ ,\ i=1,\ldots,N. \end{cases}$$

To get a simulation procedure of a trajectory of each (X_t^i), we discretize in time and we approximate μ_t by the empirical measure of the simulated particles. The Euler scheme then leads to

$$\begin{cases} Y^i_{(k+1)T/n} &= Y^i_{kT/n} + \frac{1}{N}\sum_{j=1}^{N} b(Y^i_{kT/n}, Y^j_{kT/n})\frac{T}{n} \\ &\quad + \frac{1}{N}\sum_{j=1}^{N} s(Y^i_{kT/n}, Y^j_{kT/n})\left(W^i_{(k+1)T/n} - W^i_{kT/n}\right)\ , \\ Y_0^i &= y_0^i\ ,\ i=1,\ldots,N. \end{cases} \tag{6.5}$$

In the same way, we approximate $V(t,\cdot)$ solution to (6.1) by the cumulative distribution function of μ_t:

$$\overline{V}_{kT/n}(x) := \frac{1}{N}\sum_{i=1}^{N} H(x - Y^i_{kT/n})\ ,\ \forall x\in\mathbb{R}. \tag{6.6}$$

Theorem 6.1 (Bossy and Talay [7], Bossy [5]). *Suppose (H10)-(H12). Let $V(t,x)$ be the solution of the PDE (6.1).*

There exists an increasing function $K(\cdot)$ such that, $\forall k\in\{1,\ldots,n\}$:

$$\begin{aligned} &\mathbf{E}\left\|V(kT/n,.) - \overline{V}_{kT/n}(.)\right\|_{L^1(\mathbb{R})} \\ &\quad\le K(T)\left(\|V_0 - \overline{V}_0\|_{L^1(\mathbb{R})} + \frac{1}{\sqrt{N}} + \frac{1}{\sqrt{n}}\right) \end{aligned} \tag{6.7}$$

and

$$\begin{aligned} &Var\left(\left\|V(kT/n,.) - \overline{V}_{kT/n}(.)\right\|_{L^1(\mathbb{R})}\right) \\ &\quad\le K(T)\left(\|V_0 - \overline{V}_0\|^2_{L^1(\mathbb{R})} + \frac{1}{N} + \frac{1}{n}\right). \end{aligned} \tag{6.8}$$

Besides,

$$\left\|V_0 - \overline{V}_0\right\|_{L^1(\mathbb{R})} \le \frac{C\sqrt{\log(N)}}{N}.$$

Sketch of the proof. Define $\beta\ :\ [0,T]\times\mathbb{R}\longrightarrow\mathbb{R}$ by

$$\beta(t,x) := \int_{\mathbb{R}} b(x,y)\,\mu_t(dy)\ ,$$

and $\sigma\ :\ [0,T]\times\mathbb{R}\longrightarrow\mathbb{R}$ by

$$\sigma(t,x):=\int_{\mathbb{R}}s(x,y)\,\mu_t(dy).$$

Under our hypotheses, there exists a unique strong solution to

$$\begin{cases} dz_t = \beta(t,z_t)dt + \sigma(t,z_t)\,dw_t\ , \\ z_{t=0} = z_0\ , \end{cases} \tag{6.9}$$

where z_0 is a square integrable random variable. When the law of z_0 is μ_0, the two processes (z_t) and (X_t) solution to (6.3) have the same law and

$$V(t,x) = \mathbf{E}H(x-X_t) = \mathbf{E}_{\mu_0}H(x-z_t) = \int_{\mathbb{R}}\mathbf{E}H(x-z_t(y))\,\mu_0(dy).$$

Consider the independent processes $(z_t^i)_{(i=1,..,N)}$ solutions to

$$\begin{cases} dz_t^i = \beta(t,z_t^i)\,dt + \sigma(t,z_t^i)\,dW_t^i\ , \\ z_0^i = y_0^i. \end{cases} \tag{6.10}$$

Applying the Euler scheme to (6.10), one defines the independent discrete-time processes $(\overline{z}^i_{kT/n})$:

$$\begin{cases} \overline{z}^i_{(k+1)T/n} = \overline{z}^i_{kT/n} + \beta(kT/n,\overline{z}^i_{kT/n})\,\frac{T}{n} \\ \qquad\qquad +\sigma(kT/n,\overline{z}^i_{kT/n})\left(W^i_{(k+1)/n} - W^i_{kT/n}\right)\ , \\ \overline{z}^i_0 = y^i_0. \end{cases} \tag{6.11}$$

The global error is decomposed as follows:

$$\begin{aligned} &\mathbf{E}\left\|V(kT/n,x) - \overline{V}_{kT/n}(x)\right\|_{L^1(\mathbb{R})} \\ &\leq \left\|\mathbf{E}_{\mu_0}H(x-z_{kT/n}) - \mathbf{E}_{\overline{\mu}_0}H(x-z_{kT/n})\right\|_{L^1(\mathbb{R})} \\ &+\mathbf{E}\left\|\mathbf{E}_{\overline{\mu}_0}H(x-z_{kT/n}) - \frac{1}{N}\sum_{i=1}^N H(x-z^i_{kT/n})\right\|_{L^1(\mathbb{R})} \\ &+\mathbf{E}\left\|\frac{1}{N}\sum_{i=1}^N H(x-z^i_{kT/n}) - \frac{1}{N}\sum_{i=1}^N H(x-\overline{z}^i_{kT/n})\right\|_{L^1(\mathbb{R})} \\ &+\mathbf{E}\left\|\frac{1}{N}\sum_{i=1}^N H(x-\overline{z}^i_{kT/n}) - \frac{1}{N}\sum_{i=1}^N H(x-Y^i_{kT/n})\right\|_{L^1(\mathbb{R})}. \end{aligned} \tag{6.12}$$

The first term of the right handside corresponds to the approximation error of the measure μ_0; the second term essentially is a statistical error related to the Strong Law of Large Numbers; the third term is the discretization error induced by the Euler scheme; the last term corresponds to the approximation of the coefficients $\beta(kT/n,\cdot)$ and $\sigma(kT/n,\cdot)$ by means of the empirical measure $\overline{\mu}_{kT/n}$, which introduces the family of dependent processes $(Y^i_{kT/n})$.

One successively proves:

$$\left\|\mathbf{E}_{\mu_0}H(x-z_t)-\mathbf{E}_{\overline{\mu}_0}H(x-z_t)\right\|_{L^1(\mathbb{R})}\leq C\,\|\,V_0-\overline{V}_0\,\|_{L^1(\mathbb{R})}\,,$$

$$\mathbf{E}\left\|\mathbf{E}_{\overline{\mu}_0}H(x-z_t)-\frac{1}{N}\sum_{i=1}^{N}H(x-z_t^i)\right\|_{L^1(\mathbb{R})}\leq\frac{C}{\sqrt{N}}\,,$$

$$\mathbf{E}\left\|\frac{1}{N}\sum_{i=1}^{N}H(x-z^i_{kT/n})-\frac{1}{N}\sum_{i=1}^{N}H(x-\bar{z}^i_{kT/n})\right\|_{L^1(\mathbb{R})}\leq\frac{C}{\sqrt{n}}\,,$$

$$\mathbf{E}\left\|\frac{1}{N}\sum_{i=1}^{N}H(x-\bar{z}^i_{kT/n})-\frac{1}{N}\sum_{i=1}^{N}H(x-Y^i_{kT/n})\right\|_{L^1(\mathbb{R})}$$
$$\leq C\left(\frac{1}{\sqrt{n}}+\frac{1}{\sqrt{N}}+\|V_0-\bar{V}_0\|_{L^1(\mathbb{R})}\right).$$

The three first inequalities are obtained following the guidelines presented in Section 4. and observing that an inequality of the type (4.3) holds for the density of the law of $z_t(x)$. The proof of the last inequality is based upon an induction formula with respect to k which mimics the propagation of the global error. More precisely, set

$$E_k:=\frac{1}{N}\sum_{i=1}^{N}\mathbf{E}|\bar{z}^i_{t_k}-Y^i_{t_k}|^2$$

and

$$\delta:=\|V_0-\bar{V}_0\|^2_{L^1(\mathbb{R})}+\frac{1}{N}+\frac{T}{n}.$$

A tedious computation, where the Lipschitz condition on the kernels and the estimate (2.4) play a role, shows that

$$\begin{cases} E_k\leq(1+\frac{CT}{n})E_{k-1}+\frac{CT}{n}\,(\delta+\frac{T}{n})+\frac{CT}{n}\frac{\sqrt{E_{k-1}}}{\sqrt{t_{k-1}}}\sqrt{\delta}\ \text{ for } k>1,\\ E_1\leq\frac{CT}{n}, \end{cases}$$

from which one can deduce

$$\mathbf{E}\left\|\frac{1}{N}\sum_{i=1}^{N}H(x-\bar{z}^i_{kT/n})-\frac{1}{N}\sum_{i=1}^{N}H(x-Y^i_{kT/n})\right\|_{L^1(\mathbb{R})}$$
$$\leq C\left(\frac{1}{\sqrt{n}}+\frac{1}{\sqrt{N}}+\|V_0-\bar{V}_0\|_{L^1(\mathbb{R})}\right).$$

□

Suppose now that the objective is to approximate the solution of the equation (6.4) rather than (6.1).

The above hypotheses imply that for all $t > 0$ the measure μ_t has a density $u(t,\cdot)$ w.r.t. Lebesgue's measure. To obtain an approximation of $u(kT/n,\cdot)$, we construct a regularization by convolution of the discrete measure $\overline{\mu}_{kT/n}$.

Let $\Phi_\varepsilon(\cdot)$ be the density of the Gaussian law $N(0,\varepsilon^2)$ and set

$$\overline{u}^{\varepsilon}_{kT/n}(x) := \left(\Phi_\varepsilon * \overline{\mu}_{kT/n}\right)(x) = \frac{1}{N}\sum_{i=1}^{N}\frac{1}{\sqrt{2\pi}\varepsilon}\exp\left(-\frac{(x-Y^i_{kT/n})^2}{2\varepsilon^2}\right).$$

We strengthen our hypotheses:

(H11') The kernel $b(\cdot,\cdot)$ is in $C_b^2(\mathbb{R}^2)$ and $s(\cdot,\cdot)$ is in $C_b^3(\mathbb{R}^2)$.

(H12') The initial law μ_0 has a strictly positive density $u_0(\cdot)$ in $C^2(\mathbb{R})$ satisfying: there exist strictly positive constants M, η and α such that

$$u_0(x) + |u_0'(x)| + |u_0''(x)| \leq \eta \exp\left(-\alpha\frac{x^2}{2}\right), \text{ for } |x| > M.$$

One then have the

Theorem 6.2 (Bossy and Talay [7], Bossy [5]). *Suppose (H10), (H11') and (H12'). Let $u(t,\cdot)$ be the classical solution to the PDE*

$$\begin{cases} \frac{\partial u}{\partial t}(t,x) = \frac{1}{2}\frac{\partial^2}{\partial x^2}\left[u(t,x)\left(\int_{\mathbb{R}} s(x,y)u(t,y)dy\right)^2\right] \\ \qquad - \frac{\partial}{\partial x}\left[u(t,x)\int_{\mathbb{R}} b(x,y)u(t,y)dy\right], \\ u(0,x) = u_0(x), \end{cases} \tag{6.13}$$

where $u_0(\cdot)$ is the density of μ_0.

Then there exists an increasing function $K(\cdot)$ such that, $\forall k \in \{1,\ldots,n\}$,

$$\mathbf{E}\left\|u(kT/n,.) - \overline{u}^{\varepsilon}_{kT/n}(\cdot)\right\|_{L^1(\mathbb{R})} \leq K(T)\left[\varepsilon^2 + \frac{1}{\varepsilon}\left(\|V_0 - \overline{V}_0\|_{L^1(\mathbb{R})} + \frac{1}{\sqrt{N}} + \frac{1}{\sqrt{n}}\right)\right] \tag{6.14}$$

and

$$Var\left(\left\|u(kT/n,.) - \overline{u}^{\varepsilon}_{kT/n}(\cdot)\right\|_{L^1(\mathbb{R})}\right) \leq K(T)\left[\varepsilon^4 + \frac{1}{\varepsilon^2}\left(\|V_0 - \overline{V}_0\|^2_{L^1(\mathbb{R})} + \frac{1}{N} + \frac{1}{n}\right)\right]. \tag{6.15}$$

Sketch of the proof. The first step consists in proving that the density $u(t,\cdot)$ belongs to the Sobolev space $W^{2,1}(\mathbb{R})$ and that the norm of $u(t,\cdot)$ in $W^{2,1}(\mathbb{R})$ is bounded uniformly in $t \in [0,T]$. This is done by using a criterion due to Cannarsa and Vespri [9] to check that the function $(1+x^2)u(t,x)$ belongs to $C^1([0,T];L^2(\mathbb{R}))\bigcap C([0,T];W^{2,2}(\mathbb{R}))$. Equipped with this result, one can then use the well-known estimate (cf. Raviart [41])

$$\|u(t_k,\cdot)-(u(t_k,\cdot)*\Phi_\varepsilon)\|_{L^1(\mathbb{R})} \leq C\,\varepsilon^2\,\|u(t_k,\cdot)\|_{W^{2,1}(\mathbb{R})}. \qquad (6.16)$$

The second step is easy. It consists in checking that

$$\mathbf{E}\|\,(u(t_k,\cdot)-\overline{u}_{t_k}(\cdot))*\Phi_\varepsilon\|_{L^1(\mathbb{R})} \leq \frac{C}{\varepsilon}\,\mathbf{E}\,\|V(t_k,.)-\overline{V}_{t_k}(.)\|_{L^1(\mathbb{R})}.$$

Therefore, one can conclude by applying Theorem 6.1. □

Thus, the rate of convergence depends on relations between ε, N and n. This is not estonishing: roughly speaking, if ε is too large, the smoothing by $\Phi_\varepsilon(\cdot)$ is too crude whereas, when ε is too small w.r.t. N, there may be too few particles in the windows of size ε.

7. The Burgers equation

For all the results of this section we refer to Bossy and Talay [6] and Bossy [5].

Consider the Burgers equation:

$$\begin{cases} \dfrac{\partial V}{\partial t} = \dfrac{1}{2}\sigma^2\dfrac{\partial^2 V}{\partial x^2} - V\,\dfrac{\partial V}{\partial x}\ , \text{ in } [0,T]\times\mathbb{R}\,, \\ V(0,x) = V_0(x)\,. \end{cases} \qquad (7.1)$$

This PDE can be seen as the Fokker-Planck equation for the limit law of particle systems corresponding to a kernel $b(\cdot,\cdot)$, roughly speaking, equal to a Dirac measure (see Sznitman [46]). The corresponding algorithm must involve a smoothing of this kernel. The analysis of its convergence rate is still in progress. Another stochastic particle method for the Burgers equation has been proposed by Roberts [42].

In order to construct a stochastic particle algorithm involving a kernel $b(\cdot,\cdot)$ less irregular than a Dirac measure (therefore more interesting from a numerical point of view), we interpret the solution of the Burgers equation as the distribution function of the probability measure U_t solution to the following McKean-Vlasov PDE:

$$\begin{cases} \dfrac{\partial U}{\partial t} = \dfrac{1}{2}\sigma^2\dfrac{\partial^2 U}{\partial x^2} - \dfrac{\partial}{\partial x}\left(\left(\displaystyle\int_{\mathbb{R}} H(x-y)U_t(dy)\right)U_t\right), \\ U_{t=0} = U_0\,. \end{cases} \qquad (7.2)$$

The above PDE is understood in the distribution sense. Its nonlinear part makes appear the discontinuous interaction kernel $b(x,y) = H(x-y)$.

To this McKean-Vlasov equation, is associated the nonlinear stochastic differential equation

$$\begin{cases} dX_t = \sigma dW_t + \int_{\mathbb{R}} H(X_t - y) Q_t(dy)\, dt \quad , \; Q_t(dy) \text{ is the law of } X_t \,, \\ X_{t=0} = X_0 \text{ of law } Q_0 \,. \end{cases} \tag{7.3}$$

As the kernel $H(x-y)$ is discontinuous, the "classical" results of the propagation of chaos for weakly interacting particles do not apply. Thus, one first must prove that there exists a solution to (7.3) and that the propagation of chaos holds for the corresponding particles system.

Let $\mathcal{M}(\mathbb{R})$ denote the set of probability measures on $\mathbb{R}$. For any measure $\mu \in \mathcal{M}(\mathbb{R})$ the differential operator $\mathcal{L}_{(\mu)}$ is defined by

$$\mathcal{L}_{(\mu)} f(x) = \frac{1}{2}\sigma^2 \frac{\partial^2 f}{\partial x^2}(x) + \left(\int_{\mathbb{R}} H(x-y)\mu(dy) \right) \frac{\partial f}{\partial x}(x) \,.$$

One can prove the existence and the uniqueness of a solution to the following nonlinear martingale problem (7.4) associated to the operator $\mathcal{L}_{(.)}$: for any initial distribution $Q_0 \in \mathcal{M}(\mathbb{R})$, there exists a unique $\mathcal{Q}$ in $\mathcal{M}(C([0,T];\mathbb{R}))$ (we denote by $\mathcal{Q}_t$, $t \in [0,T]$, its onedimensional distributions) such that

$$\left. \begin{array}{ll} (i) & \mathcal{Q}_0 = U_0 \,, \\ (ii) & \forall f \in C_K^2(\mathbb{R}),\, f(x(t)) - f(x(0)) - \int_0^t \mathcal{L}_{(\mathcal{Q}_s)} f(x(s)) ds, \; t \in [0,T] \\ & \text{is a } \mathcal{Q} \text{ martingale}\,, \end{array} \right\} \tag{7.4}$$

where $x(\cdot)$ denotes the canonical process on the space of continuous functions from $[0,T]$ to $\mathbb{R}$ (this is done by showing the convergence of the solutions of the martingale problems corresponding to an appropriate sequence of smoothened Heaviside functions). Equipped with this result, one can prove that there is a unique solution $\mathcal{Q}$ in the sense of probability law to (7.3). (Besides, one can show that the distribution function of $\mathcal{Q}_t$ is the *classical* solution to the Burgers equation.)

One can also prove the following

Proposition 7.1. *The propagation of chaos holds for the sequence of measures* (μ^N) *defined by*

$$\mu^N = \frac{1}{N} \sum_{i=1}^N \delta_{X^{i,N}}$$

where

$$dX_t^{i,N} = \sigma dW_t^i + \frac{1}{N} \sum_{j=1}^N H(X_t^{i,N} - X_t^{j,N})\, dt.$$

Sketch of the proof. First, one easily shows that the sequence of the laws of the μ^N's is tight. Then, let Π_1^∞ be a limit point of a convergent subsequence of $\{\mathcal{L}aw(\mu^N)\}$. Similarly to what is done in Section 4.2 of S. Méléard's contribution in this volume, set

$$F(m) := < m, \left(f(x(t)) - f(x(s)) - \int_s^t \mathcal{L}_{(m_\theta)} f(x(\theta)) d\theta \right) g(x(s_1), \ldots, x(s_k)) >$$

where $f \in C_b^2(\mathbb{R})$, $g \in C_b(\mathbb{R}^k)$, $0 < s_1 < \ldots < s_k \leq s \leq T$ and m is a probability on $C([0,T];\mathbb{R})$. Then use the two following arguments.

(a) First, $\lim_{N\to+\infty} \mathbf{E}[F(\mu^N)]^2 = 0$ since

$$\begin{aligned} \lim_{N\to+\infty} \mathbf{E}[F(\mu^N)]^2 &\leq \lim_{N\to+\infty} \frac{C}{N^2} \sum_{i=1}^N \mathbf{E}\left(\int_s^t \sigma dW_\theta^i \right)^2 \\ &= 0. \end{aligned}$$

(b) Second, one can show that the support of Π_1^∞ is the set of solutions to the nonlinear martingale problem (7.4). As the uniqueness of such a solution holds, one gets that $\Pi_1^\infty = \delta_Q$. Here, one cannot use the continuity of $F(\cdot)$ in $\mathcal{P}(\mathcal{C}([0,T];\mathbb{R}))$ endowed with the Vaserstein metric because the Heaviside function is discontinuous, but one can take advantage of the explicit form of F. The key argument is as follows. Let ν^N be defined by

$$\nu^N := \frac{1}{N^4} \sum_{i,j,k,l=1}^N \delta_{(X_\cdot^{i,N}, X_\cdot^{j,N}, X_\cdot^{k,N}, X_\cdot^{l,N})} .$$

Let $\Pi^\infty \in \mathcal{P}(\mathcal{P}(\mathcal{C}([0,T];\mathbb{R})^4))$ be the limit of a convergent subsequence of the tight family $\{\mathcal{L}aw(\nu^N)\}$. Denote by ν^1 the first marginal of a measure $\nu \in \mathcal{P}(\mathcal{C}([0,T];\mathbb{R})^4)$ (for all Borel sets A in $\mathcal{C}([0,T];\mathbb{R})$, $\nu^1(A) = \nu(A \times \mathcal{C}([0,T];\mathbb{R}) \times \mathcal{C}([0,T];\mathbb{R}) \times \mathcal{C}([0,T];\mathbb{R}))$). Then, one observes that

$$\Pi^\infty - \text{a.e,}\ \nu = \nu^1 \otimes \nu^1 \otimes \nu^1 \otimes \nu^1 .$$

Besides, one can prove that

$$\begin{aligned} &\lim_{N\to+\infty} \mathbf{E}[F(\mu^N)]^2 = \\ &\int_{\mathcal{P}(\mathcal{C}([0,T];\mathbb{R})^4)} \left\{ \int_{\mathcal{C}([0,T];\mathbb{R})^4} \left[f(x_t^1) - f(x_s^1) - \frac{\sigma^2}{2} \int_s^t f''(x_\theta^1) d\theta \right.\right. \\ &\qquad\qquad \left. - \int_s^t H(x_\theta^1 - x_\theta^2) f'(x_\theta^1) d\theta \right] \qquad (7.5) \\ &\qquad\qquad \left. g(x_{s_1}^1, \ldots, x_{s_p}^1) d\nu(x^1, x^2, x^3, x^4) \right\}^2 d\Pi^\infty(\nu) . \end{aligned}$$

One then proves that Π^∞-a.e.,

$$\int_{C([0,T];\mathbb{R})^2}\left[f(x_t^1)-f(x_s^1)-\frac{\sigma^2}{2}\int_s^t f''(x_\theta^1)d\theta-\int_s^t H(x_\theta^1-x_\theta^2)f'(x_\theta^1)d\theta\right]$$

$$g(x_{s_1}^1,\ldots,x_{s_p}^1)d\nu^1(x^1)\otimes d\nu^1(x^2)=0. \qquad (7.6)$$

Then, (7.6) and the uniqueness to the nonlinear martingale problem (7.4) imply that $\nu^1=\mathcal{Q}$ which is equivalent to

$$\lim_{N\to\infty}(Law(\mu^N))=\delta_{\mathcal{Q}}\,.$$

See [6] for details.

□

We now turn our attention to the numerical approximation of the preceding particle system. We set:

$$\begin{aligned} Y^i_{(k+1)T/n} &:= Y^i_{kT/n}+\frac{1}{N}\sum_{j=1}^N H\left(Y^i_{kT/n}-Y^j_{kT/n}\right)\frac{T}{n} \\ &\quad+\frac{1}{N}\left(W^i_{(k+1)T/n}-W^i_{kT/n}\right), \end{aligned} \qquad (7.7)$$

$$\bar{V}_{kT/n}(\cdot) := \frac{1}{N}\sum_{i=1}^N H\left(x-Y^i_{kT/n}\right). \qquad (7.8)$$

A much more complex and technical analysis than for the McKean-Vlasov equations with Lipschitz kernels (the study of the propagation from $\frac{kT}{n}$ to $\frac{(k+1)T}{n}$ of the error is very intricate when the kernels are not globally Lipschitz) leads to

Theorem 7.2 (Bossy and Talay [6], Bossy [5]). *Let $V(t,x)$ be the classical solution of the Burgers equation (7.1) with the initial condition V_0. Suppose (H12).*

Let $\overline{V}_{kT/n}(x)$ be defined as above, N being the number of particles.
There exists an increasing function $K(\cdot)$ such that for all $k\in\{1,\ldots,n\}$:

$$\begin{aligned} &\mathbf{E}\,\|V(kT/n,.)-\overline{V}_{kT/n}(.)\|_{L^1(\mathbb{R})} \\ &\quad\le K(T)\left(\|V_0-\overline{V}_0\|_{L^1(\mathbb{R})}+\frac{1}{\sqrt{N}}+\frac{1}{\sqrt{n}}\right). \end{aligned} \qquad (7.9)$$

The monotonicity of the function $V_0(\cdot)$ can be relaxed: see [7] for the modification of the algorithm when $V_0(\cdot)$ is non monotonic and for the corresponding error analysis.

In a forthcoming paper, Bossy and Talay extend this analysis to Chorin's random vortex method for the 2-D incompressible Navier-Stokes equation. The interpretation of the Navier-Stokes equation in terms of limit law of weakly interacting particles has been given by Marchioro and Pulvirenti [31] and Osada [35]. The interaction kernel is still less smooth than the Heaviside function since it is the Biot and Savart kernel, which is singular at 0. This makes the error analysis delicate.

For numerical experiments on the above stochastic particle methods related to McKean-Vlasov equations, see M. Bossy's thesis [5].

References

1. Bally, V., Talay, D., "The law of the Euler scheme for stochastic differential equations (I) : convergence rate of the distribution function", *Probability Theory and Related Fields*, 104, 43-60 (1996).
2. Bally, V., Talay, D., "The law of the Euler scheme for stochastic differential equations (II) : convergence rate of the density", Rapport de Recherche 2675, INRIA (1995). To appear in *Monte Carlo Methods and Applications.*
3. Bensoussan A., Lions, J.L., *Applications des Inéquations Variationnelles en Contrôle Stochastique*, Dunod (1978).
4. Bernard, P., Talay, D., Tubaro, L., "Rate of convergence of a stochastic particle method for the Kolmogorov equation with variable coefficients", *Math. Comp.*, 63(208), 555-587 (1994).
5. Bossy, M., *Vitesse de Convergence d'Algorithmes Particulaires Stochastiques et Application à l'Equation de Burgers*, PhD thesis, Université de Provence (1995).
6. Bossy, M., Talay, D., "Convergence rate for the approximation of the limit law of weakly interacting particles: application to the Burgers equation" (1995). To appear in *Ann. Appl. Probab.*.
7. Bossy, M., Talay, D., "A stochastic particle method for the McKean-Vlasov and the Burgers equation" (1995). To appear in *Math. Comp.*.
8. Calzolari, A., Costantini, C., Marchetti F., "A confidence interval for Monte Carlo methods with an application to simulation of obliquely reflecting Brownian motion", *Stochastic Processes and their Applications*, 29, 209-222 (1988).
9. Cannarsa, P., Vespri, C., "Generation of analytic semigroups by elliptic operators with unbounded coefficients", *SIAM J. Math. Anal.*, 18(3), 857-872 (1987).
10. Chorin, A.J., Marsden, J.E., *A Mathematical Introduction to Fluid Mechanics*, Springer Verlag (1993).
11. Chorin, A.J., "Vortex methods and Vortex Statistics – Lectures for Les Houches Summer School of Theoretical Physics", *Lawrence Berkeley Laboratory Prepublications* (1993).
12. Costantini, C., Pacchiarotti, B., SARTORETTO, F., "Numerical approximation of functionals of diffusion processes", (1995).
13. Faure, O., *Simulation du Mouvement Brownien et des Diffusions*, PhD thesis, Ecole Nationale des Ponts et Chaussées (1992).
14. Freidlin, M., *Functional Integration and Partial Differential Equations.* Annals of Mathematics Studies, Princeton University (1985).
15. Friedman, A., *Stochastic Differential Equations and Applications*, volume 1, Academic Press, New-York (1975).
16. Gaines, J.G., Lyons, T.J., "Random generation of stochastic area integrals", *SIAM Journal of Applied Mathematics*, 54(4), 1132-1146 (1994).
17. Goodman, J., "Convergence of the random vortex method", *Comm. Pure Appl. Math.*, 40, 189-220 (1987).
18. Gustafson, K.E., Sethian, J.A., (eds.), *Vortex Methods and Vortex Motions*, SIAM (1991).

19. Hald, O.H., "Convergence of random methods for a reaction diffusion equation", *SIAM J. Sci. Stat. Comput.*, 2, 85-94 (1981).
20. Hald, O.H., "Convergence of a random method with creation of vorticity", *SIAM J. Sci. Stat. Comput.*, 7, 1373-1386 (1986).
21. Has'minskii, R.Z., *Stochastic Stability of Differential Equations*, Sijthoff & Noordhoff (1980).
22. Ikeda, N, Watanabe, S., *Stochastic Differential Equations and Diffusion Processes*, North Holland (1981).
23. Kanagawa, S., "On the rate of convergence for Maruyama's approximate solutions of stochastic differential equations", *Yokohama Math. J.*, 36, 79-85 (1988).
24. Kunita, H., "Stochastic differential equations and stochastic flows of diffeomorphisms", *Ecole d'Été de Saint-Flour XII*, LNM 1097, Springer (1984).
25. Kurtz, T.G., Protter, P., "Wong–Zakai corrections, random evolutions and numerical schemes for S.D.E.'s", in *Stochastic Analysis: Liber Amicorum for Moshe Zakai*, 331-346 (1991).
26. Kusuoka S., Stroock, D., "Applications of the Malliavin Calculus, part II", *J. Fac. Sci. Univ. Tokyo*, 32, 1-76 (1985).
27. Lépingle, D., "Un schéma d'Euler pour équations différentielles réfléchies", *Note aux Comptes-Rendus de l'Académie des Sciences*, 316(I), 601-605 (1993).
28. Lépingle, D., "Euler scheme for reflected stochastic differential equations", *Mathematics and Computers in Simulation*, 38 (1995).
29. Liu, Y., "Numerical approaches to reflected diffusion processes", (submitted for publication).
30. Long, D.G., Convergence of the random vortex method in two dimensions, *J. Amer. Math. Soc.*, 1(4) (1988).
31. Marchioro, C., Pulvirenti, M., "Hydrodynamics in two dimensions and vortex theory", *Comm. Math. Phys.*, 84, 483-503 (1982).
32. Milshtein, G.N., "The solving of the boundary value problem for parabolic equation by the numerical integration of stochastic equations", (to appear).
33. Newton, N.J., "Variance reduction for simulated diffusions", *SIAM J. Appl. Math.*, 54(6), 1780-1805 (1994).
34. Nualart, D., *Malliavin Calculus and Related Topics.* Probability and its Applications, Springer-Verlag (1995).
35. Osada, H., "Propagation of chaos for the two dimensional Navier-Stokes equation", in K.Itô and N. Ikeda, editors, *Probabilistic Methods in Mathematical Physics*, pages 303–334, Academic Press (1987).
36. Pardoux, E., "Filtrage non linéaire et équations aux dérivées partielles stochastiques associées", *Cours à l'Ecole d'Eté de Probabilités de Saint-Flour XIX*, LNM 1464, Springer-Verlag (1991).
37. Protter, P., *Stochastic Integration and Differential Equations*, Springer–Verlag, Berlin (1990).
38. Protter, P., Talay, D., "The Euler scheme for Lévy driven stochastic differential equations", Rapport de Recherche 2621, INRIA, (1995). To appear in Annals Prob.
39. Puckett, E.G., A study of the vortex sheet method and its rate of convergence. *SIAM J. Sci. Stat. Comput.*, 10(2), 298-327 (1989).
40. Puckett, E.G., "Convergence of a random particle method to solutions of the Kolmogorov equation", *Math. Comp.*, 52(186), 615-645 (1989).
41. Raviart, P.A., "An analysis of particle methods", in *Numerical Methods in Fluid Dynamics*, F. Brezzi (ed.), vol. 1127 of *Lecture Notes in Math.*, 243-324, Springer-Verlag (1985).
42. Roberts, S., "Convergence of a random walk method for the Burgers equation", *Math. Comp.*, 52(186), 647-673 (1989).

43. Roynette, B., "Approximation en norme Besov de la solution d'une équation différentielle stochastique", *Stochastics and Stochastic Reports*, 49, 191-209 (1994).
44. Slominski, L., "On existence, uniqueness and stability of solutions of multidimensional SDE's with reflecting boundary conditions", *Ann. Inst. H. Poincaré*, 29 (1993).
45. Slominski, L., "On approximation of solutions of multidimensional S.D.E.'s with reflecting boundary conditions", *Stochastic Processes and their Applications*, 50(2), 197-219 (1994).
46. Sznitman, A.S., "Topics in propagation of chaos", *Ecole d'Eté de Probabilités de Saint Flour XIX*, (P.L. Hennequin, ed.), LNM 1464, Springer, Berlin, Heidelberg, New York (1989).
47. Talay, D., "Second order discretization schemes of stochastic differential systems for the computation of the invariant law", *Stochastics and Stochastic Reports*, 29(1),13-36 (1990).
48. Talay, D., "Simulation and numerical analysis of stochastic differential systems: a review", *Probabilistic Methods in Applied Physics*, (P. Krée and W. Wedig, eds.), *Lecture Notes in Physics* 451, 54-96, Springer-Verlag (1995).
49. Talay, D., Tubaro, L., *Probabilistic Numerical Methods for Partial Differential Equations*, (book in preparation).
50. Talay, D., Tubaro, L., "Expansion of the global error for numerical schemes solving stochastic differential equations", *Stochastic Analysis and Applications*, 8(4), 94-120 (1990).
51. Tweedie, R.L., "Sufficient conditions for ergodicity and recurrence of Markov chains on a general state space", *Stochastic Processes and Applications*, 3 (1975).

Weak convergence of stochastic integrals and differential equations II: Infinite dimensional case

Thomas G. Kurtz[1] * and Philip E. Protter[2] **

[1] Departments of Mathematics and Statistics University of Wisconsin - Madison, Madison, WI 53706-1388, USA

[2] Departments of Mathematics and Statistics, Purdue University, West Lafayette, IN 47907-1395, USA

1. Introduction

In Part I, we discussed weak limit theorems for stochastic integrals, with the principle result being the following (cf. Part I, Section 7):

Theorem 1.1. *For each $n = 1, 2, \ldots$, let (X_n, Y_n) be an $\{\mathcal{F}_t^n\}$-adapted process with sample paths in $D_{\mathbb{M}^{km}\times\mathbb{R}^m}[0,\infty)$ such that Y_n is an $\{\mathcal{F}_t^n\}$-semimartingale. Let $Y_n = M_n + A_n$ be a decomposition of Y_n into an $\{\mathcal{F}_t^n\}$-martingale M_n and a finite variation process A_n. Suppose that one of the following two conditions hold:*

UT (Uniform tightness.) For S_0^n, the collection of piecewise constant, $\{\mathcal{F}_t^n\}$-adapted processes

$$\mathcal{H}_t^0 = \cup_{n=1}^\infty \{|Z_- \cdot Y_n(t)| : Z \in S_0^n, \sup_{s\le t}|Z(s)| \le 1\}$$

is stochastically bounded.

UCV (Uniformly controlled variations.) $\{T_t(A_n)\}$ is stochastically bounded for each $t > 0$, and there exist stopping times $\{\tau_n^\alpha\}$ such that $P\{\tau_n^\alpha \le \alpha\} \le \alpha^{-1}$ and

$$\sup_n E[[M_n]_{t\wedge\tau_n^\alpha}] < \infty$$

for each $t > 0$.

If $(X_n, Y_n) \Rightarrow (X, Y)$ in the Skorohod topology on $D_{\mathbb{M}^{km}\times\mathbb{R}^m}[0,\infty)$, then Y is an $\{\mathcal{F}_t\}$-semimartingale for a filtration $\{\mathcal{F}_t\}$ with respect to which X is adapted and $(X_n, Y_n, X_{n-}\cdot Y_n) \Rightarrow (X, Y, X_-\cdot Y)$ in $D_{\mathbb{M}^{km}\times\mathbb{R}^m\times\mathbb{R}^k}[0,\infty)$. If $(X_n, Y_n) \to (X, Y)$ in probability, then $(X_n, Y_n, X_{n-}\cdot Y_n) \to (X, Y, X_-\cdot Y)$ in probability.

---

* Research supported in part by NSF grants DMS 92-04866 and DMS 95-04323

** Research supported in part by NSF grant INT 94-01109 and NSA grant MDR 904-94-H-2049

In this part, we consider the analogous results for stochastic integrals with respect to infinite dimensional semimartingales. We are primarily concerned with integrals with respect to semimartingale random measures, in particular, worthy martingale measures as developed by Walsh [60]. We discover, however, that the class of semimartingale random measures is not closed under the natural notion of weak limit unlike the class of finite dimensional semimartingales (Part I, Theorem 7.3). Consequently, we work with a larger class of infinite dimensional semimartingales which we call $H^{\#}$-semimartingales. This class includes semimartingale random measures, Banach-space valued martingales, and cylindrical Brownian motion.

A summary of results on semimartingale random measures is given in Section 2.. The definitions and results come primarily from Walsh [60]. $H^{\#}$-semimartingales are introduced in Section 3.. The stochastic integral is defined through approximation by finite dimensional integrands. The basic assumption on the semimartingale is essentially the "good integrator" condition that defines a semimartingale in the sense of Section 1 of Part I. This approach allows us to obtain the basic stochastic integral covergence theorems in Sections 4. and 5. as an application of Theorem 1.1.

Previous general results on convergence of infinite dimensional stochastic integrals include work of Walsh [60, Chapter 7], Cho [8, 9], and Jakubowski [25]. Walsh and Cho consider martingale random measures as distribution-valued processes converging weakly in $D_{S'(\mathbf{R}^d)}[0,\infty)$. Walsh assumes all processes are defined on the same sample space (the canonical sample space) and requires a strong form of convergence for the integrands. Cho requires $(X_n, M_n) \Rightarrow (X, M)$ in $D_{L\times S'(\mathbf{R}^d)}[0,\infty)$ where L is an appropriate function space. Both Walsh and Cho work under assumptions analogous to the UCV assumption. Jakubowski gives results for Hilbert space-valued semimartingales under the analogue of the UT condition. Our results are given under the UT condition, although estimates of the type used by Walsh and Cho are needed to verify that particular sequences satisfy the UT condition.

Section 6. contains a variety of technical results on the uniform tightness condition. Section 7. includes a uniqueness result for stochastic differential equations satisfying a Lipschitz condition with a proof that seems to be new even in the finite dimensional setting. As an example, a spin-flip model is obtained as a solution of a stochastic differential equation in sequence space. Convergence results for stochastic differential equations are given based on the results of Sections 4. and 5..

Section 8. briefly discusses stochastic differential equations for Markov processes and introduces L_1-estimates of Graham that are useful in proving existence and uniqueness, particularly for infinite systems. Infinite systems are the topic of Section 9.. Existence and uniqueness results, similar to results of Shiga and Shimizu [55] for systems of diffusions, are given for very general kinds of equations. Results of McKean-Vlasov type are given in Section 10. using the results of Section 9..

Stochastic partial differential equations are a natural area of application for the results discussed here. We have not yet developed these applications, but Section 11. summarizes some of the ideas that seem most useful in obtaining convergence theorems.

Section 12. includes several simple examples illustrating the methods of the paper. Diffusion approximations, an averaging theorem, limit theorems for jump processes, and error analysis for a simulation scheme are described.

Acknowledgements. Many people provided insight and important references during the writing of this paper. In particular, we would like to thank Don Dawson, Carl Graham, Peter Kotelenez, Jim Kuelbs, Luciano Tubaro, and John Walsh.

2. Semimartingale random measures

Let (U, r_U) be a complete, separable metric space, let $U_1 \subset U_2 \subset \cdots$ be a sequence of sets $\{U_m\} \subset \mathcal{B}(U)$ satisfying $\cup_m U_m = U$, and let $\mathcal{A} = \{B \in \mathcal{B}(U) : B \subset U_m, \text{ some } m\}$. Frequently, $U_m = U$ and $\mathcal{A} = \mathcal{B}(U)$. Let $(\Omega, \mathcal{F}, P)$ be a complete probability space and $\{\mathcal{F}_t\}$ a complete, right continuous filtration, and let Y be a stochastic process indexed by $\mathcal{A} \times [0, \infty)$ such that

- For each $B \in \mathcal{A}$, $Y(B, \cdot)$ is an $\{\mathcal{F}_t\}$-semimartingale with $Y(B, 0) = 0$.
- For each $t \geq 0$ and each disjoint sequence $\{B_i\} \subset \mathcal{B}(U)$, $Y(U_m \cap \cup_{i=1}^{\infty} B_i, t) = \sum_{i=1}^{\infty} Y(U_m \cap B_i, t)$ a.s.

Then Y is an $\{\mathcal{F}_t\}$-*semimartingale random measure*. We will say that Y is *standard* if $Y = M + V$ where $V(B, t) = \tilde{V}(B \times [0, t])$ for a random σ-finite signed measure $\tilde{V}$ on $U \times [0, \infty)$ satisfying $|\tilde{V}|(U_m \times [0, t]) < \infty$ a.s. for each $m = 1, 2, \ldots$ and $t \geq 0$ and M is a *worthy* martingale random measure in the sense of Walsh [60], that is, $M(A.\cdot)$ is locally square integrable for each $A \in \mathcal{A}$, and there exists a (positive) random measure K on $U \times U \times [0, \infty)$ such that

$$|\langle M(A), M(B)\rangle_{t+s} - \langle M(A), M(B)\rangle_t| \leq K(A \times B \times (t, t+s]) , \quad A, B \in \mathcal{A},$$

and $K(U_m \times U_m \times [0, t]) < \infty$ a.s. for each $t > 0$. K is called the dominating measure. (Merzbach and Zakai [49] define a slightly more general notion of *quasi-worthy* martingale which could be employed here. See also the defintion of *conditionally worthy* in Example 12.5.) Note that if U is finite, then every semimartingale random measure is standard.

M is *orthogonal* if $\langle M(A), M(B)\rangle_t = 0$ for $A \cap B = \emptyset$. If M is orthogonal, then $\pi(A \times (0, t]) \equiv \langle M(A)\rangle_t$ extends to a random measure on $U \times [0, \infty)$, and if we define $K(\Gamma) = \pi(f^{-1}(\Gamma))$ for $f(u, t) = (u, u, t)$, K is a dominating measure for M. In particular, if M is orthogonal, then M is worthy.

If φ is a simple function on U, that is $\varphi = \sum_{i=1}^{m} c_i I_{B_i}$ for disjoint $\{B_i\} \subset \mathcal{A}$, and $\{c_i\} \subset \mathbf{R}$, then we can define

$$Y(\varphi,t)=\sum_{i=1}^{m}c_iY(B_i,t).$$

If Y is standard and $E[K(U\times U\times[0,t])]<\infty$ for each $t>0$, then

$$E[M(\varphi,t)^2]\leq E[\int_{U\times U}|\varphi(u)\varphi(v)|K(du\times dv\times(0,t])]\leq\|\varphi\|_\infty^2E[K(U\times U\times(0,t])] \tag{2.1}$$

so $Y(\varphi,t)$ can be extended uniquely at least to all $\varphi\in B(U)$ for which the integral against $\tilde{V}$ is defined, that is

$$Y(\varphi,t)=M(\varphi,t)+\int_U\varphi(u)\tilde{V}(du\times[0,t])$$

where $M(\varphi,t)$ is defined as the limit of $M(\varphi_n,t)$ for simple φ_n using (2.1).

More generally, for each $j=0,1,\ldots$, let $\{B_i^j\}\subset\mathcal{A}$ be disjoint, let $0=t_0<t_1<t_2<\cdots$, and let C_i^j be an $\mathcal{F}_{t_j}$-measurable random variable. Define

$$X(u,t)=\sum_{i,j}C_i^jI_{B_i^j}(u)I_{(t_j,t_{j+1}]}(t) \tag{2.2}$$

and define $X\cdot Y$ by

$$X\cdot Y(t)=\sum C_i^j(Y(B_i^j,t_{j+1}\wedge t)-Y(B_i^j,t_j\wedge t))\,.$$

Again, we can estimate the martingale part

$$\begin{aligned}E[(X\cdot M(t))^2] &= E\Big[\sum_j\sum_{i_1,i_2}C_{i_1}^jC_{i_2}^j(\langle M(B_{i_1}^j),M(B_{i_2}^j)\rangle_{t_{j+1}\wedge t} && (2.3)\\ &\qquad -\langle M(B_{i_1}^j),M(B_{i_2}^j)\rangle_{t_j\wedge t})\Big] \\ &\leq E\left[\int_{U\times U\times[0,t]}|X(u,s)X(v,s)|K(du\times dv\times ds)\right] && (2.4)\end{aligned}$$

and if $E[K(U\times U\times[0,t])]<\infty$ for each $t>0$, we can extend the definition of $X\cdot M$ (and hence of $X\cdot Y$) to all bounded, $|\tilde{V}|$-integrable processes that can be approximated by simple processes of the form (2.2).

An alternative approach to defining $X_-\cdot M$ is to first consider

$$X(t,u)=\sum_{i=1}^{m}\xi_i(t)I_{B_i}(u), \tag{2.5}$$

for disjoint $\{B_i\}\subset\mathcal{A}$ and cadlag, adapted ξ_i. Set

$$X_-\cdot Y(t)=\sum_{i=1}^{m}\int_0^t\xi_i(s-)dY(B_i,s)\,,$$

where the integrals are ordinary semimartingale integrals. We then have

$$E[(X_- \cdot M(t))^2] = E[\sum_{i,j} \int_0^t \xi_i(s-)\xi_j(s-)d\langle M(B_i,\cdot), M(B_j,\cdot)\rangle_s] \tag{2.6}$$

$$\leq E[\int_{U\times U\times(0,t]} |X(s-,u)X(s-,v)|K(du \times dv \times ds)],$$

which, of course, is the same as (2.4). By Doob's inequality, we have

$$E[\sup_{s\leq t}(X_- \cdot M(s))^2] \leq 4E[\int_{U\times U\times(0,t]} |X(s-,u)X(s-,v)|K(du \times dv \times ds)]. \tag{2.7}$$

For future reference, we also note the simple corresponding inequalities for V,

$$E[\sup_{s\leq t}|X_- \cdot V(s)|] \leq E[\int_{U\times(0,t]} |X(s-,u)||\tilde{V}|(du \times ds)] \tag{2.8}$$

and

$$E[\sup_{s\leq t}(X_- \cdot V(s))^2] \leq E[|\tilde{V}|(U \times [0,t]) \int_{U\times(0,t]} |X(s-,u)|^2|\tilde{V}|(du \times ds)]. \tag{2.9}$$

Let $\mathcal{P}$ be the σ-algebra of subsets of $\Omega \times U \times [0,\infty)$ generated by sets of the form $A \times B \times (t, t+s]$ for $t, s \geq 0$, $A \in \mathcal{F}_t$, and $B \in \mathcal{B}(U)$. $\mathcal{P}$ is the σ-algebra of *predictable sets.* If $E[K(U\times U\times[0,t])] < \infty$ and $|\tilde{V}|(U\times[0,t]) < \infty$ a.s. for each $t > 0$, then the bounded $\mathcal{P}$-measurable functions gives the class of bounded processes X for which $X \cdot Y$ is defined. Of course, the estimate (2.4) also allows extension to unbounded X for which the right side is finite, provided X is also almost surely integrable with respect to $\tilde{V}$.

Note that if K satisfies

$$K(A \times B \times [0,t]) = K_1(A \cap B \times [0,t]) + K_2(A \times B \times [0,t]) \tag{2.10}$$

where K_1 is a random measure on $U \times [0,\infty)$ and we define

$$\hat{K}(A\times[0,t]) = K_1(A\times[0,t]) + \frac{1}{2}(K_2(A\times U\times[0,t]) + K_2(U\times A\times[0,t])), \tag{2.11}$$

then

$$E[\sup_{s\leq t}(X \cdot M(t))^2] \leq 4E[\int_{U\times[0,t]} |X(u,s)|^2\hat{K}(du \times ds)]. \tag{2.12}$$

For future reference, if $|\tilde{V}|(U_m \times [0,\cdot])$ is locally in L_1 for each m, that is, there is a sequence of stopping times $\tau_n \to \infty$ such that $E[|\tilde{V}|(U_m \times [0, t \wedge \tau_n])] < \infty$ for each $t > 0$, then we can define $\hat{V}(A\times[0,\cdot])$ to be the predictable projection of $|\tilde{V}|(A \times [0,\cdot])$, and we have

$$E[\int_{U\times[0,t]} X(s-,u)|\tilde{V}|(du \times ds)] = E[\int_{U\times[0,t]} X(s-,u)\hat{V}(du \times ds)] \tag{2.13}$$

for all positive, cadlag, adapted X (allowing $\infty = \infty$).

Example 2.1. Gaussian white noise.

The canonical example of a martingale random measure is given by the Gaussian process indexed by $\mathcal{A} = \{A \in \mathcal{B}(U) \times \mathcal{B}([0,\infty)) : \mu \times m(A) < \infty\}$ and satisfying $E[W(A)] = 0$ and $E[W(A)W(B)] = \mu \times m(A \cap B)$, where m denotes Lebesgue measure and μ is a σ-finite measure on U. If we define $M(A,t) = W(A \times [0,t])$ for $A \in \mathcal{B}(\bar{U})$, $\mu(A) < \infty$, and $t \geq 0$, then M is an orthogonal martingale random measure with $K(A \times B \times [0,t]) = t\mu(A \cap B)$, and for fixed A, $M(A,\cdot)$ is a Brownian motion.

Example 2.2. Poisson random measures.

Let ν be a σ-finite measure on U and let $h(u)$ be in $L^2(\nu)$. Let N be a Poisson random measure on $U \times [0,\infty)$ with mean measure $\nu \times m$, that is, for $A \in \mathcal{B}(U) \times \mathcal{B}([0,\infty))$, $N(A)$ has a Poisson distribution with expectation $\nu \times m(A)$ and $N(A)$ and $N(B)$ are independent if $A \cap B = \emptyset$. For $A \in \mathcal{B}(U)$ satisfying $\nu(A) < \infty$, define $M(A,t) = \int_A h(u)(N(du \times [0,t]) - \nu(du)t)$. Noting that $E[M(A,t)^2] = t\int_A h(u)^2\nu(du)$ and that $\{M(A_i,t)\}$ are independent for disjoint $\{A_i\}$, we can extend M to all of $\mathcal{B}(U)$ by addition.

Suppose Z is a process with independent increments with generator

$$Af(z) = \int_{\mathbf{R}} (f(z+u) - f(z) - uI_{\{|u|\leq 1\}}f'(z))\nu(du).$$

Then ν must satisfy $\int_{\mathbf{R}} u^2 \wedge 1\nu(du) < \infty$. (See, for example, Feller [18].) Let $U = \mathbf{R}$, and let N be the Poisson random measure with mean measure $\nu \times m$. Define $M(A,t) = \int_A uI_{\{|u|\leq 1\}}(N(du \times [0,t]) - \nu(du)t)$, $V(A,t) = \int_A uI_{\{|u|>1\}}N(du \times [0,t])$, and $Y(A,t) = M(A,t) + V(A,t)$. Then we can represent Z by $Z(t) = Y(\mathbf{R},t)$.

Consider a sequence of Poisson random measures with mean measures $n\nu \times m$. Define

$$M_n(A,t) = \frac{1}{\sqrt{n}}\int_A h(u)(N_n(du \times [0,t]) - nt\nu(du)). \tag{2.14}$$

Then M_n is an orthogonal martingale random measure with

$$\langle M_n(A), M_n(B)\rangle_t = t\int_{A\cap B} h(u)^2\nu(du) = K(A \times B \times [0,t]).$$

By the central limit theorem, M_n converges (in the sense of finite dimensional distributions) to the Gaussian white noise martingale random measure outlined in Example 2.1 with $\mu(A) = \int_A h(u)^2\nu(du)$.

Example 2.3. Empirical measures.

Let $\xi_1, \xi_2, \ldots$ be iid U-valued random variables with distribution μ, and define

$$M_n(A,t) = \frac{1}{\sqrt{n}} \sum_{i=1}^{[nt]} (I_A(\xi_i) - \mu(A)) . \tag{2.15}$$

Then $\langle M_n(A), M_n(B)\rangle_t = \frac{[nt]}{n}(\mu(A \cap B) - \mu(A)\mu(B))$. Note that $K_0(A \times B) = \mu(A \cap B) + \mu(A)\mu(B)$ extends to a measure on $U \times U$ and $K_n(A \times B \times (0,t]) = K_0(A \times B)\frac{[nt]}{n}$ extends to a measure on $U \times U \times [0,\infty)$ which will be a dominating measure for M_n. Of course, M_n converges to a Gaussian martingale random measure with conditional covariation $\langle M(A), M(B)\rangle_t = t(\mu(A \cap B) - \mu(A)\mu(B))$ and dominating measure $K_0 \times m$.

2.1 Moment estimates for martingale random measures

Suppose that M is an orthogonal martingale measure. If $A, B \in \mathcal{A}$ are disjoint, then $[M(A), M(B)]_t = 0$ and in particular, $M(A,\cdot)$ and $M(B,\cdot)$ have a.s. no simultaneous discontinuities. It follows that

$$\Pi(A \times [0,t]) = [M(A)]_t$$

determines a random measure on $U \times [0,\infty)$ as does

$$\Pi_k(A \times [0,t]) = \sum_{s \le t} (M(A,s) - M(A,s-))^k \tag{2.16}$$

for even $k > 2$. For odd $k > 2$, (2.16) determines a random signed measure. For X of the form (2.5), it is easy to check that

$$[X_- \cdot M]_t = \int_{U \times [0,t]} X^2(s-,u)\Pi(du \times ds),$$

and setting $Z = X_- \cdot M$ and letting $\Delta Z(s) = Z(s) - Z(s-)$, we have

$$\begin{aligned}
Z^k(t) &= \int_0^t kZ^{k-1}(s-)dZ(s) + \int_0^t \frac{k(k-1)}{2} Z^{k-2}(s)d[Z]_s \\
&\quad + \sum_{s \le t} \Big(Z^k(s) - Z^k(s-) - kZ^{k-1}(s-)\Delta Z(s) \\
&\qquad - \binom{k}{2} Z^{k-2}(s-)\Delta Z(s)^2 \Big) \\
&= \int_{U \times [0,t]} kZ^{k-1}(s-)X(s-,u)M(du \times ds) \\
&\quad + \int_{U \times [0,t]} \binom{k}{2} Z^{k-2}(s)X^2(s-,u)\Pi(du \times ds) \\
&\quad + \sum_{j=3}^{k} \binom{k}{j} \int_{U \times [0,t]} Z^{k-j}(s-)X^j(s-,u)\Pi_j(du \times ds)
\end{aligned} \tag{2.17}$$

and can be extended to more general X under appropriate conditions.

Since $M(A,\cdot)$ is locally square integrable, $[M(A)]_t$ is locally in L_1, that is, there exists a sequence of stopping times $\{\tau_n\}$ such that $\tau_n \to \infty$ and $E[[M(A)]_{t\wedge\tau_n}] < \infty$ for each $t > 0$ and each n. In addition,

$$[M(A)]_t - \langle M(A)\rangle_t = \Pi(A \times [0,t]) - \pi(A \times [0,t])$$

is a local martingale. It follows from (2.17) and L_2 approximation that

$$\begin{aligned} E[(X_- \cdot M(t))^2] &= E[\int_{U\times[0,t]} X^2(s-,u)\Pi(du \times ds)] \\ &= E[\int_{U\times[0,t]} X^2(s-,u)\pi(du \times ds)] \end{aligned} \tag{2.18}$$

whenever either the second or third expression is finite. (Note that the left side may be finite with the other two expressions infinite.) We would like to obtain similar expressions for higher moments.

A discrete time version of the following lemma can be found in Burkholder (1971), Theorem 20.2. The continuous time version was given by Lenglart, Lepingle, and Pratelli [44] (see Dellacherie, Maisonneuve and Meyer [13, page 326]). The proof we give here is from Ichikawa [22, Theorem 1].

Lemma 2.4. *For $0 < p \leq 2$ there exists a constant C_p such that for any locally square integrable martingale M with Meyer process $\langle M\rangle$ and any stopping time τ*

$$E[\sup_{s\leq\tau} |M(s)|^p] \leq C_p E[\langle M\rangle_\tau^{p/2}]$$

Proof. For $p = 2$ the result is an immediate consequence of Doob's inequality. Let $0 < p < 2$. For $x > 0$, let $\sigma_x = \inf\{t : \langle M\rangle_t > x^2\}$. Since σ_x is predictable there exists an increasing sequence of stopping times $\sigma_x^n \to \sigma_x$. Noting that $\langle M\rangle_{\sigma_x^n} \leq x^2$, we have

$$\begin{aligned} P\{\sup_{s\leq\tau} |M(s)|^p > x\} &\leq P\{\sigma_x^n \leq \tau\} + \frac{E[\langle M\rangle_{\tau\wedge\sigma_x^n}]}{x^2} \\ &\leq P\{\sigma_x^n \leq \tau\} + \frac{E[x^2 \wedge \langle M\rangle_\tau]}{x^2}, \end{aligned}$$

and letting $n \to \infty$, we have

$$P\{\sup_{s\leq\tau} |M(s)|^p > x\} \leq P\{\langle M\rangle_\tau \geq x^2\} + \frac{E[x^2 \wedge \langle M\rangle_\tau]}{x^2}. \tag{2.19}$$

Using the identity

$$\int_0^\infty E[x^2 \wedge X^2] p x^{p-3} dx = \frac{2}{2-p} E[|X|^p],$$

the lemma follows by multiplying both sides of (2.19) by px^{p-1} and integrating. □

Assume that for $2 < k \leq k_0$ and $A \in \mathcal{A}$, $|\Pi_k|(A \times [0, \infty])$ is locally in L_1 and there exist predictable random measures π_k and $\hat{\pi}_k$ such that

$$\Pi_k(A \times [0,t]) - \pi_k(A \times [0,t]) \tag{2.20}$$

and

$$|\Pi_k|(A \times [0,t]) - \hat{\pi}_k(A \times [0,t]) \tag{2.21}$$

are local martingales. Of course, for k even, $\pi_k = \hat{\pi}_k$. We define $\pi_2 = \pi$

If M is Gaussian white noise as in Example 2.1, then $\Pi_k = \pi_k = 0$ for $k > 2$. If M is as in Example 2.2, then

$$\Pi_k(A \times [0,t]) = \int_A h^k(u) N(du \times [0,t]),$$

$$\pi_k(A \times [0,t]) = t \int_A h^k(u) \nu(du),$$

and

$$\hat{\pi}_k(A \times [0,t]) = t \int_A |h|^k(u) \nu(du).$$

Theorem 2.5. *Let $k \geq 2$, and suppose that for $2 \leq j \leq k$*

$$H_{k,j} \equiv E\left[\left(\int_{U \times [0,t]} |X(s-,u)|^j \hat{\pi}_j(du \times ds)\right)^{\frac{k}{j}}\right] < \infty. \tag{2.22}$$

Then $E[\sup_{s \leq t} |Z(s)|^k] < \infty$ and

$$E[Z^k(t)] = \sum_{j=2}^{k} \binom{k}{j} E\left[\int_{U \times [0,t]} Z^{k-j}(s-) X^j(s-,u) \pi_j(du \times ds)\right] \tag{2.23}$$

Proof. For $k = 2$, the result follows by (2.18). Note that if (2.22) holds, then it holds with k replaced by $k' < k$. Consequently, proceeding by induction, suppose that $E[\sup_{s \leq t} |Z(s)|^{k-1}] < \infty$. Since

$$M_k(t) = \int_0^t k Z^{k-1}(r-) dZ(r)$$

is a local square integrable martingale with

$$\langle M_k \rangle_t = \int_0^t k^2 Z^{2k-2}(s-) d\langle Z \rangle_s,$$

by Lemma 2.4, for any stopping time τ

$$E[\sup_{s\leq t\wedge\tau}|\int_0^s kZ^{k-1}(r-)dZ(r)|]\leq C_1kE[\sqrt{\sup_{s<t\wedge\tau}|Z(s)|^{2k-k}\langle Z\rangle_{t\wedge\tau}}],$$

and letting $\tau_c=\inf\{t:|Z(t)|>c\}$, it follows that

$$\begin{aligned}E[\sup_{s\leq t\wedge\tau_c}|Z(s)|^k] &\leq C_1kE[\sqrt{\sup_{s<t\wedge\tau_c}|Z(s)|^{2k-k}\langle Z\rangle_t}] \\ &\quad+\sum_{j=2}^k\binom{k}{j}E[\sup_{s<t\wedge\tau_c}|Z(s)|^{k-j} \\ &\quad\int_{U\times[0,t]}|X(s-,u)|^j\hat{\pi}_j(du\times ds)] \quad (2.24)\end{aligned}$$

which by the Hõlder inequality implies

$$\begin{aligned}E[\sup_{s\leq t\wedge\tau_c}|Z(s)|^k] &\leq C_1kE[\sup_{s<t\wedge\tau_c}|Z(s)|^k]^{\frac{k-1}{k}} \\ &\quad E[\left(\int_{U\times[0,t]}|X(s-,u)|^2\pi_2(du\times ds)\right)^{\frac{k}{2}}]^{\frac{1}{k}} \quad (2.25) \\ &\quad+\sum_{j=2}^k\binom{k}{j}E[\sup_{s<t\wedge\tau_c}|Z(s)|^k]^{\frac{k-j}{k}} \\ &\quad E[\left(\int_{U\times[0,t]}|X(s-,u)|^j\hat{\pi}_j(du\times ds)\right)^{\frac{k}{j}}]^{\frac{j}{k}}\end{aligned}$$

where the right side is finite by (2.22) and the fact that

$$E[\sup_{s<t\wedge\tau_c}|Z(s)|^k]\leq c^k.$$

The inequality then implies that $E[\sup_{s<t\wedge\tau_c}|Z(s)|^k]\leq K_k$, where K_k is the largest number satisfying

$$K\leq C_1kK^{\frac{k-1}{k}}H_{k,2}^{\frac{1}{k}}+\sum_{j=2}^k\binom{k}{j}K^{\frac{k-j}{k}}H_{k,j}^{\frac{j}{k}}. \quad (2.26)$$

(2.23) then follows from (2.17). □

2.2 A convergence theorem for counting measures

For $n=1,2,\ldots$, let N_n be a random counting measure on $U\times[0,\infty)$ with the property that $N_n(A\times\{t\})\leq 1$ for all $A\in\mathcal{B}(U)$ and $t\geq 0$. Let ν be a σ-finite measure on U, and let $F_1\subset F_2\subset\cdots$ be closed sets such that $\nu(F_k)<\infty$, $\nu(\partial F_k)=0$, and $\nu(A)=\lim_{k\to\infty}\nu(A\cap F_k)$ for each $A\in\mathcal{B}(U)$. Let Λ_n be a random measure also satisfying $\Lambda_n(A\times\{t\})\leq 1$. Suppose that Λ_n and N_n

are adapted to $\{\mathcal{F}_t^n\}$ in the sense that $N_n(A \times [0,t])$ and $\Lambda_n(A \times [0,t])$ are $\mathcal{F}_t^n$-measurable for all $A \in \mathcal{B}(U)$ and $t \geq 0$, and suppose that

$$N_n(A \cap F_k \times [0,t]) - \Lambda_n(A \cap F_k \times [0,t])$$

is a local $\{\mathcal{F}_t^n\}$-martingale for each $A \in \mathcal{B}(U)$ and $k = 1, 2, \ldots$.

Theorem 2.6. *Let N be the Poisson random measure on $U \times [0,\infty)$ with mean measure $\nu \times m$. Suppose that for each $k = 1, 2, \ldots$, $f \in \bar{C}(U)$, and $t \geq 0$*

$$\lim_{n\to\infty} \int_{F_k} f(u)\Lambda_n(du \times [0,t]) = t \int_{F_k} f(u)\nu(du)$$

in probability. Then $N_n \Rightarrow N$ in the sense that for any $A_1, \ldots, A_m$ such that for each i, $A_i \subset F_k$ for some k and $\nu(\partial A_i) = 0$,

$$(N_n(A_1 \times [0,\cdot]), \ldots, N_n(A_m \times [0,\cdot])) \Rightarrow (N(A_1 \times [0,\cdot]), \ldots, N(A_m \times [0,\cdot])).$$

It also follows that for $f \in \bar{C}(U)$,

$$\int_{F_k} f(u)N_n(du \times [0,\cdot]) \Rightarrow \int_{F_k} f(u)N(du \times [0,\cdot]).$$

Proof. The result is essentially a theorem of Brown [6]. Alternatively, assuming $\cup_{i=1}^m A_i \subset F_k$, let

$$\tau_n = \inf\{t : \Lambda_n(F_k \times [0,t]) > t\nu(F_k) + 1\}.$$

Note that $\tau_n \to \infty$ and that

$$N_n(A_i \times [0, t \wedge \tau_n]) - \Lambda_n(A_i \times [0, t \wedge \tau_n])$$

is an $\{\mathcal{F}_t^n\}$-martingale. For $T > 0$ and $\delta > 0$, let

$$\gamma_T^n(\delta) = \sup_{t \leq T} \Lambda_n(F_k \times (t \wedge \tau_n, (t+\delta) \wedge \tau_n])$$

and observe that $\lim_{\delta\to 0} \limsup E[\gamma_T^n(\delta)] = 0$. It follows that for $0 \leq t \leq T$

$$E[N_n(A_i \times (t \wedge \tau_n, (t+\delta) \wedge \tau_n]|\mathcal{F}_t^n] \leq E[\gamma_T^n(\delta)|\mathcal{F}_t^n]$$

and the relative compactness of $\{(N_n(A_1 \times [0,\cdot]), \ldots, N_n(A_m \times [0,\cdot]))\}$ follows from Theorem 3.8.6 of Ethier and Kurtz [17]. The theorem then follows from Theorem 4.8.10 of Ethier and Kurtz [17]. □

In addition to the conditions of Theorem 2.6, we assume that there exists $h \in \bar{C}(U)$ with $0 \leq h \leq 1$ such that $\int_U (1-h(u))\nu(du) < \infty$ and for $f \in \bar{C}(U)$,

$$\int_U f(u)(1-h(u))\Lambda_n(du \times [0,t]) \to t \int_U f(u)(1-h(u))\nu(du)$$

in probability. Let D be a linear space of functions on U such that for each k and each $\varphi \in D$

$$M_n^k(\varphi,t) = \int_{F_k} \varphi(u)h(u)(N_n(du \times [0,t]) - \Lambda_n(du \times [0,t]))$$

$$\hat{M}_n^k(\varphi,t) = \int_{F_k^c} \varphi(u)h(u)(N_n(du \times [0,t]) - \Lambda_n(du \times [0,t]))$$

and

$$M_n(\varphi,t) = \int_U \varphi(u)h(u)(N_n(du \times [0,t]) - \Lambda_n(du \times [0,t]))$$

are local $\{\mathcal{F}_t^n\}$-martingales and

$$\int_U \varphi^2(u)h^2(u)\nu(du) < \infty. \tag{2.27}$$

Theorem 2.7. *Suppose that there exists $\alpha : D \to [0,\infty)$ and a sequence $m_n \to \infty$ such that for every sequence $k_n \to \infty$ with $k_n \le m_n$ and each $t \ge 0$,*

$$\lim_{n\to\infty} E[\sup_{s\le t} |\hat{M}_n^{k_n}(\varphi,s) - \hat{M}_n^{k_n}(\varphi,s-)|] = 0 \tag{2.28}$$

and

$$[\hat{M}_n^{k_n}(\varphi)]_t \to \alpha(\varphi)t \tag{2.29}$$

in probability. Then for $\varphi_1,\ldots,\varphi_m \in D$,

$$(M_n(\varphi_1,t),\ldots,M_n(\varphi_m,t)) \Rightarrow (M(\varphi_1,t),\ldots,M(\varphi_m,t))$$

for

$$M(\varphi,t) = W(\varphi,t) + \int_U \varphi(u)h(u)\tilde{N}(du \times [0,t])$$

where W is a continuous (in t), mean zero, Gaussian processes satisfying

$$E[W(\varphi_1,s)W(\varphi_2,t)] = s \wedge t\frac{1}{2}(\alpha(\varphi_1+\varphi_2) - \alpha(\varphi_1) - \alpha(\varphi_2)),$$

$\tilde{N}(A \times [0,t]) = N(A \times [0,t]) - t\nu(A)$, and W is independent of N.

Remark 2.8. Note that the linearity of D and (2.29) implies

$$[\hat{M}_n^{k_n}(\varphi_1),\hat{M}_n^{k_n}(\varphi_2)]_t \to \frac{t}{2}(\alpha(\varphi_1+\varphi_2) - \alpha(\varphi_1) - \alpha(\varphi_2)). \tag{2.30}$$

(2.28) and (2.30) verify the conditions of the martingale central limit theorem (see, for example, Ethier and Kurtz [17, Theorem 7.1.4]) and it follows that

$$(\hat{M}_n^{k_n}(\varphi_1,\cdot),\ldots,\hat{M}_n^{k_n}(\varphi_m,\cdot)) \Rightarrow (W(\varphi_1,\cdot),\ldots,W(\varphi_m,\cdot)).$$

Suppose that $A_n^k(\varphi,t)$ has the property that

$$(\hat{M}_n^k(\varphi,t))^2 - A_n^k(\varphi,t)$$

is a local $\{\mathcal{F}_t^n\}$-martingale for each $\varphi \in D$ and that for m_n and k_n as above, we replace (2.28) and (2.29) by the requirements that

$$\lim_{n\to\infty} E[\sup_{s\le t} |\hat{M}_n^{k_n}(\varphi,s) - \hat{M}_n^{k_n}(\varphi,s-)|^2] = 0 \tag{2.31}$$

$$\lim_{n\to\infty} E[\sup_{s\le t} |A_n^{k_n}(\varphi,s) - A_n^{k_n}(\varphi,s-)|] = 0$$

and

$$A_n^{k_n}(\varphi,t) \to \alpha(\varphi)t \tag{2.32}$$

in probability. Then the conclusion of the theorem remains valid. In particular, (2.31) and (2.32) verify alternative conditions for the martingale central limit theorem. Note that if $\Lambda_n(A\cap F_k\times[0,\cdot])$ is continuous for each $A\in\mathcal{B}(U)$ and $k=1,2,\ldots$, then we can take

$$A_n^k(\varphi,t) = \int_{F_k^c} \varphi^2(u)h^2(u)\Lambda_n(du\times[0,t]). \tag{2.33}$$

Proof. For simplicity, let $m=1$. For each fixed k, Theorem 2.6 implies

$$M_n^k(\varphi,\cdot) \Rightarrow \int_{F_k} \varphi(u)h(u)\tilde{N}(du\times[0,\cdot])$$

and it follows from (2.27) that

$$\lim_{k\to\infty}\int_{F_k}\varphi(u)h(u)\tilde{N}(du\times[0,t]) = \int_U \varphi(u)h(u)\tilde{N}(du\times[0,t]).$$

Consequently, for $k_n\to\infty$ sufficiently slowly,

$$M_n^{k_n}(\varphi,\cdot) \Rightarrow \int_U \varphi(u)h(u)\tilde{N}(du\times[0,t]),$$

and since we can assume that $k_n\le m_n$, for the same sequence, the martingale central limit theorem implies $\hat{M}_n^{k_n}(\varphi,\cdot)\Rightarrow W(\varphi,\cdot)$. The convergence in $D_{\mathbf{R}}[0,\infty)$ of each component implies the relative compactness of $\{(M_n^{k_n}(\varphi,\cdot),\hat{M}_n^{k_n}(\varphi,\cdot))\}$ in $D_{\mathbf{R}}[0,\infty)\times D_{\mathbf{R}}[0,\infty)$. The fact that the second component is asymptotically continuous implies relative compactness in $D_{\mathbf{R}^2}[0,\infty)$. Consequently, at least along a subsequence $(M_n^{k_n}(\varphi,\cdot),\hat{M}_n^{k_n}(\varphi,\cdot))$ converges in $D_{\mathbf{R}^2}[0,\infty)$. To see that there is a unique possible limit and hence that there is convergence along the original sequence it is enough to check that W and N are independent. To verify this assertion, check that $W(\varphi,\cdot)$, $\varphi\in D$, and $\tilde{N}(A\cap F_k\times[0,\cdot])$, $A\in\mathcal{B}(U)$, $k=1,2,\ldots$ are all martingales with respect to the same filtration. Since trivially, $[W(\varphi),\tilde{N}(A\cap F_k)]_t=0$, an application of Itô's formula verifies that W and N give a solution of the martingale problem that uniquely determines their joint distribution and implies their independence. It follows that

$$(M_n^{k_n}(\varphi,\cdot),\hat{M}_n^{k_n}(\varphi,\cdot)) \Rightarrow (\int_U \varphi(u)h(u)\tilde{N}(du\times[0,\cdot]), W(\varphi,\cdot))$$

and hence

$$M_n(\varphi,\cdot) \Rightarrow M(\varphi,\cdot).$$

□

3. $H^\#$-semimartingales.

We will, in fact, consider more general stochastic integrals than those corresponding to semimartingale random measures. As in most definitions of an integral, the first step is to define the integral for a "simple" class of integrands and then to extend the integral to a larger class by approximation. Since we already know how to define the semimartingale integral in finite dimensions, a reasonable approach is to approximate arbitrary integrands by finite-dimensional integrands.

3.1 Finite dimensional approximations

We will need the following lemma giving a partition of unity. $\bar{C}(S)$ denotes the space of bounded continuous functions on S with the sup norm.

Lemma 3.1. *Let (S,d) be a complete, separable metric space, and let $\{x_k\}$ be a countable dense subset of S. Then for each $\epsilon > 0$, there exists a sequence $\{\psi_k^\epsilon\} \subset \bar{C}(S)$ such that $supp\{\psi_k^\epsilon\} \subset B_\epsilon(x_k)$, $0 \le \psi_k^\epsilon \le 1$, $|\psi_k^\epsilon(x) - \psi_k^\epsilon(y)| \le \frac{4}{\epsilon}d(x,y)$, and for each compact $K \subset S$, there exists $N_K < \infty$ such that $\sum_{k=1}^{N_K}\psi_k^\epsilon(x) = 1$, $x \in K$. In particular, $\sum_{k=1}^{\infty}\psi_k^\epsilon(x) = 1$ for all $x \in S$.*

Proof. Fix $\epsilon > 0$. Let $\tilde{\psi}_k(x) = (1 - \frac{2}{\epsilon}d(x, B_{\epsilon/2}(x_k))) \vee 0$. Then $0 \le \tilde{\psi}_k \le 1$, $\tilde{\psi}_k(x) = 1$, $x \in B_{\epsilon/2}(x_k)$, and $\tilde{\psi}_k(x) = 0$, $x \notin B_\epsilon(x_k)$. Note also that $|\tilde{\psi}_k(x) - \tilde{\psi}_k(y)| \le \frac{2}{\epsilon}d(x,y)$. Define $\psi_1^\epsilon = \tilde{\psi}_1$, and for $k > 1$, $\psi_k^\epsilon = \max_{i\le k}\tilde{\psi}_i - \max_{i\le k-1}\tilde{\psi}_i$. Clearly, $0 \le \psi_k^\epsilon \le \tilde{\psi}_k$ and $\sum_{i=1}^k \psi_i^\epsilon = \max_{i\le k}\tilde{\psi}_i$. In particular, for compact $K \subset S$, there exists $N_K < \infty$ such that $K \subset \cup_{k=1}^{N_K} B_{\epsilon/2}(x_k)$ and hence $\sum_{k=1}^{N_K}\psi_k^\epsilon(x) = 1$ for $x \in K$. Finally,

$$|\psi_k^\epsilon(x) - \psi_k^\epsilon(y)| \le 2\max_{i\le k}|\tilde{\psi}_i(x) - \tilde{\psi}_i(y)| \le \frac{4}{\epsilon}d(x,y)$$

□

Let U be a complete, separable metric space, and let H be a Banach space of functions on U. Let $\{\varphi_k\}$ be a dense subset of H. Fix $\epsilon > 0$, and let $\{\psi_k^\epsilon\}$ be as in Lemma 3.1 with $S = H$ and $\{x_k\} = \{\varphi_k\}$. The role of the ψ_k^ϵ is quite simple. Let $x \in D_H[0,\infty)$, and define $x^\epsilon(t) = \sum_k \psi_k^\epsilon(x(t))\varphi_k$. Then

$$\|x(t) - x^{\epsilon}(t)\|_H \leq \sum_k \psi_k^{\epsilon}(x(t))\|x(t) - \varphi_k\|_H \leq \epsilon. \tag{3.1}$$

Since x is cadlag, for each $T > 0$, there exists a compact $K_T \subset H$ such that $x(t) \in K_T$, $0 \leq t \leq T$. Consequently, for each $T > 0$, there exists $N_T < \infty$ such that $x^{\epsilon}(t) = \sum_{k=1}^{N_T} \psi_k^{\epsilon}(x(t))\varphi_k$ for $0 \leq t \leq T$. This construction gives a natural way of approximating any cadlag H-valued function (or process) by cadlag functions (processes) that are essentially finite dimensional. Let Y be an $\{\mathcal{F}_t\}$-semimartingale random measure, and suppose $Y(\varphi, \cdot)$ is defined for all $\varphi \in H$ (or at least for a dense subset of φ). Let X be a cadlag, H-valued, $\{\mathcal{F}_t\}$-adapted process, and let

$$X^{\epsilon}(t) = \sum_k \psi_k^{\epsilon}(X(t))\varphi_k. \tag{3.2}$$

Then $\|X - X^{\epsilon}\|_H \leq \epsilon$, and the integral $X_-^{\epsilon} \cdot Y$ is naturally (and consistently with the previous section) defined to be

$$X_-^{\epsilon} \cdot Y(t) = \sum_k \int_0^t \psi_k^{\epsilon}(X(s-))dY(\varphi_k, s).$$

We can then extend the integral to all cadlag, adapted processes by taking the limit provided we can make the necessary estimates. This approach to the definition of the stochastic integral is similar to that taken by Mikulevicius and Rozovskii [52].

3.2 Integral estimates

Definition 3.2. *Let $\mathcal{S}$ be the collection of H-valued processes of the form*

$$Z(t) = \sum_{k=1}^{m} \xi_k(t)\varphi_k$$

where the ξ_k are **R***-valued, cadlag, and adapted.*

Suppose that $Y = M + V$ is standard and M has dominating measure K. Then for $Z \in \mathcal{S}$, we define

$$Z_- \cdot Y(t) = \sum_{k=1}^{m} \int_0^t \xi_k(s-)dY(\varphi_k, s). \tag{3.3}$$

As in the previous section, we have

$$\begin{aligned} &E[\sup_{s\leq t} |Z_- \cdot M(s)|^2] \qquad\qquad (3.4) \\ &\leq 4E\left[\int_{U\times U\times[0,t]} |Z(u, s-)||Z(v, s-))|K(du \times dv \times ds)\right], \end{aligned}$$

and, letting $|\tilde{V}|$ denote the total variation measure for the signed measure $\tilde{V}$,

$$E[\sup_{s\leq t}|Z_-\cdot V(s)|] \leq E[\int_{U\times[0,t]}|Z(s,u)||\tilde{V}|(du\times ds)]. \tag{3.5}$$

If, for example, the norm on H is the sup norm and $\|Z(s)\|_H \leq \epsilon$ for all $s \geq 0$, then

$$E[\sup_{s\leq t}|Z_-\cdot M(s)|^2] \leq 4\epsilon^2 E[K(U\times U\times[0,t])] \tag{3.6}$$

and

$$E[\sup_{s\leq t}|Z_-\cdot V(s)|] \leq \epsilon E[|\tilde{V}|(U\times[0,t])]. \tag{3.7}$$

If $H = L^p(\mu)$, for some $p \geq 2$, and $\hat{K}$ defined as in (2.11), has the representation

$$\hat{K}(du\times dt) = h(u,t)\mu(du)dt \tag{3.8}$$

and

$$\tilde{V}(du\times dt) = g(u,t)\mu(du)dt, \tag{3.9}$$

then for r satisfying $\frac{2}{p}+\frac{1}{r}=1$ and q satisfying $\frac{1}{p}+\frac{1}{q}=1$, we have for $\|Z\|_H \leq \epsilon$

$$\begin{aligned} &E[\sup_{s\leq t}|Z_-\cdot M(s)|^2] \\ &\leq 4E\left[\int_0^t\left(\int_U|Z(s,u)|^p\mu(du)\right)^{\frac{2}{p}}\left(\int_U|h(u,s)|^r\mu(du)\right)^{\frac{1}{r}}ds\right] \\ &\leq 4\epsilon^2 E\left[\int_0^t\|h(\cdot,s)\|_{L^r(\mu)}ds\right] \end{aligned} \tag{3.10}$$

and

$$\begin{aligned} &E[\sup_{s\leq t}|Z_-\cdot V(s)|] \\ &\leq E\left[\int_0^t\left(\int_U|Z(s,u)|^p\mu(du)\right)^{\frac{1}{p}}\left(\int_U|g(u,s)|^q\mu(du)\right)^{\frac{1}{q}}ds\right] \\ &\leq \epsilon E\left[\int_0^t\|g(\cdot,s)\|_{L^q(\mu)}ds\right]. \end{aligned} \tag{3.11}$$

Note that either (3.6) and (3.7) or (3.10) and (3.11) give an inequality of the form

$$E[\sup_{s\leq t}|Z_-\cdot Y(s)|] \leq \epsilon C(t) \tag{3.12}$$

which in turn implies

$$\mathcal{H}_t = \{\sup_{s\leq t}|Z_-\cdot Y(s)| \;:\; Z\in\mathcal{S},\ \sup_{s\leq t}\|Z(s)\|_H \leq 1\} \tag{3.13}$$

is stochastically bounded. The following lemma summarizes the estimates made above in a form that will be needed later.

Lemma 3.3. *a) Let $\|\cdot\|_H$ be the* sup *norm, and suppose $E[K(U\times U\times[0,t])] < \infty$ and $E[|\tilde{V}|(U\times[0,t])] < \infty$ for all $t > 0$. Then if $\sup_s \|Z(s)\|_H \leq 1$ and τ is a stopping time bounded by a constant c,*

$$E[\sup_{s\leq t}|Z_-\cdot Y(\tau+s) - Z_-\cdot Y(\tau)|] \leq 2\sqrt{E[K(U\times U\times(\tau,\tau+t])] + E[|\tilde{V}|(U\times[\tau,\tau+t])]}$$

and

$$\lim_{t\to 0}\sqrt{E[K(U\times U\times(\tau,\tau+t])] + E[|\tilde{V}|(U\times[\tau,\tau+t])]} = 0.$$

b) Let $H = L^p(\mu)$, for some $p \geq 2$, and for h and g as in (3.10) and (3.11), suppose $E[\int_0^t \|h(\cdot,s)\|_{L^r(\mu)}ds] < \infty$ and $E[\int_0^t \|g(\cdot,s)\|_{L^q(\mu)}ds] < \infty$ for all $t > 0$. Then if $\sup_s \|Z(s)\|_H \leq 1$ and τ is a stopping time bounded by a constant c,

$$E[\sup_{s\leq t}|Z_-\cdot Y(\tau+s) - Z_-\cdot Y(\tau)|] \leq 2\sqrt{E[\int_\tau^{\tau+t}\|h(\cdot,s)\|_{L^r(\mu)}ds] + E[\int_\tau^{\tau+t}\|g(\cdot,s)\|_{L^q(\mu)}ds]}$$

and

$$\lim_{t\to 0}\sqrt{E[\int_\tau^{\tau+t}\|h(\cdot,s)\|_{L^r(\mu)}ds] + E[\int_\tau^{\tau+t}\|g(\cdot,s)\|_{L^q(\mu)}ds]} = 0.$$

Proof. The probability estimates follow from the moment estimates (3.6) - (3.11), and the limits follow by the dominated convergence theorem, using the fact that $\tau \leq c$. □

We will see that for many purposes we really do not need the moment estimates of Lemma 3.3. Consequently, it suffices to assume stochastic boundedness for $|\tilde{V}|$ and to localize the estimate on K.

Lemma 3.4. *a) Let $\|\cdot\|_H$ be the sup norm. Let τ be a stopping time, and let σ be a random variable such that $P\{\sigma > 0\} = 1$, $\tau+\sigma$ is a stopping time, $E[K(U\times U\times(\tau,\tau+\sigma]] < \infty$, and $P\{|\tilde{V}|(U\times(\tau,\tau+\sigma]) < \infty\} = 1$. Then if $\sup_s\|Z(s)\|_H \leq 1$ and $\alpha > 0$,*

$$P\{\sup_{s\leq t}|Z_-\cdot Y(\tau+s) - Z_-\cdot Y(\tau)| \geq 2\alpha\} \leq \frac{\sqrt{E[K(U\times U\times(\tau,\tau+t\wedge\sigma])]}}{\alpha} + P\{|\tilde{V}|(U\times[\tau,\tau+t\wedge\sigma]) \geq \alpha\} + P\{\sigma < t\}$$

and the right side goes to zero as $t \to 0$.

b) Let $H = L^p(\mu)$, for some $p \geq 2$, and let h and g be as in (3.8) and (3.9). Let τ be a stopping time, and let σ be a random variable such that $P\{\sigma > 0\} = 1$, $\tau + \sigma$ is a stopping time, $E[\int_\tau^{\tau+\sigma} \|h(\cdot, s)\|_{L^r(\mu)} ds] < \infty$ and $P\{\int_\tau^{\tau+\sigma} \|g(\cdot, s)\|_{L^q(\mu)} ds < \infty\} = 1$. Then if $\sup_s \|Z(s)\|_H \leq 1$

$$P\{\sup_{s \leq t} |Z_- \cdot Y(\tau + s) - Z_- \cdot Y(\tau)| \geq 2\alpha\}$$
$$\leq \frac{\sqrt{E[\int_\tau^{\tau+t\wedge\sigma} \|h(\cdot, s)\|_{L^r(\mu)} ds]}}{\alpha} + P\{\int_\tau^{\tau+t\wedge\sigma} \|g(\cdot, s)\|_{L^q(\mu)} ds \geq \alpha\}$$
$$+P\{\sigma < t\}$$

and the right side goes to zero as $t \to 0$.

Proof. Observe that

$$P\{\sup_{s \leq t} |Z_- \cdot Y(\tau + s) - Z_- \cdot Y(\tau)| \geq 2\alpha\}$$
$$\leq P\{\sup_{s \leq t\wedge\sigma} |Z_- \cdot M(\tau + s) - Z_- \cdot M(\tau)| \geq \alpha\}$$
$$+P\{\sup_{s \leq t\wedge\sigma} |Z_- \cdot V(\tau + s) - Z_- \cdot V(\tau)| \geq \alpha\} + P\{\sigma < t\},$$

and note that the first two terms on the right are bounded by the corresponding terms in the desired inequalities. □

3.3 $H^\#$-semimartingale integrals

Now let H be an arbitrary, separable Banach space. With the above development in mind, we make the following definition.

Definition 3.5. *Y is an $\{\mathcal{F}_t\}$-adapted, $H^\#$*-semimartingale, *if Y is an* **R**-*valued stochastic process indexed by $H \times [0, \infty)$ such that*

- *For each $\varphi \in H$, $Y(\varphi, \cdot)$ is a cadlag $\{\mathcal{F}_t\}$-semimartingale with $Y(\varphi, 0) = 0$.*
- *For each $t \geq 0$, $\varphi_1, \ldots, \varphi_m \in H$, and $a_1, \ldots, a_m \in$* **R**, *$Y(\sum_{i=1}^m a_i\varphi_i, t) = \sum_{i=1}^m a_i Y(\varphi_i, t)$ a.s.*

The definition of the integral in (3.3) extends immediately to this more general setting. Noting (3.12), (3.13), and their relationship to the assumption that the semimartingale measure is standard, we define:

Definition 3.6. *Y is a* standard *$H^\#$-semimartingale if $\mathcal{H}_t$ defined in (3.13) is stochastically bounded for each t.*

This stochastic boundedness is implied by an apparently weaker condition.

Definition 3.7. *Let $\mathcal{S}_0 \subset \mathcal{S}$ be the collection of processes*

$$Z(t) = \sum_{k=1}^{m} \xi_k(t)\varphi_k$$

in which the ξ_k are piecewise constant, that is,

$$\xi_k(t) = \sum_{i=0}^{j} \eta_i^k I_{[\tau_i^k, \tau_{i+1}^k)}(t)$$

where $0 = \tau_0^k \leq \cdots \leq \tau_j^k$ are $\{\mathcal{F}_t\}$-stopping times and η_i^k is $\mathcal{F}_{\tau_i^k}$-measurable.

Lemma 3.8. *If*

$$\mathcal{H}_t^0 = \{|Z_- \cdot Y(t)| : Z \in \mathcal{S}_0, \sup_{s \leq t} \|Z(s)\|_H \leq 1\}$$

is stochastically bounded, then $\mathcal{H}_t$ defined in (3.13) is stochastically bounded.

Remark 3.9. If Y is real-valued, that is $H = \mathbf{R}$, then the definition of standard $H^\#$-semimartingale is equivalent to the definition of semimartingale given in Section II.1 of Protter [53], that is, the process satisfies the "good integrator" condition.

Proof. For each $\delta > 0$, there exists $K(t, \delta)$ such that

$$P\{|Z_- \cdot Y(t)| \geq K(t, \delta)\} \leq \delta \tag{3.14}$$

for all $Z \in \mathcal{S}_0$ satisfying $\sup_{s \leq t} \|Z(s)\|_H \leq 1$. We can assume, without loss of generality, that $K(t, \delta)$ is right continuous and strictly increasing in δ (so that the collection of random variables satisfying $P\{U \geq K(t, \delta)\} \leq \delta$ is closed under convergence in probability). Let $\tau = \inf\{s : |Z_- \cdot Y| \geq K(t, \delta)\}$ and $Z^\tau = I_{[0,\tau)} Z$. Then

$$\begin{aligned} P\{\sup_{s \leq t} |Z_- \cdot Y(s)| \geq K(t, \delta)\} &= P\{|Z_- \cdot Y(t \wedge \tau)| \geq K(t, \delta)\} \\ &= P\{|Z_-^\tau \cdot Y(t)| \geq K(t, \delta)\} \leq \delta\,. \end{aligned}$$

For $Z \in \mathcal{S}$ with $\sup_{s \leq t} \|Z(s)\|_H \leq 1$, there exists a sequence $\{Z_n\} \subset \mathcal{S}_0$ with $\sup_{s \leq t} \|Z_n(s)\| \leq 1$ such that $\sup_{s \leq t} \|Z(s) - Z_n(s)\|_H \to 0$. This convergence implies $Z_{n-} \cdot Y(t) \to Z_- \cdot Y$ in probability, and the lemma follows. □

The assumption of Lemma 3.8 holds if there exists a constant $C(t)$ such that for all $Z \in \mathcal{S}_0$ satisfying $\sup_{s \leq t} \|Z(s)\|_H \leq 1$,

$$E[|Z_- \cdot Y(t)|] \leq C(t).$$

The following lemma is essentially immediate. The observation it contains is useful in treating semimartingale random measures which can frequently be decomposed into a part (usually a martingale random measure) that determines an $H^\#$-semimartingale on an L^2 space and another part that determines an $H^\#$-semimartingale on an L^1-space. Note that if H_1 is a space of functions on U_1 and H_2 is a space of functions on U_2, then $H_1 \times H_2$ can be interpreted as a space of functions on $U = U_1 \cup U_2$ where, for example, $\mathbf{R} \cup \mathbf{R}$ is interpreted as the set consisting of two copies of $\mathbf{R}$.

Lemma 3.10. *Let Y_1 be a standard $H_1^\#$-semimartingale and Y_2 a standard $H_2^\#$-semimartingale with respect to the same filtration $\{\mathcal{F}_t\}$. Define $H = H_1 \times H_2$, with $\|\varphi\|_H = \|\varphi_1\|_{H_1} + \|\varphi_2\|_{H_2}$ and $Y(\varphi, t) = Y_1(\varphi_1, t) + Y_2(\varphi_2, t)$ for $\varphi = (\varphi_1, \varphi_2)$. Then Y is a standard $H^\#$-semimartingale.*

If Y is standard, then the definition of $Z_- \cdot Y$ can be extended to all cadlag, H-valued processes.

Theorem 3.11. *Let Y be a standard $H^\#$-semimartingale, and let X be a cadlag, adapted, H-valued process. Define X^ϵ by (3.2). Then*

$$X_- \cdot Y \equiv \lim_{\epsilon \to 0} X_-^\epsilon \cdot Y$$

exists in the sense that for each $t > 0$

$$\lim_{\epsilon \to 0} P\{\sup_{s \le t} |X_- \cdot Y(s) - X_-^\epsilon \cdot Y(s)| > \eta\} = 0$$

for all $\eta > 0$, and $X_- \cdot Y$ is cadlag.

Proof. Let $K(\delta, t) > 0$ be as in Lemma 3.8. Since $\|X^{\epsilon_1}(s) - X^{\epsilon_2}(s)\|_H \le \epsilon_1 + \epsilon_2$, we have that $P\{\sup_{s<t} |X_-^{\epsilon_1} \cdot Y(s) - X_-^{\epsilon_2} \cdot Y(s)| \ge (\epsilon_1 + \epsilon_2) K(\delta, t)\} \le \delta$, and it follows that $\{X_-^\epsilon \cdot Y\}$ is Cauchy in probability and that the desired limit exists. Since $X_-^\epsilon \cdot Y$ is cadlag and the convergence is uniform on bounded intervals, it follows that $X_- \cdot Y$ is cadlag. □

The following corollary is immediate from the definition of the integral.

Corollary 3.12. *Let Y be a standard $H^\#$-semimartingale, and let X be a cadlag, adapted, H-valued process. Let τ be an $\{\mathcal{F}_t\}$-stopping time and define $X^\tau = I_{[0,\tau)} X$. Then*

$$X_- \cdot Y(t \wedge \tau) = X_-^\tau \cdot Y(t) .$$

For finite dimensional semimartingale integrals, the stochastic integral for cadlag, adapted integrands can be defined as a Riemann-type limit of approximating sums

$$X_- \cdot Y(t) = \lim \sum X(t_i \wedge t)(Y(t_{i+1} \wedge t) - Y(t_i \wedge t)) \qquad (3.15)$$

where the limit is as $\max(t_{i+1} - t_i) \to 0$ for the partition of $[0, \infty)$, $0 = t_0 < t_1 < t_2 < \cdots$. Formally, the analogue for $H^\#$-semimartingale integrals would be

$$X_- \cdot Y(t) = \lim \sum (Y(X(t_i \wedge t), t_{i+1} \wedge t) - Y(X(t_i \wedge t), t_i \wedge t));$$

however, $Y(X, t)$ is not, in general, defined for random variables X. We can define an analog of the summands in (3.15) by first defining $X^{[t_i, t_{i+1})} = I_{[t_i, t_{i+1})} X(t_i)$ and then defining

$$\Delta_{[t_i, t_{i+1})} Y(X(t_i), t) = X_-^{[t_i, t_{i+1})} \cdot Y(t).$$

Similarly, we can define $\Delta_{[\tau_i, \tau_{i+1})} Y(X(\tau_i), t)$ for stopping times $\tau_i \le \tau_{i+1}$.

Proposition 3.13. *For each n let $\{\tau_i^n\}$ be an increasing sequence of stopping times $0 = \tau_0^n \le \tau_1^n \le \tau_2^n \le \cdots$, and suppose that for each $t > 0$*

$$\lim_{n \to \infty} \max\{\tau_{i+1}^n - \tau_i^n : \tau_i^n < t\} = 0.$$

If X is a cadlag, adapted H-valued process and Y is a standard $H^\#$-semimartingale, then

$$X_- \cdot Y(t) = \lim_{n \to \infty} \sum \Delta_{[\tau_i^n, \tau_{i+1}^n)} Y(X(\tau_i^n), t) \qquad (3.16)$$

where the convergence is uniform on bounded time intervals in probability.

Proof. Let

$$X_n = \sum I_{[\tau_i^n, \tau_{i+1}^n)} X(\tau_i^n).$$

Then the sum on the right of (3.16) is just $X_{n-} \cdot Y(t)$ and, with X_n^ϵ defined as above,

$$X_{n-}^\epsilon \cdot Y(t) = \sum_k \sum_i \psi_k^\epsilon(X(\tau_i^n))(Y(\varphi_k, \tau_{i+1}^n \wedge t) - Y(\varphi_k, \tau_i^n \wedge t)).$$

By the finite dimensional result, for each $\eta > 0$,

$$\lim_{n \to \infty} P\{\sup_{s \le t} |X_{n-}^\epsilon \cdot Y(s) - X_-^\epsilon \cdot Y(s)| > \eta\} = 0$$

and by standardness

$$P\{\sup_{s \le t} |X_{n-}^\epsilon \cdot Y(s) - X_{n-} \cdot Y(s)| \ge \epsilon K(\delta, t)\} + P\{\sup_{s \le t} |X_-^\epsilon \cdot Y(s) - X_- \cdot Y(s)| \ge \epsilon K(\delta, t)\} \le 2\delta$$

and the result follows. □

3.4 Predictable integrands

Definition 3.14. *Let S^* be the collection of H-valued processes of the form*

$$Z(t) = \sum_{k=1}^{m} \xi_k(t)\varphi_k$$

where the ξ_k are $\{\mathcal{F}_t\}$-predictable and bounded.

$Z_- \cdot Y$ for $Z \in S$, can be extended to $Z \in S^*$ by setting

$$Z \cdot Y(t) = \sum_{k=1}^{m} \int_0^t \xi_k(s)dY(\varphi_k, s).$$

We will show that the condition that $\mathcal{H}_t$ is stochastically bounded implies that

$$\mathcal{H}_t^* = \{\sup_{s\le t} |Z \cdot Y(s)| \ : \ Z \in S^*, \sup_{s\le t} \|Z(s)\|_H \le 1\}$$

is also stochastically bounded, and hence $Z \cdot Y$ can be extended to all predictable, H-valued processes X that satisfy a compact range condition by essentially the same argument as in the proof of Theorem 3.11.

Lemma 3.15. *If $\mathcal{H}_t$ is stochastically bounded, then $\mathcal{H}_t^*$ is stochastically bounded.*

Proof. Let $K(t,\delta)$ be as in (3.14). Fix $\varphi_1, \ldots, \varphi_m \in H$ and let $C_\varphi = \{x \in \mathbf{R}^m : \|\sum_{i=1}^m x_i\varphi_i\|_H \le 1\}$. To simplify notation, let $Y_i(t) = Y(\varphi_i, t)$. We need to show that if $(\xi_1, \ldots, \xi_m)$ is predictable and takes values in C_φ, then

$$P\{\sup_{s\le t} |\sum_{i=1}^{m} \int_0^s \xi_i(s)dY_i(s)| \ge K(t,\delta)\} \le \delta. \tag{3.17}$$

Consequently, it is enough to show that there exists cadlag, adapted ξ_i^n such that $(\xi_1^n, \ldots, \xi_m^n) \in C_\varphi$ and $\lim_{n\to\infty} \sup_{s\le t} |\int_0^s (\xi_i(u) - \xi_i^n(u-))dY_i(u)| = 0$ in probability for each i. Assume for the moment that $Y_i = M_i + A_i$ where M_i is a square integrable martingale and $E[T_t(A_i)] < \infty$, $T_t(A_i)$ denoting the total variation up to time t. Let $\Gamma(t) = \sum_{i=1}^m [M_i]_t$ and $\Lambda(t) = \sum_{i=1}^m T_t(A_i)$. Then (see Protter [53], Theorem IV.2) there exists a sequence of cadlag, adapted $\mathbf{R}^m$-valued processes $\tilde{\xi}^n$ such that

$$\lim_{n\to\infty} \left(\sqrt{E[\int_0^t |\xi(s) - \tilde{\xi}^n(s-)|^2 d\Gamma(s)]} + E[\int_0^t |\xi(s) - \tilde{\xi}^n(s-)|d\Lambda(s)] \right) = 0. \tag{3.18}$$

Letting π denote the projection onto the convex set C_φ, since $|\pi(x)-\pi(y)| \leq |x-y|$ and $\xi \in C_\varphi$, if we define $\xi^n = \pi(\tilde{\xi}^n)$, $\xi^n \in C_\varphi$ and the limit in (3.18) still holds. Finally,

$$E[\sup_{s\leq t} | \int_0^s (\xi_i(u) - \xi_i^n(u-))dY_i(u)|]$$

$$\leq 2\sqrt{E[\int_0^t |\xi_i(s) - \xi_i^n(s-)|^2 d[M_i]_s] + E[\int_0^t |\xi_i(s) - \xi_i^n(s-)|dT_s(A_i)]}$$

$$\leq 2\sqrt{E[\int_0^t |\xi(s) - \xi^n(s-)|^2 d\Gamma(s)] + E[\int_0^t |\xi(s) - \xi^n(s-)|dA(s)]}$$

so the stochastic integrals converge and the limit must satisfy (3.17).

For general Y_i, fix $\epsilon > 0$, and let $Y_i = M_i + A_i$ where M_i is a local martingale with discontinuities bounded by ϵ, that is, $\sup_t |M_i(t) - M_i(t-)| \leq \epsilon$, and A_i is a finite variation process. (Such a decomposition always exists. See Protter [53, Theorem III.13].) For $c > 0$, let $\tau_c = \inf\{t : \sum_{i=1}^m [M_i]_t + \sum_{i=1}^m T_t(A_i) \geq c\}$, and let $Y_i^{\tau_c} = Y_i(\cdot \wedge \tau_c)$. Then for cadlag, adapted ξ with values in C_φ, it still holds that

$$P\{\sup_{s\leq t} |\sum_{i=1}^m \int_0^s \xi_i(s-)dY_i^{\tau_c}(s)| \geq K(t,\delta)\} \leq \delta. \qquad (3.19)$$

(replace ξ by $I_{[0,\tau_c)}\xi$). Define $\tilde{Y}_i^{\tau_c} = M_i^{\tau_c} + A_i^{\tau_c-}$ where $A_i^{\tau_c-}(t) = A_i(t)$, for $t < \tau_c$ and $A_i^{\tau_c-}(t) = A_i(\tau_c-)$ for $t \geq \tau_c$. It follows from (3.19) and the fact that

$$|\sum_{i=1}^m \xi_i(\tau_c-)(M_i(\tau_c) - M_i(\tau_c-))| \leq \epsilon \sup_{x\in C_\varphi} \sum_{i=1}^m |x_i| \equiv \epsilon K_\varphi$$

that

$$P\{\sup_{s\leq t} |\sum_{i=1}^m \int_0^s \xi_i(s-)d\tilde{Y}_i^{\tau_c}(s)| \geq K(t,\delta) + \epsilon K_\varphi\} \leq \delta. \qquad (3.20)$$

for all cadlag, adapted processes with values in C_φ. But $M_i^{\tau_c}$ is a square integrable martingale and $T_t(A_i^{\tau_c-}) \leq c$, so it follows that (3.20) holds with $\xi(s-)$ replaced by an arbitrary predictable process with values in C_φ. Letting $c \to \infty$ and observing that ϵ is arbitrary, we see that (3.17) holds and the lemma follows. □

Proposition 3.16. *Let Y be a standard $H^\#$-semimartingale, and let X be an H-valued predictable process such that for $t, \eta > 0$ there exists compact $K_{t,\eta} \subset H$ satisfying*

$$P\{X(s) \in K_{t,\eta},\ s \leq t\} \geq 1 - \eta.$$

Then defining X^ϵ as in (3.2), $X \cdot Y \equiv \lim_{\epsilon\to 0} X^\epsilon \cdot Y$ exists.

Remark 3.17. If estimates of the form (3.4) hold, then the definition of $X \cdot Y$ can be extended to locally bounded X, that is, the compact range condition can be dropped. (Approximate X be processes of the form $XI_K(X)$ where K is compact.) We do not know whether or not the compact range condition can be dropped for every standard $H^\#$-semimartingale.

Proof. The proof is the same as for Theorem 3.11. □

3.5 Examples

The idea of an $H^\#$-semimartingale is intended to suggest, but not be equivalent to, the idea of an H^*-semimartingale, that is, a semimartingale with values in H^*. In deed, any H^*-semimartingale will be an $H^\#$-semimartingale; however, there are a variety of examples of $H^\#$-semimartingales that are not H^*-semimartingales.

Example 3.18. Poisson process integrals in L^p spaces.

Let μ be a finite measure on U, and let $H = L^p(\mu)$ for some $1 \leq p < \infty$. Let N be a Poisson point process on $U \times [0, \infty)$, and for $\varphi \in H$, define $Y(\varphi, t) = \int_{U \times [0,t]} \varphi(u) N(du \times ds)$. Of course $Y(\varphi, \cdot)$ is just a compound Poisson process whose jumps have distribution given by $\nu(A) = \int I_A(\varphi(u))\mu(du)$. Since point evaluation is not a continous linear functional on L^p, Y is an $H^\#$-semimartingale, but not an H^*-semimartingale.

Example 3.19. Cylindrical Brownian motion.

Let H be a Hilbert space and let Q be a bounded, self-adjoint, nonnegative operator on H. Then there exists a Gaussian process W with covariance

$$E[W(\varphi_1, t)W(\varphi_2, s)] = t \wedge s\langle Q\varphi_1, \varphi_2\rangle.$$

If Q is nuclear, then W will be an $H^*(= H)$-valued martingale; however, in general, W will only be an $H^\#$-semimartingale. (See, for example, Da Prato and Zabczyk [11, Section 4.3].) Note that if $X(t) = \sum_{i=1}^m \xi_i(t)\varphi_i$ is cadlag and adapted to the filtration generated by W, then

$$E[|X_- \cdot W(t)|^2] = \int_0^t E[\sum_{i,j} \xi_i(s)\xi_j(s)\langle Q\varphi_i, \varphi_j\rangle]ds \leq \|Q\| \int_0^t E[\|X(s)\|_H^2]ds$$

and it follows that W is standard.

4. Convergence of stochastic integrals

Let H be a separable Banach space, and for each $n \geq 1$, let Y_n be an $\{\mathcal{F}_t^n\}$-$H^{\#}$-semimartingale. Note that the Y_n need not all be adapted to the same filtration nor even defined on the same probablity space. The minimal convergence assumption that we will consider is that for $\varphi_1, \ldots, \varphi_m \in H$, $(Y_n(\varphi_1, \cdot), \ldots, Y_n(\varphi_m, \cdot)) \Rightarrow (Y(\varphi_1, \cdot), \ldots, Y(\varphi_m, \cdot))$ in $D_{\mathbf{R}^m}[0, \infty)$ with the Skorohod topology.

Let $\{X_n\}$ be cadlag, H-valued processes. We will say that $(X_n, Y_n) \Rightarrow (X, Y)$, if $(X_n, Y_n(\varphi_1, \cdot), \ldots, Y_n(\varphi_m, \cdot)) \Rightarrow (X, Y(\varphi_1, \cdot), \ldots, Y(\varphi_m, \cdot))$ in $D_{H \times \mathbf{R}^m}[0, \infty)$ for each choice of $\varphi_1, \ldots, \varphi_m \in H$. We are interested in conditions on $\{(X_n, Y_n)\}$, under which $X_{n-} \cdot Y_n \Rightarrow X_- \cdot Y$. In the finite dimensional setting of Kurtz and Protter [42], convergence was obtained by first approximating by piecewise constant processes. This approach was also taken by Cho [8, 9] for integrals driven by martingale random measures. Here we take a slightly different approach, approximating the H-valued processes by finite dimensional H-valued processes in a way that allows us to apply the results of Kurtz and Protter [42] and Jakubowski, Mémin, and Pagès [27].

Lemma 4.1. *Suppose that for each $\varphi \in H$, the sequence $\{Y_n(\varphi, \cdot)\}$ of $\mathbf{R}$-valued semimartingales satisfies the conditions of Theorem 1.1. Let $X_n^\epsilon(t) \equiv \sum_k \psi_k^\epsilon(X_n(t))\varphi_k$. If $(X_n, Y_n) \Rightarrow (X, Y)$, then $(X_n, Y_n, X_{n-}^\epsilon \cdot Y_n) \Rightarrow (X, Y, X_-^\epsilon \cdot Y)$. If $(X_n, Y_n) \to (X, Y)$ in probability, then $(X_n, Y_n, X_{n-}^\epsilon \cdot Y_n) \to (X, Y, X_-^\epsilon \cdot Y)$ in probability.*

Proof. By tightness, there exists a sequence of compact $K_m \subset H$ such that $P\{X_n(t) \in K_m, t \leq m\} \geq 1 - \frac{1}{m}$. Let $\tau_n^m = \inf\{t : X_n(t) \notin K_m\}$. Then $P\{\tau_n^m \geq m\} \geq 1 - \frac{1}{m}$, and

$$X_{n-}^\epsilon \cdot Y_n(t) = Z_n^m(t) \equiv \sum_{k=1}^{N_{K_m}} \int_0^t \psi_k^\epsilon(X_n(s-))dY_n(\varphi_k, s)$$

for $t < \tau_n^m$. Theorem 1.1 implies $(X_n, Y_n, Z_n^m) \Rightarrow (X, Y, Z^m)$ for each m, where $Z^m(t) \equiv \sum_{k=1}^{N_{K_m}} \int_0^t \psi_k^\epsilon(X(s-))dY(\varphi_k, s)$. Since $Z_n^m(t) = X_{n-}^\epsilon \cdot Y_n(t)$ for $t \leq \tau_n^m$, using the metric of Ethier and Kurtz [17, Chapter 3, Formula (5.2)], we have $d(X_{n-}^\epsilon \cdot Y_n, Z_n^m) \leq e^{-\tau_n^m}$, and the convergence of $\{Z_n^m\}$ for each m implies the desired convergence for $X_{n-}^\epsilon \cdot Y_n$. □

In order to prove the convergence for $X_{n-} \cdot Y_n$, by Lemma 4.1, it is enough to show that $X_{n-}^\epsilon \cdot Y_n$ is a good approximation of $X_{n-} \cdot Y_n$, that is, we need to estimate $(X_{n-} - X_{n-}^\epsilon) \cdot Y_n$. If the Y_n correspond to semimartingale random measures, then (3.6) and (3.7) or (3.10) and (3.11) give estimates of the form

$$E[\sup_{s \leq t} |(X_{n-} - X_{n-}^\epsilon) \cdot Y(s)|] \leq \epsilon C_n(t).$$

If $\sup_n C_n(t) < \infty$ for each t, then defining

$$\mathcal{H}_{n,t} = \{\sup_{s\leq t} |Z_- \cdot Y_n(s)| \ : \ Z \in \mathcal{S}^n, \sup_{s\leq t} \|Z(s)\|_H \leq 1\}, \tag{4.1}$$

$\hat{\mathcal{H}}_t = \cup_n \mathcal{H}_{n,t}$ is stochastically bounded for each t. This last assertion is essentially the *uniform tightness* (UT) condition of Jakubowski, Mémin, and Pagès [27]. As in Lemma 3.8, the condition that $\cup_n \mathcal{H}_{n,t}$ be stochastically bounded is equivalent to the condition that

$$\hat{\mathcal{H}}_t^0 = \cup_n \mathcal{H}_{n,t}^0 = \cup_n \{|Z_- \cdot Y_n(t)| : Z \in \mathcal{S}_0^n, \sup_{s\leq t} \|Z(s)\|_H \leq 1\} \tag{4.2}$$

be stochastically bounded.

For finite dimensional semimartingales, the uniform tightness condition, Condition UT of Theorem 1.1, implies uniformly controlled variations, Condition UCV of Theorem 1.1. In the present setting, the relationship of the UT condition and some sort of "uniform worthiness" is not clear and certainly not so simple. Consequently, the following theorem is really an extension of the convergence theorem of Jakubowski, Mémin, and Pagès [27] rather than the results of Kurtz and Protter [42], although in the finite dimensional setting of those results, the conditions of the two theorems are equivalent.

Theorem 4.2. *For each $n = 1, 2, \ldots$, let Y_n be a standard $\{\mathcal{F}_t^n\}$-adapted, $H^\#$-semimartingale. Let $\mathcal{H}_{n,t}^0$ and $\hat{\mathcal{H}}_t^0$ be defined as in (4.2), and suppose that for each $t > 0$, $\hat{\mathcal{H}}_t^0$ is stochastically bounded. If $(X_n, Y_n) \Rightarrow (X, Y)$, then there is a filtration $\{\mathcal{F}_t\}$, such that Y is an $\{\mathcal{F}_t\}$-adapted, standard, $H^\#$-semimartingale, X is $\{\mathcal{F}_t\}$-adapted and $(X_n, Y_n, X_{n-} \cdot Y_n) \Rightarrow (X, Y, X_- \cdot Y)$. If $(X_n, Y_n) \to (X, Y)$ in probability, then $(X_n, Y_n, X_{n-} \cdot Y_n) \to (X, Y, X_- \cdot Y)$ in probability.*

Remark 4.3. a) One of the motivations for introducing $H^\#$-semimartingales rather than simply posing the above result in terms of semimartingale random measures is that the Y_n may be given by standard semimartingale random measures while the limiting Y is not.

b) Jakubowski [25] proves the above theorem in the case of Hilbert space-valued semimartingales.

Proof. The stochastic boundedness condition implies that for each $t, \delta > 0$ there exists $K(\delta, t)$ such that for all $R \in \hat{\mathcal{H}}_t$, $P\{|R| \geq K(\delta, t)\} \leq \delta$. Without loss of generality, we can assume that $K(\delta, t)$ is a nondecreasing, right continuous function of t. Note that this inequality will hold for $R = \sup_{s\leq t} |Z_{n-} \cdot Y_n(s)|$ for any cadlag, $\{\mathcal{F}_t^n\}$-adapted Z_n satisfying $\sup_{s\leq t} \|Z_n(s)\|_H \leq 1$.

Let $\mathcal{F}_t = \sigma(X(s), Y(\varphi, s) : s \leq t, \varphi \in H)$. Define

$$Z_n(t) = \sum_{i=1}^m f_i(X_n, Y_n(\hat{\varphi}_1, \cdot), \ldots, Y_n(\hat{\varphi}_d, \cdot), t)\varphi_i$$

where $(f_1, \ldots, f_m)$ is a continous function mapping $D_{H\times \mathbf{R}^d}[0,\infty)$ into $C_{A(\varphi_1,\ldots,\varphi_m)}[0,\infty)$, $A(\varphi_1,\ldots,\varphi_m) = \{\alpha \in \mathbf{R}^m : \|\sum_i \alpha_i\varphi_i\|_H \le 1\}$, in such a way that $f_i(x, y_1, \ldots, y_d, t)$ depends only on $(x(s), y_1(s), \ldots, y_d(s))$ for $s \le t$ (ensuring that Z_n is $\{\mathcal{F}_t^n\}$-adapted and $Z = \sum_{i=1}^m f_i(X, Y(\hat\varphi_1,\cdot), \ldots, Y(\hat\varphi_d,\cdot))\varphi_i$ is $\{\mathcal{F}_t\}$-adapted). Theorem 1.1 implies $Z_{n-}\cdot Y_n \Rightarrow Z_-\cdot Y$, and it follows (using the right continuity of $K(\delta,t)$) that

$$P\{\sup_{s\le t}|Z_-\cdot Y| \ge K(\delta,t)\} \le \delta. \tag{4.3}$$

By approximation, one can see that (4.3) holds for any process Z of the form

$$Z(t) = \sum_{i=1}^m \xi_i(t)\varphi_i$$

where $\xi = (\xi_1, \ldots, \xi_m)$ is $\{\mathcal{F}_t\}$-adapted, cadlag and has values in $A(\varphi_1, \ldots, \varphi_m)$. By (4.3), it follows that $Y(\varphi,\cdot)$ is an $\{\mathcal{F}_t\}$-semimartingale for each φ and hence that Y is an $H^\#$-semimartingale. It also follows from (4.3) that Y is standard.

Finally, observing that $\|X_n(s) - X_n^\epsilon(s)\|_H/\epsilon$ is bounded by 1, we have that

$$P\{\sup_{s\le t}|X_{n-}\cdot Y_n - X_{n-}^\epsilon\cdot Y_n| \ge \epsilon(K(\delta,t)\} \le \delta$$

and similarly for X and Y. Consequently, the Theorem follows from Lemma 4.1. □

Example 4.4. Many particle random walk.

For each n let X_k^n, $k = 1, \ldots, n$, be independent, continous-time, reflecting random walks on $E_n = \{\frac{i}{n} : i = 0, \ldots, n\}$ with generator

$$B_n f(x) = \begin{cases} \frac{n^2}{2}(f(x+\frac{1}{n}) + f(x-\frac{1}{n}) - 2f(x)), & 0 < x < 1 \\ n^2(f(\frac{1}{n}) - f(0)), & x = 0 \\ n^2(f(1-\frac{1}{n}) - f(1), & x = 1 \end{cases}$$

and $X_k^n(0)$ uniformly distributed on E_n. Let $H = C^1([0,1])$ with $\|\varphi\|_H = \sup_{0\le x\le 1}(|\varphi(x)| + |\varphi'(x)|)$, and define

$$Y_n(\varphi,t) = \frac{1}{\sqrt{n}}\sum_{k=1}^n \left(\varphi(n^{-1}X_k^n(t)) - \varphi(n^{-1}X_k^n(0)) - \int_0^t B_n\varphi(X_k^n(s))ds\right).$$

Note that Y_n corresponds to a martingale random measure and that

$$E[(Y_n(\varphi,t_2) - Y_n(\varphi,t_1))^2|\mathcal{F}_{t_1}^n] = \frac{1}{n}\sum_{k=1}^n$$

$$E\left[\int_{t_1}^{t_2} n^2 \frac{\left(\varphi(X_k^n(s)+\frac{1}{n})-\varphi(X_k^n(s))\right)^2+\left(\varphi(X_k^n(s)-\frac{1}{n})-\varphi(X_k^n(s))\right)^2}{2}\, ds \middle| \mathcal{F}_t^n\right].$$

It follows that for $Z \in \mathcal{S}_0^n$,

$$E[(Z_- \cdot Y_n(t))^2] \leq t \sup_{0\leq s\leq t} \sup_{0\leq x\leq 1} |Z'(s,x)|^2 \leq t \sup_{0\leq s\leq t} \|Z(s)\|_H^2,$$

and hence $\{Y_n\}$ is uniformly tight. The martingale central limit theorem gives $Y_n \Rightarrow Y$ where Y is Gaussian and satisfies

$$\langle Y(\varphi_1,\cdot), Y(\varphi_2,\cdot)\rangle_t = t\int_0^1 \varphi_1'(x)\varphi_2'(x)dx.$$

It follows that Y does not correspond to a standard martingale random measure.

5. Convergence in infinite dimensional spaces

Theorem 4.2 extends easily to integrals with range in $\mathbf{R}^d$. The interest in semimartingales in infinite dimensional spaces, however, is frequently in relation to stochastic partial differential equations. Consequently, extension to function-valued integrals is desirable. For semimartingale random measures, this extension would be to integrals of the form

$$Z(t,x) = \int_{U\times[0,t]} X(s-,x,u)Y(du\times ds)$$

where X is a process with values in a function space on $E \times U$. We will take (E,r) to be a complete, separable metric space. More generally, let X be an H-valued process indexed by $[0,\infty)\times E$. If X is $\{\mathcal{F}_t\}$-adapted and $X(\cdot,x)$ is cadlag for each x, then

$$Z(t,x) = X(\cdot-,x)\cdot Y(t)$$

is defined for each x; however, the properties of Z as a function of x (even the measurability) are not immediately clear. Consequently, we construct the desired integral more carefully.

5.1 Integrals with infinite-dimensional range

Let (E,r_E) and (U,r_U) be complete, separable metric spaces, let L be a separable Banach space of $\mathbf{R}$-valued functions on E, and let H be a separable Banach space of $\mathbf{R}$-valued functions on U. (We restrict our attention to function spaces so that for $f \in L$ and $\varphi \in H$, $f\varphi$ has the simple interpretation as a pointwise product. The restriction to function spaces could be dropped with the introduction of an appropriate definition of product.) Let $G_L = \{f_i\} \subset L$

be a sequence such that the finite linear combinations of the f_i are dense in L, and let $G_H = \{\varphi_j\}$ be a sequence such that the finite linear combinations of the φ_j are dense in H.

Definition 5.1. *Let $\hat{H}$ be the completion of the linear space $\{\sum_{i=1}^{l}\sum_{j=1}^{m} a_{ij} f_i\varphi_j : f_i \in G_L, \varphi_i \in G_H\}$ with respect to some norm $\|\cdot\|_{\hat{H}}$.*

For example, if

$$\|\sum_{i=1}^{l}\sum_{j=1}^{m} a_{ij} f_i\varphi_j\|_{\hat{H}}$$

$$= \sup\{\sum_{i=1}^{m} a_{ij}\langle\lambda, f_i\rangle\langle\eta, \varphi_i\rangle : \lambda \in L^*, \eta \in H^*, \|\lambda\|_{L^*} \leq 1, \|\eta\|_{H^*} \leq 1\}$$

then we can interpret $\hat{H}$ as a subspace of bounded linear operators mapping H^* into L. Metivier and Pellaumail [50] develop the stochastic integral in this setting.

We say that a norm $\|\cdot\|_G$ on a function space G is monotone, if $g \in G$ implies $|g| \in G$ and $|g_1| \leq |g_2|$ implies $\|g_1\|_G \leq \|g_2\|_G$. If $\|\cdot\|_L$ and $\|\cdot\|_H$ are both monotone, we may take

$$\|\sum_{i=1}^{l}\sum_{j=1}^{m} a_{ij} f_i\varphi_j\|_{\hat{H}} = \|\|\sum_{i=1}^{l}\sum_{j=1}^{m} a_{ij} f_i\varphi_j\|_L\|_H. \tag{5.1}$$

Note that in the above examples, the mapping $(f,\varphi) \in L \times H \to f\varphi \in \hat{H}$ is continuous, although in general, we do not require this continuity.

Let $\zeta_k = \sum a_{kij} f_i\varphi_j$, $k = 1, 2, \ldots$ be a dense sequence in $\hat{H}$, where each sum is finite, and let $\{\hat{\psi}_k^\epsilon\}$ be as in Lemma 3.1 with $\{x_k\}$ replaced by $\{\zeta_k\}$. Then for each $v \in D_{\hat{H}}[0,\infty)$, we can define $v^\epsilon(t) = \sum_k \hat{\psi}_k^\epsilon(v(t))\zeta_k$, and we have $\|v(t) - v^\epsilon(t)\|_{\hat{H}} \leq \epsilon$. Furthermore, if we define

$$c_{ij}^\epsilon(v) = \sum_k \hat{\psi}_k^\epsilon(v)a_{kij}\,, \tag{5.2}$$

then $v^\epsilon(t) = \sum c_{ij}^\epsilon(v(t)) f_i\varphi_j$, and only finitely many of the $\{c_{ij}(v(t))\}$ are non-zero on any bounded interval $0 \leq t \leq T$.

With the above approximation in mind, let

$$X(t) = \sum_{i,j} \xi_{ij}(t) f_i\varphi_j$$

where the ξ_{ij} are R-valued, cadlag, adapted processes and only finitely many of the ξ_{ij} are non-zero. If Y is an $H^{\#}$-semimartingale, we can define

$$X_- \cdot Y(t) = \sum_i f_i \sum_j \int_0^t \xi_{ij}(s-)dY(\varphi_j, s).$$

Then $X_- \cdot Y$ is in $D_L[0,\infty)$.

Definition 5.2. *Let $S^0_{\hat{H}}$ be the collection of simple $\hat{H}$-valued processes of the form*

$$X = \sum \xi_{kij} I_{[t_k,t_{k+1})} f_i \varphi_j$$

where ξ_{kij} is an **R***-valued, $\mathcal{F}_{t_k}$-measurable random variable and all but finitely many of the ξ_{kij} are zero, and let $S_{\hat{H}}$ be the collection of $\hat{H}$-valued processes of the form*

$$X(t) = \sum_{ij} \xi_{ij}(t) f_i \varphi_j,$$

where the ξ_{ij} are cadlag and adapted and all but finitely many are zero.

For $X \in S^0_{\hat{H}}$

$$X_- \cdot Y(t) = \sum \xi_{kij} f_i (Y(\varphi_j, t \wedge t_{k+1}) - Y(\varphi_j, t \wedge t_k)),$$

and for $X \in S_{\hat{H}}$

$$X_- \cdot Y(t) = \sum_{ij} f_i \int_0^t \xi_{ij}(s-)dY(\varphi_j, s).$$

As in the **R**-valued case, we make the following definition.

Definition 5.3. *An $H^\#$-semimartingale Y is a* standard *$(L, \hat{H})^\#$-semimartingale if*

$$\mathcal{H}^0_t = \{\|X_- \cdot Y(t)\|_L \;:\; X \in S^0_{\hat{H}},\ \sup_{s\le t} \|X(s)\|_{\hat{H}} \le 1\} \tag{5.3}$$

is stochastically bounded for each t, or equivalently (as in Lemma 3.8)

$$\mathcal{H}_t = \{\sup_{s\le t} \|X_- \cdot Y(s)\|_L \;:\; X \in S_{\hat{H}},\ \sup_{s\le t} \|X(s)\|_{\hat{H}} \le 1\}$$

is stochastically bounded for each t.

As before, this assertion holds if there exists a constant $C(t)$ such that for all $X \in S^0_{\hat{H}}$ satisfying $\sup_{s\le t} \|X(s)\|_{\hat{H}} \le 1$,

$$E[\|X_- \cdot Y(t)\|_L] \le C(t).$$

If Y is standard, then, as in the **R**-valued case, the definition of $X_- \cdot Y$ can be extended to all cadlag, $\hat{H}$-valued processes X by approximating X by

$$X^{\epsilon}(t) = \sum_k \hat{\psi}_k^{\epsilon}(X(t))\zeta_k = \sum_{ij} c_{ij}^{\epsilon}(X(t)) f_i \varphi_j. \tag{5.4}$$

In fact, Lemma 3.15 and Proposition 3.16 extend to the present setting and $X \cdot Y$ can be defined for all predictable $\hat{H}$-valued processes X satisfying the compact range condition. Note that $X \cdot Y$ will be cadlag.

Remark 5.4. Other approaches. Metivier and Pellaumail [50] develop an integral for, in our notation, H^*-semimartingales with integrands whose values are bounded linear operators from H^* to L. (Take $\|\cdot\|_{\hat{H}}$ to be given by (5.1).) They assume moment conditions that imply that the integrator is a standard $(L, \hat{H})^{\#}$-semimartingale. These conditions are sufficient to extend the integral to all locally bounded predictable integrands. (See Remark 3.17.) Da Prato and Zabczyk [11] is a recent account of the theory of stochastic integration and stochastic ordinary and partial differential equations driven by infinite dimensional Brownian motions of the form described in Example 3.19. Ustunel [59] develops stochastic integration in nuclear spaces. Mikulevicius and Rozovskii [52] consider integrals in more general linear topological spaces considering integrands in continously embedded Hilbert subspaces determined by the covariance operator of the martingale.

5.2 Convergence theorem

Let $\{Y_n\}$ be a sequence of standard $(L, \hat{H})^{\#}$-semimartingales and define

$$\mathcal{H}_{n,t}^0 = \{\|X_{n-} \cdot Y_n(t)\|_L \; : \; X \in \mathcal{S}_{n,\hat{H}}^0, \; \sup_{s \le t} \|X_n(s)\|_{\hat{H}} \le 1\} \tag{5.5}$$

If $\hat{\mathcal{H}}_t^0 = \cup \mathcal{H}_{n,t}^0$ is stochastically bounded for each t, we will again say that $\{Y_n\}$ is uniformly tight. (Since bounded sets in L are not, in general, compact, this terminology is not entirely appropriate; however, see Lemma 6.14.) As in Lemma 3.8, uniform tightness implies

$$\hat{\mathcal{H}}_t = \cup_n \{\sup_{s \le t} \|X_{n-} \cdot Y_n(s)\|_L \; : \; X_n \in \mathcal{S}_{n,\hat{H}}, \sup_{s \le t} \|X_n(s)\|_{\hat{H}} \le 1\}$$

is stochastically bounded for each t.

If $L = \mathbf{R}^k$ and $\hat{H} = H^k$ with $\|(\varphi_1, \ldots, \varphi_k)\|_{\hat{H}} = \sum_{i=1}^k \|\varphi_i\|_H$, then any uniformly tight sequence of $H^{\#}$-semimartingales is a uniformly tight sequence of $(L, \hat{H})^{\#}$-semimartingales.

Theorem 5.5. *For each $n = 1, 2, \ldots$, let Y_n be a standard $\{\mathcal{F}_t^n\}$-adapted, $(L, \hat{H})^{\#}$-semimartingale, and assume that $\{Y_n\}$ is uniformly tight.*

If $(X_n, Y_n) \Rightarrow (X, Y)$, then there is a filtration $\{\mathcal{F}_t\}$, such that Y is an $\{\mathcal{F}_t\}$-adapted, standard, $(L, \hat{H})^{\#}$-semimartingale and X is $\{\mathcal{F}_t\}$-adapted, and $(X_n, Y_n, X_{n-} \cdot Y_n) \Rightarrow (X, Y, X_- \cdot Y)$. If $(X_n, Y_n) \to (X, Y)$ in probability, then $(X_n, Y_n, X_{n-} \cdot Y_n) \to (X, Y, X_- \cdot Y)$ in probability.

Proof. Noting that Lemma 4.1 extends to this setting, the proof is exactly the same as the proof of Theorem 4.2. □

5.3 Verification of standardness

If $L = C[0,1])$ and $H = \mathbf{R}$ and we identify $\hat{H}$ with L, then a scalar semimartingale Y defines a standard $(L, L)^{\#}$-semimartingale if and only if Y is a finite variation process. In particular, for any $t_0 < t_1 < \cdots < t_m = t$ we can define $Z = \sum_{i=0}^{m-1} \xi_i I_{[t_i, t_{i+1})}$ so that

$$\|Z_- \cdot Y(t)\|_L = \sum_{i=0}^{m-1} |Y(t_{i+1}) - Y(t_i)|,$$

simply by ensuring that for each of the 2^m choices of $\theta_i = \pm 1, i = 0, \ldots, m-1$, there is some value of $x \in [0,1]$ such that $\xi_i(x) = \theta_i$.

Fortunately, more interesting examples of $(L, \hat{H})^{\#}$-semimartingales exist in other spaces. Let $L = L_2(\nu)$, and assume that $Y = M + V$ is a standard semimartingale random measure with dominating measure K. Then, as is (2.7),

$$E\left[\sup_{s \le t}\left(\int_{U \times [0,s]} X(s-, x, u) M(du \times ds)\right)^2\right] \tag{5.6}$$

$$\le 4E\left[\int_{U \times U \times [0,t]} |X(s-, x, u)||X(s-, x, v)| K(du \times dv \times ds)\right],$$

and, since the integral of the sup is greater than the sup of the integral, it follows that

$$E\left[\sup_{s \le t}\left\|\int_{U \times [0,s]} X(s-, \cdot, u) M(du \times ds)\right\|_L^2\right]$$

$$\le 4E\left[\int_{U \times U \times [0,t]} \int_E |X(s-, x, u)||X(s-, x, v)| \nu(dx) K(du \times dv \times ds)\right]$$

$$\le 4E\left[\int_{U \times U \times [0,t]} \|X(s-, \cdot, u\|_L \|X(s-, \cdot, v\|_L K(du \times dv \times ds)\right].$$

If the norm on H is the sup norm and $\|X(t)\|_{\hat{H}} \equiv \|\|X(t, \cdot, \cdot)\|_L\|_H \le 1$,

$$E\left[\sup_{s \le t}\left\|\int_{U \times [0,s]} X(s-, \cdot, u) M(du \times ds)\right\|_L^2\right]$$

$$\le 4E[K(U \times U \times [0,t])]$$

The analogous inequality for V is simply

$$E\left[\sup_{s\le t}\left\|\int_{U\times[0,s]} X(s-,\cdot,u)V(du\times ds)\right\|_L\right]$$
$$\le E\left[\int_{U\times[0,t]} \|X(s-,\cdot,u)\|_L|\tilde{V}|(du\times ds)\right]$$
$$\le E[(|\tilde{V}|(U\times[0,t]))]\,.$$

With $\hat{K}$ defined as in (2.11) and $\hat{V}$ as in (2.13), suppose that $\hat{K}(du\times dt) = h(u,t)\mu(du)dt$ and $\hat{V}(du\times dt) = g(u,t)\mu(du)dt$. Let $H = L_p(\mu)$ for $2\le p<\infty$ and again assume that $\|X(t)\|_{\hat{H}} \equiv \|\|X(t,\cdot,\cdot)\|_L\|_H \le 1$. Then with $\frac{2}{p}+\frac{1}{r}=1$ and $\frac{1}{p}+\frac{1}{q}=1$

$$E\left[\sup_{s\le t}\left\|\int_{U\times[0,s]} X(s-,\cdot,u)M(du\times ds)\right\|_L^2\right]$$
$$\le 4E\left[\int_0^t\int_U \|X(s-,\cdot,u)\|_L^2 h(u,s)\mu(du)ds\right]$$
$$\le 4E\left[\int_0^t\left(\int_U \|X(s-,\cdot,u)\|_L^p\mu(du)\right)^{\frac{2}{p}}\left(\int_U |h(u,s)|^r\mu(du)\right)^{\frac{1}{r}} ds\right]$$
$$\le 4E\left[\int_0^t\left(\int_U |h(u,s)|^r\mu(du)\right)^{\frac{1}{r}} ds\right]$$

(with the obvious modification if $r=\infty$) and for V

$$E\left[\sup_{s\le t}\left\|\int_{U\times[0,s]} X(s-,\cdot,u)V(du\times ds)\right\|_L\right]$$
$$\le E\left[\int_{U\times[0,t]} \|X(s-,\cdot,u)\|_L|\tilde{V}|(du\times ds)\right]$$
$$\le E\left[\int_{U\times[0,t]} \|X(s-,\cdot,u)\|_L\hat{V}(du\times ds)\right]$$
$$\le E\left[\int_0^t \|\|X(s,x,\cdot)\|_L\|_H\left(\int_U |g(u,s)|^q\mu(du)\right)^{\frac{1}{q}} ds\right]$$
$$\le E\left[\int_0^t\left(\int_U |g(u,s)|^q\mu(du)\right)^{\frac{1}{q}} ds\right].$$

From the above inequalities, we see that, at least in the Hilbert space setting, we can give conditions under which a semimartingale random measure gives a standard $(L,\hat{H})^{\#}$-semimartingale and conditions under which a sequence of such $(L, \hat{H})^{\#}$-semimartingales satisfies a uniform tightness condition. In particular, we have the following analog of Lemma 3.3. An analog of Lemma 3.4 will also hold.

Lemma 5.6. *Let* $L = L_2(\nu)$ *and* $\|\cdot\|_{\hat{H}} = \|\|\cdot\|_L\|_H$.

a) Let $\|\cdot\|_H$ *be the* sup *norm, and suppose* $E[K(U \times U \times [0,t])] < \infty$ *and* $E[|\tilde{V}|(U \times [0,t])] < \infty$ *for all* $t > 0$. *Then if* $\sup_s \|Z(s)\|_{\hat{H}} \leq 1$ *and* τ *is a stopping time bounded by a constant* c,

$$E[\sup_{s \leq t} \|Z_- \cdot Y(\tau + s) - Z_- \cdot Y(\tau)\|_L] \leq 2\sqrt{E[K(U \times U \times (\tau, \tau + t])]} + E[|\tilde{V}|(U \times [\tau, \tau + t])]$$

and

$$\lim_{t \to 0} E[K(U \times U \times (\tau, \tau + t])] + E[(|\tilde{V}|(U \times [\tau, \tau + t]))^2] = 0. \qquad (5.7)$$

b) Let $H = L_p(\mu)$, *for some* $p \geq 2$, *and for* h *and* g *as in (3.10) and (3.11), suppose* $E[\int_0^t \|h(\cdot, s)\|_{L_r(\mu)} ds] < \infty$ *and* $E[(\int_0^t \|g(\cdot, s)\|_{L_q(\mu)} ds)^2] < \infty$ *for all* $t > 0$. *Then if* $\sup_s \|Z(s)\|_{\hat{H}} \leq 1$ *and* τ *is a stopping time bounded by a constant* c,

$$E[\sup_{s \leq t} \|Z_- \cdot Y(\tau + s) - Z_- \cdot Y(\tau)\|_L] \leq 2\sqrt{E[\int_\tau^{\tau+t} \|h(\cdot, s)\|_{L_r(\mu)} ds]} + E[(\int_\tau^{\tau+t} \|g(\cdot, s)\|_{L_q(\mu)} ds)^2]\alpha^2$$

and

$$\lim_{t \to 0} E[\int_\tau^{\tau+t} \|h(\cdot, s)\|_{L_r(\mu)} ds] + E[(\int_\tau^{\tau+t} \|g(\cdot, s)\|_{L_q(\mu)} ds)^2] = 0.$$

Now let $L = C([0,1]^d)$ with the sup norm. It is clear from the discussion at the beginning of this subsection, that we cannot let $\|f\|_{\hat{H}} = \sup_x \|f(x, \cdot)\|_H$ if we want interesting standard $(L, \hat{H})^{\#}$-semimartingales. It is sufficient, however, to give $\hat{H}$ some kind of Hölder norm. In particular, let M be an orthogonal martingale random measure and suppose the $\hat{\pi}_j$ defined as in (2.21) satisfy

$$\hat{\pi}_j(du \times ds) = h_j(u, s)\nu_j(du)ds.$$

Let $\frac{1}{p} + \frac{1}{q} = 1$, and suppose for each $t > 0$,

$$C_{k,j}(t) \equiv E\left[\left(\int_0^t \left(\int_U |h_j(u, s)|^q \nu_j(du)\right)^{\frac{1}{q}}\right)^{\frac{k}{j}}\right] < \infty.$$

For k and α satisfying $0 < \alpha \le 1$ and $k\alpha > d$, define

$$\|f\|_H = \sum_{j=2}^{k} \left(\int_U |f(u)|^{jp} \nu_j(du) \right)^{\frac{1}{jp}},$$

and

$$\|g\|_{\hat{H}} = \sup_x \|g(x,\cdot)\|_H + \sup_{x,y} \frac{\|g(x,\cdot) - g(y,\cdot)\|_H}{|x-y|^\alpha}.$$

Recall that if M is continuous, then $\nu_j = 0$ for $j > 2$ and H is just $L_{2p}(\nu_2)$. Now suppose $X \in S^0_{\hat{H}}$ and $\|X\|_{\hat{H}} \le 1$. Fix $x, y \in [0,1]^d$, and set

$$\begin{aligned} Z(t,x,y) &= X(\cdot,x,\cdot)_- \cdot M(t) - X(\cdot,y,\cdot)_- \cdot M(t) \\ &= \int_{U\times[0,t]} (X(s-,x,u) - Y(s-,y,u)) M(du \times ds). \end{aligned}$$

With reference to (2.22),

$$\begin{aligned} H_{k,j}(x,y) &= E\left[\left(\int_{U\times[0,t]} |X(s-,x,u) - X(s-,y,u)|^j \hat{\pi}_j(du \times ds) \right)^{\frac{k}{j}} \right] \\ &\le E\left[\left(\int_0^t \left(\int_U |X(s,x,u) - X(s,y,u)|^{jp} \nu_j(du) \right)^{\frac{1}{p}} \right.\right. \\ &\qquad \left.\left. \left(\int_U |h_j(u,s)|^q \nu_j(du) \right)^{\frac{1}{q}} \right)^{\frac{k}{j}} \right] \\ &\le |x-y|^{\alpha k} E\left[\left(\int_0^t \left(\int_U |h_j(u,s)|^q \nu_j(du) \right)^{\frac{1}{q}} \right)^{\frac{k}{j}} \right] \\ &= |x-y|^{\alpha k} C_{k,j}(t). \end{aligned}$$

Then with reference to (2.26),

$$E[|Z(t,x,y)|^k] \le K|x-y|^{\alpha k} \tag{5.8}$$

where K is the largest number satisfying

$$K \le C_1 k C_{k,2}^{\frac{1}{k}} K^{\frac{k-1}{k}} + \sum_{j=2}^{k} \binom{k}{j} C_{k,j}^{\frac{1}{k}} K^{\frac{k-j}{k}}.$$

The remainder of the proof of standardness is essentially the same as for the proof of the Kolmogorov continuity criterion. (Recall that the exponent αk on the right side of (5.8) is greater than d.) Note that any $x \in [0,1]^d$ can be represented as

$$x = \sum_{i=1}^{\infty} \frac{1}{2^i}(\theta_1^i(x), \ldots, \theta_d^i(x))$$

where $\theta_k^i(x)$ is 0 or 1. Let $x^0 = 0$ and

$$x^m = \sum_{i=1}^{m} \frac{1}{2^i}\theta^i(x)$$

It follows that

$$X(\cdot, x, \cdot)_- \cdot M(t) = X(\cdot, 0, \cdot) \cdot M(t) + \sum_{m=0}^{\infty} Z(t, x^{m+1}, x^m).$$

For each $\theta \in \{0,1\}^d$ and $m = 1, 2, \ldots$, let

$$\eta_m(\theta) = \sum_{\{y \in [0,1)^d : 2^m y \in \mathbf{Z}^d\}} |Z(t, y + \frac{1}{2^m}\theta, y)|^k.$$

Then

$$\sup_{x \in [0,1]^d} |X(\cdot, x, \cdot)_- \cdot M(t)| \le |X(\cdot, 0, \cdot) \cdot M(t)| + \sum_{m=1}^{\infty} \sum_{\theta \in \{0,1\}^d} \eta_m(\theta)^{\frac{1}{k}},$$

and since by (5.8)

$$\begin{aligned} E\left[\sum_{m=1}^{\infty} \sum_{\theta \in \{0,1\}^d} \eta_m(\theta)^{\frac{1}{k}}\right] &\le \sum_{m=1}^{\infty} \sum_{\theta \in \{0,1\}^d} E[\eta_m(\theta)]^{\frac{1}{k}} \\ &\le \sum_{m=1}^{\infty} 2^d \left[K\left(\frac{\sqrt{d}}{2^m}\right)^{\alpha k} 2^{md}\right]^{\frac{1}{k}} \\ &\le 2^d K^{\frac{1}{k}} d^{\frac{\alpha}{2}} \sum_{m=1}^{\infty} \left(\frac{1}{2^m}\right)^{\frac{\alpha k - d}{k}} < \infty, \end{aligned}$$

the stochastic boundedness of $\mathcal{H}_t^0$ follows.

5.4 Equicontinuity of stochastic integrals

The fact that the estimates in Lemma 5.6 along with those in Lemmas 3.3 and 3.4 tend to zero as $t \to 0$ will be needed for the uniqueness theorem in Section 7.. This fact holds very generally for standard $(L, \hat{H})^{\#}$-semimartingales, although we have not been able to prove that it always holds. The following lemma follows easily from Araujo and Giné [1], Theorem 3.2.8.

Lemma 5.7. *Let $\{\theta_i\}$ be iid with $P\{\theta_i = 1\} = P\{\theta_i = -1\} = \frac{1}{2}$, let $\{x_i\} \subset L$, and define $S_k = \sum_{i=1}^{k} \theta_i x_i$. Then for $p \geq 1$,*

$$2P\{\|S_n\|_L > a\} \geq P\{\max_{k\leq n} \|S_k\|_L > a\} \geq 2^{1-p}\left(1 - \frac{a^p(1+3^p)}{E[\|S_n\|_L^p]}\right).$$

Let $f_{p,n}(x_1, \ldots, x_n) = E[\|S_n\|_L^p]$. If $X_1, X_2, \ldots$ are L-valued random variables, independent of the $\{\theta_i\}$, and $\tilde{S}_k = \sum_{i=1}^{k} \theta_i X_i$, then

$$P\{\max_{k\leq n} \|\tilde{S}_k\|_L > a\} \geq 2^{-p} P\{f_{p,n}(X_1, \ldots, X_n) \geq 2(1+3^p)a^p\}.$$

Remark 5.8. By definition, if L is cotype p, then

$$f_{p,n}(x_1, \ldots, x_n) \geq c \sum_{i=1}^{n} \|x_i\|_L^p.$$

If L is uniformly convex, for each $\epsilon > 0$ there exists $\alpha(\epsilon) > 0$ such that

$$f_{2,n}(x_1, \ldots, x_n) \geq \alpha(\epsilon) \sum_{i=1}^{n} I_{[\epsilon,\infty)}(\|x_i\|_L). \tag{5.9}$$

Note that the cotype inequality implies (5.9) (with 2 replaced by p). Since $L_1(\nu)$ is cotype 2 (see Araujo and Giné [1], page 188) and $L_r(\nu), 1 < r < \infty$ is uniformly convex, (5.9) holds for all $L_r(\nu)$, $1 \leq r < \infty$. The inequality fails for $L_\infty(\nu)$.

Proof. If $\max_{1\leq i\leq n} \|x_i\|_L > 2a$, then $P\{\max_{k\leq n} \|S_k\|_L > a\} = 1$, so the bound c in Araujo and Giné [1], Theorem 3.2.8, can, in the present setting, be replaced by $2a$, and the first inequality follows. The second inequality is an immediate consequence of the first. □

Lemma 5.9. *Suppose that L satisfies (5.9) (possibly with $f_{2,n}$ replaced by $f_{p,n}$). Let Y be a standard $(L, \hat{H})^{\#}$-semimartingale, and define*

$$K(t,\delta) = \inf\{K : P\{\sup_{s\leq t} \|Z_- \cdot Y(s)\|_L \geq K\} \leq \delta, Z \in S_{\hat{H}}, \|Z\|_{\hat{H}} \leq 1\}.$$

Then for each $\delta > 0$, $\lim_{t\to 0} K(t,\delta) = 0$. More generally, let τ be a stopping time bounded by a constant and define

$$\begin{aligned} &K_\tau(t,\delta) \\ &\quad = \inf\{K : P\{\sup_{s\leq t} \|Z_- \cdot Y(\tau+s) - Z_- \cdot Y(\tau)\|_L \geq K\} \leq \delta, \\ &\qquad\qquad Z \in S_{\hat{H}}, \|Z\|_{\hat{H}} \leq 1\}. \end{aligned}$$

Then $\lim_{t\to 0} K_\tau(t,\delta) = 0$.

Proof. Consider the case $\tau = 0$. The general case is similar. Suppose that $\lim_{t\to 0} K(t,\delta) > 0$. Then there exists $t_n \to 0$, $K > 0$, and $Z_n \in S_{\hat{H}}$ with $\|Z_n\|_{\hat{H}} \leq 1$, such that

$$P\{\sup_{s\leq t_n} \|Z_{n-}\cdot Y(s)\|_L \geq K\} \geq \frac{2}{3}\delta.$$

Since $Z_{n-}\cdot Y$ is right continuous and vanishes at zero, we can select $0 < r_n < t_n$ such that

$$P\{\sup_{r_n\leq s\leq t_n} \|Z_{n-}\cdot Y(s) - Z_{n-}\cdot Y(r_n)\|_L \geq \frac{K}{2}\} \geq \frac{\delta}{2}.$$

Let

$$\sigma_n = \inf\{s > r_n : \|Z_{n-}\cdot Y(s) - Z_{n-}\cdot Y(r_n)\|_L \geq \frac{K}{2}\},$$

and note that

$$P\{\|Z_{n-}\cdot Y(\sigma_n \wedge t_n) - Z_{n-}\cdot Y(r_n)\|_L \geq \frac{K}{2}\} \geq \frac{\delta}{2}.$$

Without loss of generality, we can assume that there is a sequence $\{\theta_i\}$, as in Lemma 5.7, that is independent of the Z_n and is $\mathcal{F}_0$-measurable. (If not, enlarge the sample space and the filtration to include such a sequence and note that the stochastic boundedness of $\mathcal{H}_t^0$ is unaffected.)

Select a subsequence satisfying $t_{n_1} = t_1$ and $t_{n_{k+1}} < r_{n_k}$, and define

$$Z^{(m)} = \sum_{k=1}^{m} \theta_k I_{[r_{n_k}, t_{n_k}\wedge\sigma_{n_k})} Z_{n_k}.$$

Then $Z^{(m)}$ is cadlag and adapted, $\|Z\|_{\hat{H}} \leq 1$, and setting $X_k = Z_{n_k-}\cdot Y(\sigma_{n_k}\wedge t_{n_k}) - Z_{n_k-}\cdot Y(r_{n_k})$

$$Z_-^{(m)}\cdot Y(t_1) = \sum_{k=1}^{m} \theta_k X_k.$$

By Lemma 5.7 and (5.9)

$$\lim_{m\to\infty} P\{\|Z_-^{(m)}\cdot Y(t_1)\|_L > a\} \geq 2^{-p} P\{\|X_k\|_L > \frac{K}{4}, i.o.\} \geq 2^{-p-1}\delta.$$

Since a is arbitrary, this estimate violates the stochastic boundedness of $\mathcal{H}_t^0$, and the lemma follows. □

The following variation on the above lemma may be useful in proving uniqueness in spaces in which (5.9) fails.

Lemma 5.10. *Let Y be a standard $(L,\hat{H})^{\#}$-semimartingale, let τ be a stopping time bounded by a constant, and let $\Gamma \subset \hat{H}$ be compact. Define*

$$\begin{aligned} K_\tau(t,\delta,\Gamma) &\\ = \inf\{K : P\{\sup_{s\le t} \|Z_- \cdot Y(\tau+s) - Z_- \cdot Y(\tau)\|_L \ge K\} \le \delta, &\\ Z \in S_{\hat{H}}, Z(s) \in \Gamma, s \le t\}. \end{aligned}$$

Then for each $\delta > 0$, $\lim_{t\to 0} K_\tau(t,\delta,\Gamma) = 0$.

Proof. Take $\tau = 0$. The general case is similar. Let Z_n, t_{n_k}, and τ_{n_k} be as in the proof of the previous lemma, and define

$$Z = \sum_{k=1}^{\infty} I_{(\tau_{n_k}, t_{n_k} \wedge \sigma_{n_k}]} Z_{n_k}.$$

Then Z is predictable and takes values in the compact set Γ. Consequently, $Z \cdot Y$ is defined and cadlag. In particular, $\lim_{t\to 0} Z \cdot Y(t) = 0$. But for $\tau_{n_k} \le t < t_{n_k}$, $Z \cdot Y(t) - Z \cdot Y(\tau_{n_k}) = Z_{n_k -} \cdot Y(t) - Z_{n_k -} \cdot Y(\tau_{n_k})$ so

$$\limsup_{t\to 0} \sup_{s \le t} \|Z \cdot Y(t) - Z \cdot Y(s)\|_L \ge \frac{K}{2}$$

with probability at least $\frac{\delta}{2}$, contradicting the right continuity at 0 and giving the result. □

6. Consequences of the uniform tightness condition

Let $\{Y_n\}$ be a sequence of $H^{\#}$-semimartingales, and let $\hat{\mathcal{H}}_t^0$ be as in Theorem 4.2, that is,

$$\hat{\mathcal{H}}_t^0 = \cup_n \{|X_{n-} \cdot Y_n(t)| : X_n \in S_0^n, \sup_{s\le t} \|X_n(s)\|_H \le 1\},$$

where S_0^n is the collection of simple, finite dimensional H-valued, $\{\mathcal{F}_t^n\}$-adapted processes. The sequence $\{Y_n\}$ is *uniformly tight (UT)* if $\hat{\mathcal{H}}_t^0$ is stochastically bounded for each $t > 0$. For real-valued semimartingales, this condition appears first in Stricker [56] where it is shown to imply a type of relative compactness for the sequence of semimartingales previously studied by Meyer and Zheng [51] under somewhat different conditions. Jakubowski [26] develops the topological properties of the corresponding convergence, and Kurtz [39] characterizes the convergence in terms of convergence in the Skorohod topology of a time-changed sequence. Uniform tightness is the basic condition in the stochastic integral convergence results of Jakubowski, Mémin and Pagès [27].

As noted previously, uniform tightness is equivalent to the stochastic boundedness of

$$\hat{\mathcal{H}}_t = \cup_n \{\sup_{s \le t} |X_{n-} \cdot Y_n(s)| : X_n \in \mathcal{S}^n, \sup_{s \le t} \|X_n(s)\|_H \le 1\}, \tag{6.1}$$

for each $t > 0$. In particular, if $\{Y_n\}$ is uniformly tight, for each $t > 0$ and $\delta > 0$ there exists a $K(t, \delta)$ such that

$$P\{\sup_{s \le t} |X_{n-} \cdot Y_n(s)| \ge K(t, \delta)\} \le \delta \tag{6.2}$$

for each n and all $X_n \in \mathcal{S}^n$ satisfying $\|X_n(s)\|_H \le 1$. The following lemma is an immediate consequence of (6.2).

Lemma 6.1. *Let $\{Y_n\}$ be as above and let X_n be an H-valued, $\{\mathcal{F}_t^n\}$-adapted, cadlag process. Let $\tau_n^M = \inf\{t : \|X_n\|_H \ge M\}$. Then for all $t, \delta > 0$,*

$$P\{\sup_{s \le t \wedge \tau_n^M} |X_{n-} \cdot Y_n(s)| \ge MK(t, \delta)\} \le \delta.$$

The next result gives conditions under which stochastic integrals of uniformly tight sequences define uniformly tight sequences.

Proposition 6.2. *Let H_0 be a Banach space of functions on U with the property that $g \in H_0$ and $h \in H$ implies $gh \in H$ and $\|gh\|_H \le \|g\|_{H_0} \|h\|_H$. Let $\{Y_n\}$ be a uniformly tight sequence of $H^\#$-semimartingales, and for each n, let X_n be an H-valued, cadlag, $\mathcal{F}_t^n$-adapted process such that for each t, the sequence $\{\sup_{s \le t} \|X_n(s)\|_H\}$ is stochastically bounded. Define*

$$Z_n(g, \cdot) = gX_{n-} \cdot Y_n$$

for $g \in H_0$. Then $\{Z_n\}$ is a uniformly tight sequence of $H_0^\#$-semimartingales. In particular (taking $H_0 = \mathbf{R}$), $\{X_{n-} \cdot Y_n\}$ is a uniformly tight sequence of $\mathbf{R}$-valued semimartingales.

Proof. Let $\tilde{X}_n$ be a simple, $\{\mathcal{F}_t^n\}$-adapted, H_0-valued process satisfying $\|\tilde{X}_n(s)\|_{H_0} \le 1$. Then $\tilde{X}_{n-} \cdot Z_n = \tilde{X}_n X_{n-} \cdot Y_n$, and

$$P\{\sup_{s \le t} |\tilde{X}_{n-} \cdot Z_n(s)| \ge MK(t, \delta)\} \le \delta + P\{\sup_{s \le t} \|X_n(s)\|_H \ge M\}.$$

In particular, select M such that $P\{\sup_{s \le t} \|X_n(s)\|_H \ge M\} \le \delta$, and define $\tilde{K}(t, 2\delta) = MK(t, \delta)$. □

We have the following analogue of Lemma 3.10.

Lemma 6.3. *Let $\{Y_n^1\}$ be a uniformly tight sequence of $H_1^\#$-semimartingales and $\{Y_n^2\}$ a uniformly tight sequence of $H_2^\#$-semimartingales, where the conditions on Y_n^1 and Y_n^2 are with respect to the same filtration $\{\mathcal{F}_t^n\}$. Define $H = \{(\varphi_1, \varphi_2) : \varphi_1 \in H_1, \varphi_2 \in H_2\}$, with $\|\varphi\|_H = \|\varphi_1\|_{H_1} + \|\varphi_2\|_{H_2}$ and $Y_n(\varphi, t) = Y_n^1(\varphi_1, t) + Y_n^2(\varphi_2, t)$ for $\varphi = (\varphi_1, \varphi_2)$. Then $\{Y_n\}$ is a uniformly tight sequence of $H^\#$-semimartingales.*

We will need a number of technical lemmas regarding convergence in distribution in the Skorohod topology. For any nonnegative, nondecreasing function a defined on $[0,\infty)$, we define $a^{-1}(t) = \inf\{u : a(u) > t\}$. Let $T_c[0,\infty)$ be the collection of continuous, nondecreasing functions γ satisfying $\gamma(0) = 0$ and $\lim_{t\to\infty} \gamma(t) = \infty$. (Note that γ^{-1} is right continuous and strictly increasing and that $\gamma(t) = \inf\{u : \gamma^{-1}(u) > t\}$.) If $\{X_n\}$ is a sequence of processes in $D_E[0,\infty)$ and $\{\gamma_n\}$ is a sequence of processes in $T_c[0,\infty)$, then we say that $\{\gamma_n\}$ *regularizes* $\{X_n\}$ if $\{\gamma_n^{-1}(t)\}$ is stochastically bounded for each $t > 0$ and the sequence $\{(X_n \circ \gamma_n, \gamma_n)\}$ is relatively compact in $D_{E\times\mathbf{R}}[0,\infty)$. Note that if $Y_n = X_n \circ \gamma_n$, then $X_n(t) = Y_n(\gamma_n^{-1}(t))$. The following lemma is a restatement of Theorem 1.1b,c of Kurtz [39] using the above terminology.

Lemma 6.4. *Suppose that $\{\gamma_n\}$ regularizes $\{X_n\}$ and that $(X_n \circ \gamma_n, \gamma_n) \Rightarrow (Y, \gamma)$. Then (by the Skorohod representation theorem) there exists a proability space on which are defined processes $\{(\hat{Y}_n, \hat{\gamma}_n)\}$ converging almost surely to a process $(\hat{Y}, \hat{\gamma})$ in the Skorohod topology on $D_{E\times\mathbf{R}}[0,\infty)$ such that $(\hat{Y}_n, \hat{\gamma}_n)$ has the same distribution as $(X_n \circ \gamma_n, \gamma_n)$ (and hence $\hat{X}_n = \hat{Y}_n \circ \hat{\gamma}_n^{-1}$ has the same distribution as X_n) and with probability 1, $\hat{X}_n(t) \to \hat{X}(t) \equiv \hat{Y} \circ \hat{\gamma}^{-1}(t)$ for all but countably many $t \geq 0$.*

If γ is strictly increasing, then $X_n \Rightarrow Y \circ \gamma^{-1}$.

Remark 6.5. Under the conditions of Lemma 6.4, there exists a countable set $D \subset [0,\infty)$ such that for $(t_1, \ldots, t_m) \subset [0,\infty) - D$, $(X_n(t_1), \ldots, X_n(t_m)) \Rightarrow (X(t_1), \ldots, X(t_m))$, for $X = Y \circ \gamma^{-1}$.

It follows from results of Stricker [56] and Kurtz [39] that any uniformly tight sequence of real-valued semimartingales can be regularized. We will prove a corresponding result for $H^\#$-semimartingales. For a given $\mathbf{R}$-valued, cadlag process X and $u < v$, $N(u,v,t)$ will denote the number of upcrossings of the interval (u,v) by X before time t, and for $u > v$, $N(u,v,t)$ will denote the number of downcrossings of the interval (v,u) before time t. Note, for example, that $|N(u,v,t) - N(v,u,t)| \leq 1$.

Lemma 6.6. *Let $\{X_n\}$ be a sequence of cadlag processes with X_n adapted to $\{\mathcal{F}_t^n\}$, and let $N_n(u,v,t)$ count down/upcrossings for X_n. Suppose that for each $(u,v,t) \in \mathbf{R}^2 \times [0,\infty)$, $u \neq v$, $\{N_n(u,v,t)\}$ is stochastically bounded and that for each $t \geq 0$, $\{\sup_{s\leq t} |X_n(s)|\}$ is stochastically bounded. Then for each n there exists a strictly increasing, $\{\mathcal{F}_t^n\}$-adapted process C_n such that for each $t > 0$, $\{C_n(t)\}$ is stochastically bounded and $\{\gamma_n\}$ defined by $\gamma_n = C_n^{-1}$ regularizes $\{X_n\}$, that is $\{(X_n \circ \gamma_n, \gamma_n)\}$ is relatively compact in the Skorohod topology.*

Proof. Let $\{(u_l, v_l), l \geq 1\}$ be some ordering of $\{(u,v) : u, v \in \mathbf{Q}, u \neq v\}$. By the stochastic boundedness assumptions, for $k, l = 1, 2, \ldots$ there exist $0 < c_{kl} \leq 1$ such that

$$\sup_n P\{c_{kl} N_n(u_l, v_l, k) > 2^{-(k+l)}\} \leq 2^{-(k+l)}.$$

Define

$$C_n(t) = 1 + t + \sum_{l=1}^{\infty} \int_0^t c_{[s]+1,l} dN_n(u_l, v_l, s).$$

Note that

$$H_{n,L}(t) = \sum_{l=L+1}^{\infty} \int_0^t c_{[s]+1,l} dN_n(u_l, v_l, s) \leq \sum_{l=L+1}^{\infty} \sum_{k=1}^{[t]+1} c_{kl} N_n(u_l, v_l, k)$$

and

$$P\{H_{n,L}(t) > 2^{-L}\} \leq 2^{-L},$$

so

$$P\{C_n(t) > t + a + 2\} \leq 2^{-L} + P\{\sum_{l=1}^{L} N_n(u_l, v_l, t) > a\},$$

and the stochastic boundedness of $\{C_n(t)\}$ follows from the assumed stochastic boundedness of $\{N_n(u_l, v_l, t)\}$.

Let $\gamma_n = C_n^{-1}$, and define $Z_n(t) = X_n(\gamma_n(t))$. Let $\epsilon > 0$ be rational. Note that each time X_n crosses an interval (u_l, v_l), C_n jumps and that each jump of C_n corresponds to an interval on which γ_n, and hence Z_n, is constant.

Define $\tau_0^{n,\epsilon} = 0$,

$$\tau_{k+1}^{n,\epsilon} = \inf\{t > \tau_k^{n,\epsilon} : Z_n(t) \notin [[Z_n(\tau_k^{n,\epsilon})/\epsilon]\epsilon - \epsilon, [Z_n(\tau_k^{n,\epsilon})/\epsilon] + 2\epsilon]\}$$

and

$$Z_n^\epsilon(t) = Z_n(\tau_k^{n,\epsilon}), \quad \tau_k^{n,\epsilon} \leq t < \tau_{k+1}^{n,\epsilon}$$

so that $|Z_n(t) - Z_n^\epsilon(t)| \leq 2\epsilon$. To show that $\{Z_n\}$ is relatively compact, it is enough to show that $\{Z_n^\epsilon\}$ is relatively compact. Since the Z_n^ϵ are piecewise constant, it is enough to show that $\{\sup_{s \leq t} |Z_n^\epsilon(s)|\}$ is stochastically bounded (which follows from the fact that $\sup_{s \leq t} |Z_n^\epsilon(s)| \leq \sup_{s \leq t} |X_n(s)|$) and that $\{\min\{\tau_{k+1}^{n,\epsilon} - \tau_k^{n,\epsilon} : \tau_k^{n,\epsilon} < t\}, n = 1, 2, \ldots\}$ is stochastically bounded away from 0, that is, for each $\delta > 0$ there exists an $\eta > 0$ such that $P\{\min\{\tau_{k+1}^{n,\epsilon} - \tau_k^{n,\epsilon} : \tau_k^{n,\epsilon} < t\} \leq \eta\} \leq \delta$. Let $c(u, v, k) = c_{kl}$ if $(u, v) = (u_l, v_l)$. Define $\eta(a, t) = \min\{c(i\epsilon, j\epsilon, [t] + 1) : |i\epsilon|, |j\epsilon| \leq a, j = i \pm 1\}$. Then

$$P\{\min\{\tau_{k+1}^{n,\epsilon} - \tau_k^{n,\epsilon} : \tau_k^{n,\epsilon} < t\} < \eta(a,t)\} \leq P\{\sup_{s \leq t} |Z_n(s)| > a\}$$

since $\tau_1^{n,\epsilon} \geq 1$ and at each time $\tau_k^{n,\epsilon}$ with $k > 0$, Z_n finishes a downcrossing or an upcrossing of an interval of the form $[i\epsilon, (i+1)\epsilon]$ and is constant for an interval of length at least $c(i\epsilon,(i+1)\epsilon,[\tau_k^{n,\epsilon}]+1)$ (in the case of an upcrossing) after time $\tau_k^{n,\epsilon}$ implying $\tau_{k+1}^{n,\epsilon}-\tau_k^{n,\epsilon}\geq c(i\epsilon,(i+1)\epsilon,[\tau_k^{n,\epsilon}]+1)$. □

Following the proof of Theorem 2 of Stricker [56], we have the following lemma.

Lemma 6.7. *Let $\{Y_n\}$ be a uniformly tight sequence of* $\mathbf{R}$*-valued semimartingales, and let $\{N_n\}$ be as in Lemma 6.6. Then for $u\neq v\in\mathbf{R}$ and $t>0$, $\{N_n(u,v,t)\}$ is stochastically bounded.*

Proof. Suppose $u<v$. Define $\tau_1^n=\inf\{s\geq 0: X_n(s)\leq u\}$, $\sigma_k^n=\inf\{s>\tau_k^n: X_n(s)\geq v\}$, and for $k>1$, $\tau_k^n=\inf\{s>\sigma_k^n: X_n(s)\leq u\}$. Note that $N_n(u,v,t)=\max\{k:\sigma_k^n\leq t\}$. Define $X_n=\sum_k I_{[\tau_k^n,\sigma_k^n)}$. Then $X_{n-}\cdot Y_n(t)\geq(v-u)N_n(u,v,t)-2\sup_{s\leq t}|Y_n(s)|$ and hence, $(v-u)N_n(u,v,t)\leq|X_{n-}\cdot Y_n(t)|+2\sup_{s\leq t}|Y_n(s)|$. Since $\{|X_{n-}\cdot Y_n(t)|\}$ is stochastically bounded by the definition of the uniform tightness of $\{Y_n\}$ and $\{\sup_{s\leq t}|Y_n(s)|\}$ is stochastically bounded by (6.1), the stochastic boundedness of $\{N_n(u,v,t)\}$ follows. The proof for $u>v$ is essentially the same. □

Lemmas 6.6 and 6.7 immediately give the following:

Lemma 6.8. *For each n, let Y_n be an* $\mathbf{R}$*-valued $\{\mathcal{F}_t^n\}$-semimartingale, and let the sequence $\{Y_n\}$ be uniformly tight. Then there exist $\{\mathcal{F}_t^n\}$-adapted C_n satisfying $C_n(t)-C_n(s)\geq t-s$, $t>s\geq 0$, and $\{C_n(t)\}$ stochastically bounded for each $t>0$, such that for $\gamma_n=C_n^{-1}$, $\{Y_n\circ\gamma_n\}$ is relatively compact.*

The next lemma gives conditions under which a process with values in a product space can be regularized.

Lemma 6.9. *For each $k=1,2,\ldots$, let (E_k,r_k) be a complete, separable metric space, and for each $n=1,2,\ldots$, let X_k^n be an $\{\mathcal{F}_t^n\}$-adapted process in $D_{E_k}[0,\infty)$. Let E denote the product space $E_1\times E_2\times\cdots$. Suppose, for each k, that $\{X_k^n\}$ is relatively compact (in the sense of convergence in distribution in the Skorohod topology). (This assumption implies that the sequence of E-valued processes $\{(X_1^n,X_2^n,\ldots)\}$ is relatively compact in $D_{E_1}[0,\infty)\times D_{E_2}[0,\infty)\times\cdots$ but not necessarily in $D_E[0,\infty)$). Then there exist strictly increasing processes $\{C_n\}$, such that $C_n(0)\geq 0$; for $t>s$, $C_n(t)-C_n(s)\geq t-s$; for each $t>0$, $\{C_n(t)\}$ is stochastically bounded; for each n, C_n is $\{\mathcal{F}_t^n\}$-adapted; and defining $\gamma_n=C_n^{-1}$, the sequence $\{(\hat{X}_1^n,\hat{X}_2^n,\ldots)\}$ obtained by setting $\hat{X}_k^n=X_k^n\circ\gamma_n$ is relatively compact in $D_E[0,\infty)$.*

Proof. Recall that a sequence converging in $D_{E_1}[0,\infty)\times D_{E_2}[0,\infty)\times\cdots$ fails to converge in $D_E[0,\infty)$ if discontinuities in two of the components "coalesce". With that in mind, C_n should be constructed to slow down the time scale after a jump in such a way that coalescence of discontinuities is prevented. Let $N_k^n(t,r)$, $t,r>0$ be the cardinality of the set $\{s: s\leq t, r_k(X_k^n(s),X_k^n(s-))\geq$

$r\}$. The relative compactness of $\{X_k^n\}$ ensures the stochastic boundedness of $\{N_k^n(t,r)\}$ for each choice of t, r and k. For $m = 0,1,2,\ldots$ and $l = 1,2,\ldots$, let $c_k(l, 2^{-m}) = \sup\{c : c \le 1, \sup_n P\{cN_k^n(l,2^{-m}) > 2^{-m}\} \le 2^{-m}\}$ and for $l-1 \le s < l$, define $c_k(s,r) = c_k(l,2^{-m})$ for $2^{-m} \le r < 2^{-(m-1)}$, $m \ge 1$, and $c_k(s,r) = c_k(l,1)$ for $r \ge 1$. Then $C_n^k(t) = \sum_{s\le t} c_k(s, r_k(X_k^n(s), X_k^n(s-))$ converges and

$$P\{C_n^k(l) \ge x\} \le P\{N_k^n(l,2^{-m}) \ge x - l\} + l2^{-m}. \tag{6.3}$$

It follows from the stochastic boundedness of $\{N_k^n(l,2^{-m})\}$ and the fact that m is arbitrary, that $\lim_{x\to\infty} \sup_n P\{C_n^k(l) \ge x\} = 0$. Finally, let $0 < a_{kl} \le 1$ satisfy

$$\sup_n P\{a_{kl}(C_n^k(l) - C_n^k(l-1)) \ge 2^{-k-l}\} \le 2^{-k-l}.$$

Define $a_k(s) = a_{kl}$, $l-1 < s \le l$ and

$$C_n(t) = t + \sum_k \sum_{s\le t} a_k(s) c_k(s, r_k(X_k^n(s), X_k^n(s-)). \tag{6.4}$$

Noting that $C_n(l) = l + \sum_{m=1}^{l} \sum_k a_{km}(C_n^k(m) - C_n^k(m-1))$, the stochastic boundedness of $\{C_n(t)\}$ follows, as in (6.3), from the stochastic boundedness of the $\{C_n^k(t)\}$ and the definition of the a_k.

Note that $\gamma_n = C_n^{-1}$ is continuous, in fact, absolutely continuous with $\gamma_n' \le 1$. Setting $\hat{X}_k^n = X_k^n \circ \gamma_n$, we have $\{(\gamma_n, \hat{X}_1^n, \hat{X}_2^n, \ldots)\}$ relatively compact in $S \equiv D_{\mathbf{R}}[0,\infty) \times D_{E_1}[0,\infty) \times D_{E_2}[0,\infty) \times \cdots$. The proof of the lemma follows by showing that any subsequence that converges in S also converges in $D_{\mathbf{R}\times E}[0,\infty)$. The Skorohod representation theorem (for example, Ethier and Kurtz [17, Theorem 2.1.8]) and the characterization of convergence in the Skorohod topology given in Proposition 2.6.5 of Ethier and Kurtz [17] can be used to complete the proof. □

Lemma 6.10. *Suppose that $\{X_n\}$ is relatively compact in $D_E[0,\infty)$ and that C_n is nonnegative and satisfies $C_n(t) - C_n(s) \ge \alpha(t-s)$, $t > s \ge 0$ for some $\alpha > 0$. Define $\gamma_n = C_n^{-1}$ and $\hat{X}_n = X_n \circ \gamma_n$. Then $\{\hat{X}_n\}$ is relatively compact.*

Proof. Note that γ_n is differentiable with $\gamma_n' \le \frac{1}{\alpha}$. Let $w'(x,\delta,T)$ denote the modulus of continuity for $D_E[0,\infty)$ defined in (3.6.2) of Ethier and Kurtz [17]. Then

$$w'(\hat{X}_n, \delta, T) \le w'(X_n, \delta/\alpha, T/\alpha).$$

This estimate and the relative compactness of $\{X_n\}$ implies the relative compactness of $\{\hat{X}_n\}$ by Theorem 7.2 of Ethier and Kurtz [17]. □

The next lemma says that if a sequence of time changes regularized a sequence of processes then a sequence of time changes that grows more slowly will also regularize the sequence of processes.

Lemma 6.11. *Let X_n be cadlag, E-valued, and $\{\mathcal{F}_t^n\}$-adapted. Let C_n^1 and C_n^2 be strictly increasing and $\{\mathcal{F}_t^n\}$-adapted. Define $\gamma_n^1(t) = \inf\{u : C_n^1(u) > t\}$ and $\gamma_n(t) = \inf\{u : C_n^1(u) + C_n^2(u) > t\}$. Suppose $\{X_n \circ \gamma_n^1\}$ is relatively compact. Then $\{X_n \circ \gamma_n\}$ is relatively compact.*

Proof. Let $0 = t_0 < t_1 < \cdots$. Then there exist $0 = s_0 < s_1 < \cdots$ satisfying $\gamma_n(s_k) = \gamma_n^1(t_k)$ and $t_{k+1} - t_k \leq s_{k+1} - s_k$. It follows that $w'(X_n \circ \gamma_n, \delta, t) \leq w'(X_n \circ \gamma_n^1, \delta, t)$, where w' is the modulus of continuity defined in (3.6.2) in Ethier and Kurtz [17]. Since $\gamma_n(t) \leq \gamma_n^1(t)$, for any compact set $K \subset E$, $P\{X_n \circ \gamma_n(s) \in K, s \leq t\} \geq P\{X_n \circ \gamma_n^1(s) \in K, s \leq t\}$. The lemma follows by Theorem 3.7.2 and Remark 3.7.3 of Ethier and Kurtz [17]. □

Lemma 6.12. *If $\gamma_n = C_n^{-1}$ regularizes $\{X_n\}$, then for $a > 0$, γ_n^a given by $\gamma_n^a(t) = \inf\{s : aC_n(s) > t\}$ regularizes $\{X_n\}$.*

Proof. Note that $\gamma_n^a(t) = \gamma_n(t/a)$ and that if $\{Z_n\}$ is relatively compact, then $\{Z_n(\cdot/a)\}$ is relatively compact. □

Lemma 6.13. *Suppose for each $k = 1, 2, \ldots$ and $n = 1, 2, \ldots$ C_n^k is $\{\mathcal{F}_t^n\}$-adapted, $C_n^k(t) - C_n^k(s) \geq t - s$, $t > s \geq 0$, for each $t > 0$, $\{C_n^k(t)\}$ is stochastically bounded, and $\gamma_n^k(t) = \inf\{u : C_n^k(u) > t\}$ regularizes $\{X_n^k\}$ in $D_{E^k}[0,\infty)$. Then there exists $\{\mathcal{F}_t^n\}$-adapted C_n such that $C_n(t) - C_n(s) \geq t - s$, $t > s \geq 0$, for each $t > 0$, $\{C_n(t)\}$ is stochastically bounded, and $\gamma_n = C_n^{-1}$ regularizes $\{(X_n^1, X_n^2, \ldots)\}$ in $D_E[0,\infty)$, $E = E_1 \times E_2 \times \cdots$.*

Proof. First construct a $\tilde{C}_n$ such that for $\tilde{\gamma}_n = \tilde{C}_n^{-1}$, $\{(X_n^1 \circ \tilde{\gamma}_n, X_n^2 \circ \tilde{\gamma}_n, \ldots)\}$ is relatively compact in $D_{E_1}[0,\infty) \times D_{E_2}[0,\infty) \times \cdots$ and then perturb $\tilde{C}_n$ by a process constructed as in (6.4) to obtain the desired C_n. □

The next result extends Lemma 6.8 to sequences of $H^\#$ semimartingales.

Proposition 6.14. *Let $\{Y_n\}$ be a uniformly tight sequence of $H^\#$-semimartingales, Y_n adapted to $\{\mathcal{F}_t^n\}$, and for each n, let U_n be an $\{\mathcal{F}_t^n\}$-adapted process in $D_E[0,\infty)$ where E is a complete, separable metric space. Suppose $\{U_n\}$ is relatively compact (in the sense of convergence in distribution in the Skorohod topology). Then there exist strictly increasing, $\{\mathcal{F}_t^n\}$-adapted processes C_n, with $C_n(0) \geq 0$, $C_n(t+h) - C_n(t) \geq h$, $t, h \geq 0$, and $\{C_n(t)\}$ stochastically bounded for all $t \geq 0$, such that, defining $\gamma_n = C_n^{-1}$ and $\hat{Y}_n(\varphi, t) = Y_n(\varphi, \gamma_n(t))$, $\{\hat{Y}_n\}$ is uniformly tight and $\{(U_n \circ \gamma_n, \hat{Y}_n)\}$ is relatively compact.*

Proof. For C_n with the desired properties, the uniform tightness of $\{\hat{Y}_n\}$ follows from the fact that $\gamma_n' \leq 1$. By the uniform tightness of $\{\hat{Y}_n\}$, to prove relative compactness, it is enough to prove relative compactness of $\{(U_n \circ \gamma_n, \hat{Y}_n(\varphi_1, \cdot), \hat{Y}_n(\varphi_2, \cdot), \ldots)\}$ in $D_{E \times \mathbb{R}^\infty}[0,\infty)$ for a dense sequence $\{\varphi_k\}$ or for a sequence whose finite linear combinations are dense. By

Lemma 6.8, for each $k = 1, 2, \ldots$ there exists a strictly increasing, $\{\mathcal{F}_t^n\}$-adapted process $\tilde{C}_n^k$ such that $\tilde{\gamma}_n^k$ defined by $\tilde{\gamma}_n^k(t) = \inf\{s : \tilde{C}_n^k(s) > t\}$ regularizes $\{Y_n(\varphi_k, \cdot)\}$. Letting $\iota(t) = t$, there exist $a_k > 0$ such that $\tilde{C}_n = \iota + \sum_k a_k \tilde{C}_n^k$ exists and has the property that for each $t > 0$, $\{\tilde{C}_n(t)\}$ is stochastically bounded. Setting $\tilde{\gamma}_n = \tilde{C}_n^{-1}$, it follows from Lemmas 6.11 and 6.12, that $\{(U_n \circ \tilde{\gamma}_n, Y_n(\varphi_1, \tilde{\gamma}_n(\cdot)), Y_n(\varphi_2, \tilde{\gamma}_n(\cdot)), \ldots)\}$ is relatively compact in $D_E[0,\infty) \times D_{\mathbf{R}}[0,\infty) \times D_{\mathbf{R}}[0,\infty) \times \cdots$. As in Lemma 6.9, we must modify $\tilde{C}_n$ to ensure that discontinuities for different components do not coalesce. Setting, $X_0^n = U_n$ and $X_k^n = Y_n(\varphi_k, \cdot)$, let $N_k^n(t,r)$ be as in the proof of Lemma 6.9. Note that $\{N_0^n(t,r)\}$ is stochastically bounded by the assumption that $\{U_n\}$ is relatively compact and that for $k > 0$, $\{N_k^n(t,r)\}$ is stochastically bounded by the assumption that $\{Y_n\}$ is uniformly tight. (In particular, this assertion follows from the fact that $\{[Y_n]_t\}$ is stochastically bounded for each $t > 0$.) Consequently, we can add the analogue of the right side of (6.4) to $\tilde{C}_n$ defined above to obtain a nonnegative, strictly increasing, $\{\mathcal{F}_t^n\}$-adapted process C_n such that for $\gamma_n = C_n^{-1}$, $\{(U_n \circ \gamma_n, Y_n(\varphi_1, \gamma_n(\cdot)), Y_n(\varphi_2, \gamma_n(\cdot)), \ldots)\}$ is relatively compact in $D_E[0,\infty) \times D_{\mathbf{R}}[0,\infty) \times D_{\mathbf{R}}[0,\infty) \times \cdots$ (by Lemma 6.11) and discontinuities do not coalesce, so that, in fact, relative compactness is in $D_{E\times\mathbf{R}^\infty}[0,\infty)$ as desired. □

Lemma 6.15. *For each n, let Z_n be an $\{\mathcal{F}_t^n\}$-adapted process in $D_L[0,\infty)$, and suppose that the compact containment condition holds, that is, for each $\epsilon > 0$ and $t > 0$, there exists a compact $K_{\epsilon,t} \subset L$ such that $P\{Z_n(s) \in K_{\epsilon,t}, s \le t\} \ge 1 - \epsilon$. Let $\{\lambda_i\} \subset L^*$ satisfy $\sup_i \langle \lambda_i, x\rangle = \|x\|_L$ for all $x \in L$, and suppose for each i, the sequence $\{\langle \lambda_i, Z_n\rangle\}$ is uniformly tight. Then there exist $\{\mathcal{F}_t^n\}$-adapted C_n satisfying $C_n(t) - C_n(s) \ge t - s$, $t > s \ge 0$, and $\{C_n(t)\}$ stochastically bounded for each $t > 0$, such that for $\gamma_n = C_n^{-1}$, $\{Z_n \circ \gamma_n\}$ is relatively compact.*

Proof. By Lemma 6.8 there is a C_n^i such that the corresponding γ_n^i regularizes $\{\langle \lambda_i, Z_n\rangle\}$, and by Lemma 6.13, there exists a C_n such that $\{(\langle \lambda_1, Z_n \circ \gamma_n\rangle, \langle \lambda_2, Z_n \circ \gamma_n\rangle, \ldots)\}$ is relatively compact in $D_{\mathbf{R}^\infty}[0,\infty)$. The relative compactness of $\{Z_n \circ \gamma_n\}$ then follows from Theorem 3.9.1 of Ethier and Kurtz [17]). □

The following lemma generalizes Proposition 4.3 of Kurtz and Protter [42].

Lemma 6.16. *For each $n = 1, 2, \ldots$, let U_n be an $\{\mathcal{F}_t^n\}$-adapted process in $D_L[0,\infty)$ and let Y_n be an $\{\mathcal{F}_t^n\}$-$(L, \hat{H})^\#$-semimartingale. Suppose that $\{Y_n\}$ is uniformly tight and that $\{(U_n, Y_n)\}$ is relatively compact in the sense that $\{(U_n, Y_n(\varphi_1, \cdot), \ldots, Y_n(\varphi_m, \cdot))\}$ is relatively compact in $D_{L\times\mathbf{R}^m}[0,\infty)$ for any finite collection $\varphi_1, \ldots, \varphi_m \in H$. Let X_n, be an $\{\mathcal{F}_t^n\}$-adapted process in $D_{\hat{H}}[0,\infty)$. Define*

$$Z_n(t) = U_n(t) + X_{n-} \cdot Y_n(t)\,.$$

Let C_n be a strictly increasing, $\{\mathcal{F}_t^n\}$-adapted process with $C_n(0) \geq 0$ and $C_n(t+h) - C_n(t) \geq h$, $t, h \geq 0$. Suppose that $\{C_n(t)\}$ is stochastically bounded for all $t \geq 0$. Define $\gamma_n = C_n^{-1}$, $\hat{U}_n = U_n \circ \gamma_n$, $\hat{Y}_n = Y_n \circ \gamma_n$, and $\hat{X}_n = X_n \circ \gamma_n$, and suppose that $\{(\hat{U}_n, \hat{X}_n, \hat{Y}_n, \gamma_n)$ is relatively compact in the sense that

$$\{(\hat{U}_n, \hat{X}_n, (\hat{Y}_n(\varphi_1, \cdot), \ldots, \hat{Y}_n(\varphi_m, \cdot), \gamma_n)\}$$

is relatively compact in $D_{L\times\hat{H}\times\mathbf{R}^m\times\mathbf{R}}[0,\infty)$ for $\varphi_1, \ldots, \varphi_m \in H$. Then $\{(Z_n, U_n, Y_n)\}$ is relatively compact.

Proof. First replace X_n by X_n^ϵ defined by $X_n^\epsilon(t) = \sum c_{ij}^\epsilon(X_n(t)) f_i \varphi_j$ where the c_{ij}^ϵ are defined as in (5.2) giving

$$Z_n^\epsilon(t) = U_n(t) + X_{n-}^\epsilon \cdot Y_n(t) = U_n(t) + \sum_{i,j} f_i \int_0^t c_{ij}^\epsilon(X_n(t)) dY_n(\varphi_j, s)$$

and apply Kurtz and Protter [42, Lemma 4.3], to obtain the relative compactness of $\{(Z_n^\epsilon, U_n, Y_n)\}$. The lemma then follows from the uniform approximation of X_n by X_n^ϵ and the uniform tightness of $\{Y_n\}$. □

The following lemma shows that the "uniform tightness" terminology is appropriate even in the infinite dimensional setting, that is, if we restrict our attention to integrands taking values in a compact set, then the distributions of the values of the integrals are "tight" in the usual weak convergence sense.

Lemma 6.17. *Let $\{Y_n\}$ be a uniformly tight sequence of standard $(L, \hat{H})^\#$-semimartingales. Then for each compact $\Gamma_0 \subset \hat{H}$ and $\eta > 0$, there exists a compact $\Gamma_1 \subset L$ such that for any Γ_0-valued, $\mathcal{F}_t^n$-adapted, cadlag process X_n, $P\{X_{n-} \cdot Y_n(s) \in \Gamma_1, s \leq t\} \geq 1 - \eta$.*

Proof. With $\{\zeta_k\}$ as in Section 5., let

$$X_n^\epsilon(t) = \sum_k \hat{\psi}_k^\epsilon(X_n(t))\zeta_k = \sum_{i,j} c_{ij}^\epsilon(X_n(t)) f_i \varphi_j.$$

Then since $\sup_{x\in K_1} c_{ij}^\epsilon(x) = 0$ except for finitely many values of i and j, say $1 \leq i \leq I$ and $1 \leq j \leq J$, and the coefficient of f_i in the integral $X_{n-}^\epsilon \cdot Y_n(t)$ will be of the form

$$\sum_{j=1}^J \int_0^t c_{ij}^\epsilon(X_n(s-)) dY_n(\varphi_j, s)$$

where the c_{ij}^ϵ are uniformly bounded on Γ_0, for any $\eta_0 > 0$ there will exist a compact set of the form

$$\Gamma = \{\sum_{i=1}^I \alpha_i f_i : |\alpha_i| \leq a_i\} \tag{6.5}$$

such that

$$P\{X^{\epsilon}_{n-} \cdot Y_n(s) \in \Gamma, s \leq t\} \geq 1 - \eta_0$$

for all Γ_0-valued, $\mathcal{F}^n_t$-adapted, cadlag processes X_n. Let

$$\Gamma^a = \{x : \|x - \Gamma\|_L < a\}.$$

Then for $\delta > 0$, Then for $\delta > 0$,

$$\begin{aligned} P\{X_{n-} \cdot Y_n(s) \notin \Gamma^{\epsilon K(t,\delta)}, \text{ some } s \leq t\} \\ \leq \ \eta_0 + P\{\sup_{s \leq t} \|(X_{n-} - X^{\epsilon}_{n-}) \cdot Y_n(s)\|_L \geq \epsilon K(t,\delta)\} \\ \leq \ \eta_0 + \delta \,. \end{aligned}$$

For $m \geq 2$, let $\delta_m = \eta 2^{-m}$, let $\epsilon_m = 1/(mK(t,\delta_m))$, and select Γ_m of the form (6.5) so that

$$P\{X^{\epsilon_m}_{n-} \cdot Y_n(s) \in \Gamma_m, s \leq t\} \geq 1 - \eta 2^{-m}$$

for all Γ_0-valued, $\mathcal{F}^n_t$-adapted, cadlag processes X_n. Then

$$P\{X_{n-} \cdot Y_n(s) \notin \Gamma^{1/m}_m, \text{ some } s \leq t\} \leq \eta 2^{-(m-1)},$$

and letting Γ_1 denote the closure of $\cap_{m=2}^{\infty} \Gamma^{1/m}_m$, we have

$$P\{X_{n-} \cdot Y_n(s) \notin \Gamma_1, \text{ some } s \leq t\} \leq \sum_{m=2}^{\infty} \eta 2^{-(m-1)} = \eta.$$

Note that Γ_1 is compact since it is complete and totally bounded. □

The following lemma is an immediate consequence of Lemma 6.17.

Lemma 6.18. *Let $\{Y_n\}$ be a uniformly tight sequence of standard $(L, \hat{H})^{\#}$-semimartingales. For each $n = 1, 2, \ldots$, let X_n be an $\{\mathcal{F}^n_t\}$-adapted process in $D_{\hat{H}}[0,\infty)$, and suppose that $\{X_n\}$ satisfies the compact containment condition. Then $\{X_{n-} \cdot Y_n\}$ satisfies the compact containment condition, that is, for $\eta > 0$, there exists a compact $\Gamma \subset L$ such that $P\{X_{n-} \cdot Y_n(s) \in \Gamma, s \leq t\} \geq 1 - \eta$.*

7. Stochastic differential equations

In this section we consider stochastic differential equations of the form

$$X(t) = U(t) + F(X, \cdot-) \cdot Y(t) \,, \tag{7.1}$$

where $F : D_L[0,\infty) \to D_{\hat{H}}[0,\infty)$ and F is nonanticipating in the sense that for $x \in D_L[0,\infty)$ and $x^t(\cdot) = x(\cdot \wedge t)$, $F(x,t) = F(x^t,t)$ for all $t \geq 0$. The usual Lipschitz condition on F implies uniqueness for equations driven by $(L, \hat{H})^{\#}$-semimartingales satisfying a condition slightly stronger than the assumption that Y is standard. Weak existence, under certain continuity assumptions, follows from a convergence theorem given below (see Corollary 7.7), and strong existence follows from weak existence and strong uniqueness.

7.1 Uniqueness for stochastic differential equations

Theorem 7.1. *Let Y be an $(L, \hat{H})^\#$-semimartingale adapted to $\{\mathcal{F}_t\}$. Suppose that for each $\{\mathcal{F}_t\}$-stopping time τ, bounded by a constant, and $t, \delta > 0$, there exists $K_\tau(t,\delta)$ such that*

$$P\{\|Z_- \cdot Y(\tau + t) - Z_- \cdot Y(\tau)\|_L \geq K_\tau(t,\delta)\} \leq \delta \tag{7.2}$$

for all $Z \in S_{\hat{H}}$ satisfying $\sup_s \|Z(s)\|_{\hat{H}} \leq 1$, and that $\lim_{t\to 0} K_\tau(t,\delta) = 0$. Suppose that there exists $M > 0$ such that

$$\sup_{s\leq t} \|F(x,s) - F(y,s)\|_{\hat{H}} \leq M \sup_{s\leq t} \|x(s) - y(s)\|_L$$

for all $x, y \in D_L[0,\infty)$. Then there is at most one solution of (7.1).

Remark 7.2. Note that if L and H are finite dimensional and Y is a finite dimensional semimartingale, then the hypothesized estimate (7.2) holds.

More generally, if L satisfies the conditions of Lemma 5.9, then (7.2) will hold for any standard $(L, \hat{H})^\#$-semimartingale.

Finally, one can apply Lemma 5.10 to prove uniqueness for any standard $(L, \hat{H})^\#$-semimartingale under the additional condition on F that for each compact $\Gamma \subset L$ there exists a compact $K_\Gamma \subset \hat{H}$ such that $x(s) \in \Gamma$, $s \leq t$, implies $F(x,t) \in K_\Gamma$. (See Theorem 7.6 for another application of a related condition.) In particular, if $F(x,t) = f(x(t))$ for a Lipschitz continuous $f : L \to \hat{H}$, then uniqueness holds for all standard $(L, \hat{H})^\#$-semimartingales.

Proof. Without loss of generality, we can assume that $K_\tau(t,\delta)$ is a nondecreasing function of t.

Suppose X and $\tilde{X}$ satisfy (7.1). Let $\tau_0 = \inf\{t : \|X(t) - \tilde{X}(t)\|_L > 0\}$, and suppose $P\{\tau_0 < \infty\} > 0$. Select $r, \delta, t > 0$, such that $P\{\tau_0 < r\} > \delta$ and $MK_{\tau_0\wedge r}(t,\delta) < 1$. Note that if $\tau_0 < \infty$, then

$$X(\tau_0) - \tilde{X}_0(\tau_0) = (F(X,\cdot-) - F(\tilde{X},\cdot-)) \cdot Y(\tau_0) = 0. \tag{7.3}$$

Define

$$\tau_\epsilon = \inf\{s : \|X(s) - \tilde{X}(s)\|_L \geq \epsilon\}.$$

Noting that $\|X(s) - \tilde{X}(s)\|_L \leq \epsilon$ for $s < \tau_\epsilon$, we have

$$\|F(X,s) - F(\tilde{X},s)\|_{\hat{H}} \leq \epsilon M,$$

for $s < \tau_\epsilon$, and

$$\|F(X,\cdot-) \cdot Y(\tau_\epsilon) - F(\tilde{X},\cdot-) \cdot Y(\tau_\epsilon)\|_L = \|X(\tau_\epsilon) - \tilde{X}(\tau_\epsilon)\|_L \geq \epsilon.$$

Consequently, for $r > 0$, letting $\tau_0^r = \tau_0 \wedge r$, we have

$$
\begin{aligned}
&P\{\tau_\epsilon - \tau_0^r \le t\} \\
&\le P\{\sup_{s \le t \wedge (\tau_\epsilon - \tau_0^r)} \|F(X,\cdot-)\cdot Y(\tau_0^r + s) - F(\tilde{X},\cdot-)\cdot Y(\tau_0^r + s)\|_L \\
&\qquad\qquad \ge \epsilon M K_{\tau_0^r}(t,\delta)\} \\
&\le \delta.
\end{aligned}
$$

Since the right side does not depend on ϵ and $\lim_{\epsilon\to 0}\tau_\epsilon = \tau_0$, it follows that $P\{\tau_0 - \tau_0 \wedge r < t\} \le \delta$ and hence that $P\{\tau_0 < r\} \le \delta$, contradicting the assumption on δ and proving that $\tau_0 = \infty$ a.s.

□

Example 7.3. Equation for spin-flip models.

A spin-flip model, for example on the lattice $\mathbf{Z}^d$, is a stochastic process whose state $\eta = \{\eta_i : i \in \mathbf{Z}^d\}$ assigns to each lattice point $i \in \mathbf{Z}^d$ the value ± 1. The model is prescribed by specifying for each $i \in \mathbf{Z}^d$ a flip rate c_i which determines the rate at which the associated state variable η_i changes sign. The rates c_i may depend on the full configuration η. We formulate a slightly more general model by tracking the cumulative number of sign changes X_i rather than just the current sign. Of course, if the initial configuration is known, then the current configuration can be recovered by the formula

$$\eta_i(t) = \eta_i(0)(-1)^{X_i(t)-X_i(0)}.$$

The model, then, consists of a collection of counting processes $X = \{X_i : i \in \mathbf{Z}^d\}$, and the specification of the rates c_i corresponds to the requirement that there exist a filtration $\{\mathcal{F}_t\}$ such that for each $i \in \mathbf{Z}^d$,

$$X_i(t) - \int_0^t c_i(X(s))ds. \tag{7.4}$$

is an $\{\mathcal{F}_t\}$-martingale. To completely specify the martingale problem corresponding to the $\{c_i\}$ we must also require that no two X_i have simultaneous jumps (that is, the martingales are orthogonal).

We formulate a corresponding system of stochastic differential equations for the X_i. Let $\{N_i : i \in \mathbf{Z}^d\}$ be independent Poisson random measures on $[0,\infty) \times [0,\infty)$ with Lebesgue mean measure. Then we require

$$X_i(t) = X_i(0) + \int_{[0,\infty)\times[0,t]} \sigma_i(X(s-),z)N_i(dz \times ds) \tag{7.5}$$

where

$$\sigma_i(x,z) = \begin{cases} 1 & 0 \le z \le c_i(x) \\ 0 & otherwise \end{cases}$$

Note that any solution of this system will be a family of counting process without simultaneous jumps (by the independence of the N_i). Letting $\tilde{N}_i(A) = N_i(A) - m(A)$, we can rewrite (7.5) as

$$X_i(t) = X_i(0) + \int_{[0,\infty)\times[0,t]} \sigma_i(X(s-),z)\tilde{N}_i(dz\times ds) + \int_0^t c_i(X(s))ds\,. \quad (7.6)$$

The second term on the right will be a martingale (at least, for example, if the c_i are bounded), and hence (7.4) will be a martingale.

We now give conditions under which the solution of (7.5) is unique. Let $J = \{\{k_i, i \in \mathbf{Z}^d\} : k_i \in \mathbf{Z}_+\}$. Assume

$$|c_i(x+e_j) - c_i(x)| \le a_{ij} \qquad |c_i(x)| \le b_i$$

for all $x, y \in J$ and $i, j \in \mathbf{Z}^d$ where e_j is the element of J such $e_{ji} = 0$ for $i \neq j$ and $e_{jj} = 1$. In addition, assume that there exist $\alpha_i > 0$ and $C > 0$ such that

$$\sum_i \alpha_i a_{ij} \le C\alpha_j \qquad \sum_i \alpha_i b_i < \infty. \quad (7.7)$$

These conditions are similar to those under which Liggett [45] (see also Liggett [46, Chapter III]) proves uniqueness of the martingale problem for the spin-flip model. To give a complete proof of Liggett's theorem, we would need to show that any solution of the martingale problem is a weak solution of (7.5). We do not pursue this issue here, but see Kurtz [38] for a closely related approach involving a different system of stochastic equations. Uniqueness of (7.5) can actually be proved under the same conditions as used in Kurtz [38] using different methods; however, our point here is to illustrate the breadth of applicability of the theorem above.

In order to interpret the system as a solution of a single equation of the form (7.1), let $U = [0,\infty) \times \mathbf{Z}^d$ and $E = \mathbf{Z}^d$, and define $F_i(x,u) = \sigma_i(x,z)\delta_{ij}$ for $u = (z,j)$. Define

$$\begin{aligned} \|f\|_L &= \sum_i \alpha_i |f(i)| \\ \|\varphi\|_H &= \sum_j \int_0^\infty |\varphi(z,j)|dz \\ \|g\|_{\hat{H}} &= \sum_i \sum_j \alpha_i \int_{\mathbf{R}} |g(i,z,j)|du = \|\|g\|_H\|_L \end{aligned}$$

Then

$$\|F.(x,\cdot)\|_{\hat{H}} \le \sum_i \alpha_i \int_0^\infty |\sigma_i(x,z)|dz = \sum_i \alpha_i c_i(x) \le \sum_i \alpha_i b_i$$

and

$$\begin{aligned} \|F.(x,\cdot) - F.(y,\cdot)\|_{\hat{H}} &= \textstyle\sum_i \alpha_i \int_0^\infty |\sigma_i(x,z) - \sigma_i(y,z)|dz \\ &= \textstyle\sum_i \alpha_i |c_i(x) - c_i(y)| \\ &\le \textstyle\sum_i \alpha_i \sum_j a_{ij}|x_j - y_j| \\ &\le \textstyle\sum_j C\alpha_j |x_j - y_j| = C\|x-y\|_L \end{aligned}$$

so F is bounded and Lipschitz.

For many examples $a_{ij} = \rho(i-j)$, where ρ has bounded support and $b_i \equiv b$. One then can take $\alpha_i = \frac{1}{1+|i|^{d+1}}$ so that

$$\sum_i \frac{\rho(i-j)}{1+|i|^{d+1}} \leq \left(\sup_k \sum_i \frac{1+|k|^{d+1}}{1+|i|^{d+1}} \rho(i-k)\right) \frac{1}{1+|j|^{d+1}}$$

which gives (7.7).

To verify (7.2), suppose

$$\|X(s)\|_{\hat{H}} = \sum_i \sum_j \alpha_i \int_0^\infty |X(i,z,j)| dz \leq 1 .$$

Then

$$\begin{aligned} E[\|X_- \cdot N(t)\|_L] &= E[\sum_i \sum_j \alpha_i \int_{[0,\infty)\times[0,t]} X(s-,i,z,j) N_j(dz \times ds)] \\ &\leq E[\sum_i \sum_j \alpha_i \int_{[0,\infty)\times[0,t]} |X(s-,i,z)| dz ds] \\ &\leq t \end{aligned}$$

7.2 Sequences of stochastic differential equations

Consider a sequence of equations of the above form

$$X_n(t) = U_n(t) + F_n(X_n, \cdot -) \cdot Y_n(t). \tag{7.8}$$

The following analogue of Proposition 5.1 of Kurtz and Protter [42] is an immediate consequence of Theorems 4.2 and 5.5.

Proposition 7.4. *Suppose (U_n, X_n, Y_n) satisfies (7.8), that $\{(U_n, X_n, Y_n)\}$ is relatively compact, that $(U_n, Y_n) \Rightarrow (U, Y)$, and that $\{Y_n\}$ is uniformly tight. Assume that F_n and F satisfy the following continuity condition:*

C.1 *If $(x_n, y_n) \to (x, y)$ in $D_{L\times \mathbf{R}^m}[0,\infty)$, then*

$$(x_n, y_n, F_n(x_n, \cdot)) \to (x, y, F(x, \cdot))$$

in $D_{L\times \mathbf{R}^m \times \hat{H}}[0,\infty)$.

Then any limit point of $\{X_n\}$ satisfies (7.1).

In many situations, the relative compactness of $\{X_n\}$ is easy to verify; however, with somewhat stronger conditions on the sequence $\{F_n\}$ (satisfied, for example, if $F_n(x,t) = f_n(x(t))$ for a sequence of continous mappings $f_n : \mathbf{R}^k \to H$ converging uniformly on compact subsets of $\mathbf{R}^k$ to $f : \mathbf{R}^k \to H$) we can drop the a priori assumption of relative compactness of $\{X_n\}$. Let

$T_1[0,\infty)$ be the collection of nondecreasing mappings λ of $[0,\infty)$ onto $[0,\infty)$ satisfying $|\lambda(t)-\lambda(s)| \le |t-s|$. Let the topology on $T_1[0,\infty)$ be given by the metric

$$d_1(\lambda_1,\lambda_2) = \sup_{0\le t<\infty} \frac{|\lambda_1(t)-\lambda_2(t)|}{1+|\lambda_1(t)-\lambda_2(t)|}\,.$$

ι will denote the identity map, $\iota(t)=t$. We assume the following:

C.2a There exist $G_n, G : D_L[0,\infty)\times T_1[0,\infty) \to D_{\hat{H}}[0,\infty)$ such that $F_n(x)\circ\lambda = G_n(x\circ\lambda,\lambda)$ and $F(x)\circ\lambda = G(x\circ\lambda,\lambda)$, $x\in D_L[0,\infty)$, $\lambda\in T_1[0,\infty)$.
C.2b For each compact $\mathcal{K}\subset D_L[0,\infty)\times T_1[0,\infty)$ and $t>0$,

$$\sup_{(x,\lambda)\in\mathcal{K}} \sup_{s\le t} \|G_n(x,\lambda,s)-G(x,\lambda,s)\|_{\hat{H}} \to 0.$$

C.2c For $\{(x_n,\lambda_n)\}\in D_L[0,\infty)\times T_1[0,\infty)$, $\sup_{s\le t}\|x_n(s)-x(s)\|_L \to 0$ and $\sup_{s\le t}|\lambda_n(s)-\lambda(s)|\to 0$ for each $t>0$ implies

$$\sup_{s\le t}\|G(x_n,\lambda_n,s)-G(x,\lambda,s)\|_{\hat{H}} \to 0.$$

Note that if $F(x,t)=f(x(t))$, then $G(x,\lambda,t)=f(x(t))$; if $F(x,t)=f(\int_0^t h(x(s))ds)$, then $G(x,\lambda,t)=f(\int_0^t h(x(s))\lambda'(s)ds)$. (See Kurtz and Protter [42] for additional examples.)

The following theorem is the analogue of Theorem 5.4 of Kurtz and Protter [42]. For simplicity, we assume that F_n and F are uniformly bounded. The localization argument used in the earlier theorem can again be applied here to extend the result to the unbounded case.

Theorem 7.5. *Let $L=\mathbf{R}^k$. Suppose (U_n,X_n,Y_n) satisfies (7.8), that $(U_n,Y_n)\Rightarrow(U,Y)$, and that $\{Y_n\}$ is uniformly tight. Assume that $\{F_n\}$ and F satisfy Condition C.2 and that $\sup_n\sup_x\|F_n(x,\cdot)\|_{H^k}<\infty$. Then $\{(U_n,X_n,Y_n)\}$ is relatively compact and any limit point satisfies (7.1).*

Proof. Since Condition C.2 implies Condition C.1, it is enough to show the relative compactness of (U_n,X_n,Y_n). By Proposition 6.2 $\{F_n(X_n,\cdot-)\cdot Y_n\}$ is uniformly tight. By Proposition 6.14, there exists a γ_n such that $\{(U_n\circ\gamma_n, F_n(X_n,\cdot-)\cdot Y_n\circ\gamma_n)\}$ is relatively compact, which in turn implies $\{(X_n\circ\gamma_n, U_n\circ\gamma_n, F_n(X_n,\cdot-)\cdot Y_n\circ\gamma_n)\}$ is relatively compact. Setting $\hat{X}_n = X_n\circ\gamma_n$, etc., Condition C.2a implies

$$\hat{X}_n(t) = \hat{U}_n(t) + G_n(\hat{X}_n,\gamma_n,\cdot-)\cdot\hat{Y}_n(t)$$

and the relative compactness of $\{(\hat{X}_n,\hat{U}_n,\hat{Y}_n)\}$ implies the relative compactness of

$$\{(\hat{X}_n,\hat{U}_n,\hat{Y}_n,G_n(\hat{X}_n,\gamma_n,\cdot),\gamma_n)\}\,.$$

By Lemma 6.16, $\{(X_n,U_n,Y_n)\}$ is relatively compact, and the theorem follows from Proposition 7.4. □

Theorem 7.6. *Suppose (U_n, X_n, Y_n) satisfies (7.8), that $(U_n, Y_n) \Rightarrow (U, Y)$, and that $\{Y_n\}$ is uniformly tight. Assume that $\{F_n\}$ and F satisfy Condition C.2, that $\sup_n \sup_x \|F_n(x, \cdot)\|_{\hat{H}} < \infty$, and that for each $\kappa > 0$ there exists a compact $K_\kappa \subset \hat{H}$ such that $\sup_{s \le t} \|x(s)\|_L \le \kappa$ implies $F_n(x, t) \in K_\kappa$ for all n. Then $\{(U_n, X_n, Y_n)\}$ is relatively compact and any limit point satisfies (7.1).*

Proof. Again it is sufficient to verify the relative compactness of $\{(U_n, X_n, Y_n)\}$. The uniform tightness of $\{Y_n\}$ and the boundedness of the F_n imply the stochastic boundedness of $\{\sup_{s<t} \|X_n(s)\|_L\}$. The compactness condition on the F_n then ensures that $\{F_n(X_n, \cdot)\}$ satisfies the compact containment condition. The sequence $\{F_n(X_n, \cdot-) \cdot Y_n\}$ satisfies the compact containment condition by Lemma 6.18 which in turn implies $\{X_n\}$ satisfies the compact containment condition. Lemma 6.15 then ensures the existence of γ_n as in the proof of Theorem 7.5, and the remainder of the proof is the same. □

Corollary 7.7. *a) Let $L = \mathbf{R}^k$. Suppose that F and G satisfy C.2a and C.2c, that*

$$\sup_x \|F(x, \cdot)\|_{H^k} < \infty,$$

and that Y is a standard $H^\#$-semimartingale. Then weak existence holds for (7.1).

b) For general L, suppose that F, G satisfy C.2a and C.2c, that $\sup_x \|F(x, \cdot)\|_{\hat{H}} < \infty$, and that F satisfies the compactness condition of Theorem 7.6, and that Y is a standard $(L, \hat{H})^\#$-semimartingale. Then weak existence holds for (7.1).

Proof. Let $F_n \equiv F$, and define $U_n(t) = U(\frac{[nt]}{n})$ and $Y_n(\varphi, t) = Y(\varphi, \frac{[nt]}{n})$. Let X_n be the solution of

$$X_n(t) = U_n(t) + F(X_n, \cdot-) \cdot Y_n(t)$$

which is easily seen to exist since U_n and Y_n are constant except for a discrete set of jumps. Then X_n, U_n, Y_n, and F_n satisfy the conditions of Theorem 7.5 in part (a) and of Theorem 7.6 in part (b). Consequently, a subsequence of X_n will converge in distribution to a process $\tilde{X}$ satisfying $\tilde{X}(t) = \tilde{U}(t) + F(\tilde{X}, \cdot-) \cdot \tilde{Y}(t)$, where $(\tilde{U}, \tilde{Y})$ has the same distribution as (U, Y), that is, $\tilde{X}$ is a weak solution of (7.1). □

The following corollary is the analogue, in the present setting, of a result of Yamada and Watanabe [61] stating that weak existence and strong uniqueness imply strong existence. (See Engelbert [16] for a more recent discussion.) The proof of the corollary is the same as that of Lemma 5.5 of Kurtz and Protter [42]. We say that strong uniqueness holds for (7.1) if any two solutions X_1 and X_2 satisfy $X_1 = X_2$ a.s.

Corollary 7.8. *Suppose, in addition to the conditions of Corollary 7.7, that strong uniqueness holds for (7.1) for any version of (U,Y) for which Y is a standard $H^\#$ (or $(L,\hat{H})^\#$)-semimartingale. Then any solution of (7.1) is a measurable function of (U,Y), that is, if X satisfies (7.1) and the finite linear combinations of $\{\varphi_k\}$ are dense in H, there exists a measurable mapping*

$$g : D_{L\times\mathbf{R}^\infty}[0,\infty) = D_L[0,\infty)$$

such that $X = g(U, Y(\varphi_1,\cdot), Y(\varphi_2,\cdot),\ldots)$. In particular, there exists a strong solution of (7.1).

Remark 7.9. Note that under the conditions of Theorem 7.1 (see Remark 7.2), the strong uniqueness hypothesis of the present Corollary will hold.

The proof of the following corollary is essentially the same as that of Corollary 5.6 of Kurtz and Protter [42].

Corollary 7.10. *Suppose, in addition to the conditions of Theorem 7.5 or Theorem 7.6, that strong uniqueness holds for any version of (U,Y) for which Y is a standard $H^\#$ $(L,\hat{H})^\#$)-semimartingale and that $(U_n,Y_n)\to(U,Y)$ in probability. Then $(U_n,Y_n,X_n)\to(U,Y,X)$ in probability.*

8. Markov processes

An $H^\#$-semimartingale has ***stationary independent increments*** if $(Y(\varphi_1,\cdot),\ldots,Y(\varphi_m,\cdot))$ has stationary independent increments for each choice of $\varphi_1,\ldots,\varphi_m\in H$. If Y is a standard $(L,\hat{H})^\#$-semimartingale with stationary independent increments, $F : L\to\hat{H}$, and the equation

$$X(t) = x_0 + F(X(\cdot-))\cdot Y(t) \tag{8.1}$$

has a unique solution for each $x_0\in L$, then X is a temporally homogeneous Markov process. If Y is given by a standard semimartingale random measure, then (8.1) can be written

$$X(t) = x_0 + \int_{U\times[0,t]} F(X(s-),u)Y(du\times ds)\,.$$

For $\varphi_1,\ldots,\varphi_m\in H$, $(Y(\varphi_1,\cdot),\ldots,Y(\varphi_m,\cdot))$ has a generator of the form (see, for example, Ethier and Kurtz [17, Theorem 8.3.4])

$$\begin{aligned} B(\varphi_1,\ldots,\varphi_m)f(x) &= \frac{1}{2}\sum_{i,j=1}^m a_{ij}(\varphi_1,\ldots,\varphi_m)\partial_i\partial_j f(x) \\ &\quad + \sum_{i=1}^m b_i(\varphi_1,\ldots,\varphi_m)\partial_i f(x) \\ &\quad + \int_{\mathbf{R}^m}\left(f(x+y)-f(x)-\frac{y\cdot\nabla f(x)}{1+|y|^2}\right) \\ &\qquad\qquad \mu(\varphi_1,\ldots,\varphi_m,dy) \end{aligned}$$

where it is enough to consider $f \in C_c^\infty(\mathbf{R}^m)$, that is, the closure of $B(\varphi_1,\dots,\varphi_m)$ defined on $C_c^\infty(\mathbf{R}^m)$, the space of infinitely differentiable functions with compact support, generates the strongly continuous contraction semigroup on $\hat{C}(\mathbf{R}^m)$ corresponding to the process $(Y(\varphi_1,\cdot),\dots,Y(\varphi_m,\cdot))$. The assumption that Y is standard $H^\#$-semimartingale implies $B(\varphi_1,\dots,\varphi_m)f$ is a continuous function of $(\varphi_1,\dots,\varphi_m)$. (Note, however, that this continuity does not imply the continuity of a_{ij}.)

Suppose $L = \mathbf{R}^k$. Formally, at least, the solution of (8.1) should have a generator given by

$$Af(x) = B(F_1(x),\dots,F_k(x))f(x),$$

that is, the solution of (8.1) is a solution of the martingale problem for A.

Consider a sequence of stochastic differential equations

$$X_n(t) = X_n(0) + \int_{U\times[0,t]} F_n(X(s-),u)Y_n(du\times ds)$$

where Y_n is an $H^\#$-semimartingale with stationary, independent increments, and let the generator for $(Y_n(\varphi_1,\cdot),\dots,Y_n(\varphi_m,\cdot))$ be denoted $B_n(\varphi_1,\dots,\varphi_m)$. The following theorem is an immediate consequence of Theorem 7.5 and Theorems 1.6.1 and 4.2.5 of Ethier and Kurtz [17].

Theorem 8.1. *Let $L = \mathbf{R}^k$. Let F in (8.1) be bounded and Lipschitz. Assume that $\{Y_n\}$ is uniformly tight, Y_n is independent of $X_n(0)$, and $X_n(0) \Rightarrow X(0)$. Suppose for each $\varphi_1,\dots,\varphi_m \in H$ and each $f \in C_c^\infty(\mathbf{R}^m)$*

$$\lim_{n\to\infty} \sup_x |B_n(\varphi_1,\dots,\varphi_m)f(x) - B(\varphi_1,\dots,\varphi_m)f(x)| = 0, \tag{8.2}$$

and for each compact $K \subset \mathbf{R}^k$

$$\lim_{n\to\infty} \sup_{x\in K} \|F_n(x) - F(x)\|_{H^k} = 0.$$

Then $(X_n, Y_n) \Rightarrow (X, Y)$, where Y is a standard $H^\#$-semimartingale, $(Y(\varphi_1,\cdot),\dots,Y(\varphi_m,\cdot))$ has generator $B(\varphi_1,\dots,\varphi_m)$, and X is the unique solution of

$$X(t) = X(0) + F(X(\cdot-))\cdot Y(t).$$

Proof. The convergence of B_n to B implies the convergence

$$(Y_n(\varphi_1,\cdot),\dots,Y(\varphi_m,\cdot)) \Rightarrow (Y(\varphi_1,\cdot),\dots,Y(\varphi_m,\cdot)),$$

and the theorem follows from Theorem 7.5 □

For a specific form for (8.1), let W denote Gaussian white noise on $U_0 \times [0,\infty)$ with $E[W(A,t)^2] = \mu(A)t$, let N_1 and N_2 be Poisson random measures on $U_1\times[0,\infty)$ and $U_2\times[0,\infty)$ with σ-finite mean-measures $\nu_1\times m$ and $\nu_2\times m$, respectively. Let $M_1(A,t) = N_1(A,t) - \nu_1(A)t$, and note that

$$\langle M_1(A,\cdot), M_1(B,\cdot)\rangle_t = t\nu_1(A\cap B).$$

Consider the equation for an $\mathbf{R}^k$-valued process X

$$\begin{aligned} X(t) &= X(0) + \int_{U_0\times[0,t]} F_0(X(s-),u)W(du\times ds) \\ &\quad + \int_{U_1\times[0,t]} F_1(X(s-),u)M_1(du\times ds) \\ &\quad + \int_{U_2\times[0,t]} F_2(X(s-),u)N_2(du\times ds) + \int_0^t F_3(X(s-))ds \end{aligned} \quad (8.3)$$

Note that the above equation is essentially the same as that originally considered by Itô [24]. (The diffusion term in Itô's equation was driven by a finite dimensional standard Brownian motion rather than the Gaussian white noise.) The generality of this equation is demonstrated in the work of Çinlar and Jacod [10].

Theorem 8.2. *Suppose that there exists $M > 0$ such that*

$$\begin{aligned} &\sqrt{\int_{U_0} |F_0(x,u)-F_0(y,u)|^2\mu(du)} + \sqrt{\int_{U_1} |F_1(x,u)-F_1(y,u)|^2\nu_1(du)} \\ &\quad + \int_{U_2} |F_2(x,u)-F_2(y,u)|\nu_2(du) + |F_3(x)-F_3(y)| \\ &\quad \le M|x-y|\,. \end{aligned}$$

Then there exists a unique solution of (8.3).

Remark 8.3. This result is essentially Theorem 1.2 of Graham [20], the only difference being that we allow general Gaussian white noise and Graham only considers finite dimensional Brownian motion. His methods, however, which employ an L_1-martingale inequality instead of the typical application of Doob's L_2-martingale inequality (see Lemma 2.4 and additional discussion below), could also be used in the present setting. As Graham points out, the ability to use L_1 estimates is crucial for many applications. For example, consider the equation corresponding to the generator

$$Af(x) = \int_{\mathbf{R}^d} \lambda(x,u)(f(x+u)-f(x))\nu(du)$$

given by

$$X(t) = X(0) + \int_{\mathbf{R}^d\times[0,\infty)} uI_{[0,\lambda(X(s-),u)]}(z)N(du\times dz\times ds),$$

where N is the Poisson random measure on $\mathbf{R}^d\times[0,\infty)$ with mean measure $\eta = \nu\times m$. The $L_1(\eta)$ Lipschitz condition becomes

$$\int_{\mathbf{R}^d} u|\lambda(x,u)-\lambda(y,u)|\nu(du) \le K|x-y|$$

which will be satisfied under reasonable conditions on λ. Since the square of an indicator is the indicator, the corresponding L_2 condition would require

$$\int_{\mathbf{R}^d} u^2|\lambda(x,u)-\lambda(y,u)|\nu(du) \le K|x-y|^2,$$

which essentially says that λ(x,u) is constant in x. The classical conditions of Itô [24] (see also Ikeda and Watanabe [23]) as well as more recent work (for example, Kallianpur and Xiong [31]) based on L_2 estimates do not cover this example. Roughly speaking, L_2 estimates will work if the jump sizes vary smoothly with the location of the process and the jump rates are constant, but fail if the jump rates vary.

Proof. Let $H_0 = L_2(\mu)$, $H_1 = L_2(\nu_1)$, $H_2 = L_1(\nu_2)$, and $H_3 = \mathbf{R}$, and let $Y_0(\varphi,t) = \int_{U_0} \varphi(u)W(du,t)$ for $\varphi \in L_2(\mu)$, $Y_1(\varphi,t) = \int_{U_1} \varphi(u)M_1(du,t)$ for $\varphi \in H_1$, $Y_2(\varphi,t) = \int_{U_2} \varphi(u)N_2(du,t)$ for $\varphi \in H_2$, and $Y_3(\varphi,t) = \varphi t$ for $\varphi \in \mathbf{R}$. Y_0 and Y_1 are standard $H_0^{\#}$- and $H_1^{\#}$-semimartingales, respectively, by Lemma 3.3. For $Z \in \mathcal{S}_{H_2}$, we have

$$\begin{aligned} E[|Z_- \cdot Y_2(t)|] &\le E[\int_{U_2\times[0,t]} |Z(s-,u)|N_2(du\times ds)] \\ &= E[\int_0^t \int_{U_2} |Z(s-,u)|\nu_2(du)ds], \end{aligned}$$

which implies Y_2 is a standard $H_2^{\#}$-semimartingale. Trivially, Y_3 is a standard $H_3^{\#}$-semimartingale. Setting $H = H_0 \times H_1 \times H_2 \times H_3$, Y defined as in Lemma 3.10 is an $H^{\#}$-semimartingale, and the theorem follows by Theorem 7.1 and Corollaries 7.7 and 7.8. □

Let X and $\tilde{X}$ be solutions of (8.3). Graham's approach (cf. Remark 8.3) depends on the inequality

$$\begin{aligned} &E[\sup_{s\le t}|X(s)-\tilde{X}(s)|] \\ &\le E[|X(0)-\tilde{X}(0)|] \\ &\quad +C_1E\left[\left(\int_0^t\int_{U_0}|F_0(X(s-),u)-F_0(\tilde{X}(s-),u)|^2\mu(du)ds\right)^{1/2}\right] \\ &\quad +C_1E\left[\left(\int_0^t\int_{U_1}|F_1(X(s-),u)-F_1(\tilde{X}(s-),u)|^2\nu_1(du)ds\right)^{1/2}\right] \\ &\quad +E\left[\int_0^t\int_{U_2}|F_2(X(s-),u)-F_2(\tilde{X}(s-),u)|\nu_2(du)ds\right] \end{aligned}$$

$$+E\left[\int_0^t |F_3(X(s-)) - F_3(\tilde{X}(s-))|ds\right]$$
$$\leq E[|X(0) - \tilde{X}(0)|] + D(\sqrt{t}+t)E[\sup_{s\leq t}|X(s) - \tilde{X}(s)|] \quad (8.4)$$

where C_1 is the constant in Lemma 2.4 and D depends on C_1 and the Lipschitz constants of F_0 - F_3. Uniqueness follows by selecting t so that $D(\sqrt{t}+t) < 1$. We will need estimates like this one in Section 9.

9. Infinite systems

Most uniqueness proofs for stochastic differential equations are based on L_2-estimates. Graham [20] is one exception (see Remark 8.3), and as he notes, L_2-estimates may be completely inadequate in treating jump processes. Example 7.3 illustrates the problem. For that example, $E[(\sigma_i(X(t),z)-\sigma_i(\tilde{X}(t),z))^2] = E[|\sigma_i(X(t),z) - \sigma_i(\tilde{X}(t),z)|]$, since σ_i is an indicator function, and any attempt to estimate $E[(X_i(t) - \tilde{X}_i(t))^2]$ by the usual Gronwall argument will fail. One of the main advantages of the techniques developed in Theorem 7.1 is that a mixture of L_2 and L_1 (or other) estimates can be used to to obtain the fundamental estimate (7.2).

For finite dimensional problems or single equations in a Banach space, the techniques of Theorem 7.1 should work any time either L_2 or L_1 estimates work. (One exception appears to be the results of Yamada and Watanabe [61] involving non-Lipschitz F.) For infinite systems, however, there are several examples for which L_2 or L_1 methods are effective but for which we have not found an analogue of the approach of Theorem 7.1. Example 7.3 is in fact one. The conditions under which we have applied Theorem 7.1, although they cover most of the standard examples of spin-flip models, are not the usual conditions employed. For example, the conditions in Liggett [46, Chapter III] require $\sup_i \sum_j a_{ij} < \infty$. Kurtz [38] considers the general L_p in i, L_q in j analogs of the L_1 in i, and L_∞ in j conditions covered in Example 7.3 and the L_∞ in i, L_1 in j conditions of Liggett.

9.1 Systems driven by Poisson random measures

We consider a system more general than (7.5)

$$X_i(t) = X_i(0) + \int_{V\times[0,t]} F_i(X(s-),v)N(dv\times ds) \quad (9.1)$$

where N is a Poisson random measures on $V\times[0,\infty)$ with mean measure $\nu\times m$. Note that to represent (7.5) as an equation of the form (9.1), let $V = \mathbf{Z}^d\times[0,\infty)$, $\nu = \gamma\times m$ where γ is the measure with mass 1 at each point in $\mathbf{Z}^d$, and

$$F_i(x,(k,z)) = \sigma_i(x,z)\delta_{ik}.$$

We assume that $E[|X_i(0)|] < \infty$ and

$$\int_V |F_i(x,v)|\nu(dv) \leq b_i$$

which ensures that $E[|X_i(t)|] < E[|X_i(0)|] + b_i t < \infty$. Observe that if X and Y are solutions of (9.1) with $X(0) = Y(0)$, then

$$E[|X_i(t) - Y_i(t)|] \leq \int_0^t E[\int_V |F_i(X(s-),v) - F_i(Y(s-),v)|\nu(dv)]ds\,.$$

With this inequality and Example 7.3 in mind, we assume that

$$\int_V |F_i(x,v) - F_i(y,v)|\nu(dv) \leq \sum_j a_{ij}|x_j - y_j|\,, \tag{9.2}$$

which implies

$$\begin{aligned} \alpha_i E[|X_i(t) - Y_i(t)|] &\leq \int_0^t \sum_j \frac{\alpha_i a_{ij}}{\alpha_j} \alpha_j E[|X_j(s) - Y_j(s)|] \qquad (9.3) \\ &\leq \left\| \frac{\alpha_i a_{ij}}{\alpha_j} \right\|_{p,j} \int_0^t \|\alpha_j E[|X_j(s) - Y_j(s)|]\|_{q,j}\, ds \end{aligned}$$

where

$$\|f(i,j)\|_{p,j} = \left[\sum_j |f(i,j)|^p\right]^{1/p}$$

and similarly for $\|\cdot\|_{q,j}$. If

$$\|\alpha_j b_j\|_{q,j} < \infty\,, \tag{9.4}$$

then $\|\alpha_j E[|X_j(s) - Y_j(s)|]\|_{q,j} < \infty$, and (9.3) implies

$$\|\alpha_i E[|X_i(t) - Y_i(t)|]\|_{q,i} \leq \left\| \left\| \frac{\alpha_i a_{ij}}{\alpha_j} \right\|_{p,j} \right\|_{q,i} \int_0^t \|\alpha_j E[|X_j(s) - Y_j(s)|]\|_{q,j}\, ds\,. \tag{9.5}$$

Consequently, if

$$\left\| \left\| \frac{\alpha_i a_{ij}}{\alpha_j} \right\|_{p,j} \right\|_{q,i} < \infty, \tag{9.6}$$

by Gronwall's inequality,

$$\|\alpha_i E[|X_i(t) - Y_i(t)|]\|_{q,i} = 0$$

and hence $X = Y$.

Note that with the appropriate definition of V, a system of the form

$$X_i(t) = X_i(0) + \sum_{k,l} \int_{[0,\infty)\times[0,t]} F_{ikl}(X(s-), z) N_{kl}(dz \times ds) \tag{9.7}$$

where the N_{kl} are independent Poisson random measures with mean measure $m \times m$, can be written in the form (9.1). Let

$$F_{ikl}(x, z) = (1 - x_i)x_k \delta_{li} I_{[0,\lambda(k,l)]}(z) - x_i(1 - x_l)\delta_{ki} I_{[0,\lambda(k,l)]}(z),$$

and assume that $X_i(0)$ is 0 or 1 for each i. Then (9.7) gives a simple exclusion model with jump rates $\lambda(k, l)$, that is, a particle at k attempts a jump to l at rate $\lambda(k, l)$; however, if l is occupied, the jump is rejected. Note that

$$\begin{aligned}\sum_{k,l} \int |F_{ikl}(x, z) - F_{ikl}(y, z)| dz & \\ \leq \sum_k |(1 - x_i)x_k - (1 - y_i)y_k| \lambda(k, i) & \\ + \sum_l |x_i(1 - x_l) - y_i(1 - y_l)| \lambda(i, l) &\end{aligned}$$

We see that (9.2) is satisfied for

$$a_{ii} = \sum_{k\neq i} \lambda(k, i) + \sum_{l\neq i} \lambda(i, l) \qquad a_{ij} = \lambda(i, j) + \lambda(j, i),\ i \neq j\,.$$

If

$$\sup_i \sum_{j\neq i} (\lambda(i, j) + \lambda(j, i)) < \infty$$

then (9.6) is satisfied for $p = 1$ and $q = \infty$.

9.2 Uniqueness for general systems

We now consider a general system of the form

$$X_i(t) = U_i(t) + F_i(X, \cdot-) \cdot Y_i(t) \tag{9.8}$$

where for each i, Y_i is an $(L, \hat{H})^\#$-semimartingale and $F_i : D_{L^\infty}[0, \infty) \to \hat{H}$ and is nonanticipating. Recall that while the product space L^∞ is not a Banach space, it is metrizable with a complete metric, for example,

$$d_{L^\infty}(x, y) = \sum_{k=1}^{\infty} \frac{\|x_k - y_k\|_L \wedge 1}{2^k}.$$

We could allow L and $\hat{H}$ to depend on i, but the notation is bad enough already. With (8.4) in mind, the following theorem employs L_1-estimates. A similar result could be stated using L_2-estimates.

Theorem 9.1. *Suppose that for each $T > 0$, there exists a nonnegative function C_T such that for every cadlag, adapted, $\hat{H}$-valued process Z and each stopping time $\tau \leq T$ and positive number $t \leq 1$*

$$E[\sup_{s\leq t} \|Z_- \cdot Y_i(\tau+s) - Z_- \cdot Y_i(\tau)\|_{\hat{H}}] \leq C_T(t) E[\sup_{s\leq t} \|Z(\tau+s)\|_{\hat{H}}] \quad (9.9)$$

and $\lim_{t\to 0} C_T(t) = 0$ (cf. (8.4)). Assume (cf. (9.2)) that for $x, y \in D_{L^\infty}[0,\infty)$ and all $t \geq 0$, $\|F_i(x,t)\| \leq b_i$ and

$$\|F_i(x,t) - F_i(y,t)\|_{\hat{H}} \leq \sum_j a_{ij} \sup_{s\leq t} \|x_j(s) - y_j(s)\|_L \quad (9.10)$$

and that for some positive sequence $\{\alpha_i\}$ and some p and q satisfying $p^{-1} + q^{-1} = 1$

$$\|\alpha_j b_j\|_{q,j} < \infty\,, \quad (9.11)$$

and

$$\left\| \left\| \frac{\alpha_i a_{ij}}{\alpha_j} \right\|_{p,j} \right\|_{q,i} < \infty. \quad (9.12)$$

Then there exists a unique solution of the system (9.8).

Remark 9.2. a) Shiga and Shimizu [55] give a similar result for systems of diffusions.

b) One advantage of L_1 and L_2 estimates over the probability estimates used in Theorem 7.1 is that a direct, iterative approach to existence is possible.

Proof. To show uniqueness, let X and $\tilde{X}$ be solutions of (9.8) and let $\tau = \inf\{t : X(t) \neq \tilde{X}(t)\}$. Fix $T > 0$. As in (7.3), $X(\tau) = \tilde{X}(\tau)$, so by (9.9) and (9.10),

$$\begin{aligned} &E[\sup_{s\leq t} \|X_i(\tau\wedge T+s) - \tilde{X}_i(\tau\wedge T+s)\|_L] \\ &\quad = E[\sup_{s\leq t} \| \left(F_i(X,\cdot-) - F_i(\tilde{X},\cdot-)\right) \cdot Y_i(\tau\wedge T+s)\|_L] \\ &\quad \leq C_T(t) \sum_j a_{ij} E[\sup_{s\leq t} \|X_j(\tau\wedge T+s) - \tilde{X}_j(\tau\wedge T+s)\|_L], \end{aligned}$$

and hence, as in (9.5),

$$\begin{aligned} &\left\| \alpha_i E[\sup_{s\leq t} \|X_i(\tau\wedge T+s) - \tilde{X}_i(\tau\wedge T+s)\|_L] \right\|_{q,i} \quad (9.13) \\ &\leq \; C_T(t) \left\| \left\| \frac{\alpha_i a_{ij}}{\alpha_j} \right\|_{p,j} \right\|_{q,i} \left\| \alpha_j E[\sup_{s\leq t} \|X_j(\tau\wedge T+s) - \tilde{X}_j(\tau\wedge T+s)\|_L] \right\|_{q,j} \end{aligned}$$

and selecting $t > 0$ such that

$$C_T(t)\left\|\left\|\frac{\alpha_i a_{ij}}{\alpha_j}\right\|_{p,j}\right\|_{q,i} < 1, \tag{9.14}$$

we see that $\tau \geq \tau \wedge T + t$ a.s, which, in particular, implies $\tau > T$ a.s. Since T is arbitrary, $\tau = \infty$ a.s. and the uniqueness follows.

If t satisfies (9.14), then existence on the time interval $[0,t]$ follows by iteration using (9.13), that is, let $X^0 \equiv x^0$ and define X^n recursively by

$$X_i^{n+1}(t) = U_i(t) + F_i(X^n, \cdot-)\cdot Y_i(t).$$

Then as in (9.13),

$$\left\|\alpha_i E[\sup_{s\leq t}\|X_i^{n+1}(s) - X_i^n(s)\|_L]\right\|_{q,i}$$

$$\leq C_T(t)\left\|\left\|\frac{\alpha_i a_{ij}}{\alpha_j}\right\|_{p,j}\right\|_{q,i}\left\|\alpha_j E[\sup_{s\leq t}\|X_j^n(s) - X_j^{n-1}(s)\|_L]\right\|_{q,j} \tag{9,15}$$

and it follows that $\{X^n\}$ is Cauchy. For $s < T$, if existence is known on $[0,s]$, then the solution can be extended to $[0,s+t]$ by the same interation argument. Consequently, existence holds on $[0,T]$, and since T is arbitrary, we have global existence of the solution. □

9.3 Convergence of sequences of systems

We now consider a sequence of systems of the form

$$X_{n,i}(t) = U_{n,i}(t) + F_{n,i}(X_n, \cdot-)\cdot Y_{n,i}(t) \tag{9.16}$$

and extend the convergence theorems of Section 7. to this settings. (Note that the extension to finite systems is immediate.) We view $X_n = (X_{n,1}, X_{n,2},\ldots)$ and $U_n = (U_{n,1}, U_{n,2},\ldots)$ as processes in $D_{L^\infty}[0,\infty)$ and $F_n = (F_{n,1}, F_{n,2},\ldots)$ as a mapping from $D_{L^\infty}[0,\infty)$ into $D_{\hat{H}^\infty}[0,\infty)$, and by $(U_n, Y_n) \Rightarrow (U,Y)$, we mean that

$$\begin{aligned}(U_n, Y_{n,1}(\varphi_{1,1}),\ldots,Y_{n,1}(\varphi_{1,m_1}),\ldots,Y_{n,l}(\varphi_{l,1}),\ldots,Y_{n,l}(\varphi_{l,m_l}))\\ \Rightarrow (U, Y_1(\varphi_{1,1}),\ldots,Y_1(\varphi_{1,m_1}),\ldots,Y_l(\varphi_{l,1}),\ldots,Y_l(\varphi_{l,m_l}))\end{aligned}$$

in $D_{L^\infty\times\mathbb{R}^{m_1+\cdots+m_l}}[0,\infty)$ for all choices of $\varphi_{i,j} \in H$. Recall that convergence in $D_{L^\infty}[0,\infty)$ is equivalent to convergence of the first k components in $D_{L^k}[0,\infty)$ for each k and that a sequence of processes $\{X^n\}$ in $D_{L^\infty}[0,\infty)$ satisfies the compact containment condition if and only if each component satisfies the compact containment condition. We need the following modification of Condition C2.

C.3a There exist $G_n, G : D_{L^\infty}[0,\infty)\times T_1[0,\infty) \to D_{\hat{H}^\infty}[0,\infty)$ such that $F_n(x)\circ\lambda = G_n(x\circ\lambda,\lambda)$ and $F(x)\circ\lambda = G(x\circ\lambda,\lambda)$, $x \in D_{L^\infty}[0,\infty)$, $\lambda \in T_1[0,\infty)$.

C.3b For each compact $\mathcal{K} \subset D_{L^\infty}[0,\infty) \times T_1[0,\infty)$ and $t > 0$,

$$\sup_{(x,\lambda)\in\mathcal{K}} \sup_{s\leq t} d_{\hat{H}^\infty}(G_n(x,\lambda,s), G(x,\lambda,s)) \to 0.$$

C.3c For $\{(x_n,\lambda_n)\} \in D_{L^\infty}[0,\infty) \times T_1[0,\infty)$, $\sup_{s\leq t} d_{L^\infty}(x_n(s), x(s)) \to 0$ and $\sup_{s\leq t} |\lambda_n(s) - \lambda(s)| \to 0$ for each $t > 0$ implies

$$\sup_{s\leq t} d_{\hat{H}^\infty}(G(x_n,\lambda_n,s), G(x,\lambda,s)) \to 0.$$

Theorem 9.3. *Let $L = \mathbf{R}^k$. Suppose that (U_n, X_n, Y_n) satisfies (9.16), that $(U_n, Y_n) \Rightarrow (U, Y)$, and that $\{F_n\}$ and F satisfy Condition C.3. For each i, assume that $\{Y_{n,i}\}$ is uniformly tight and that* $\sup_n \sup_x \|F_{n,i}(x,\cdot)\|_{H^k} < \infty$. *Then $\{(U_n, X_n, Y_n)\}$ is relatively compact and any limit point satisfies (9.8).*

Proof. As in the proof of Theorem 7.5, it is enough to show the relative compactness of (U_n, X_n, Y_n). By Proposition 6.2, $\{F_{n,i}(X_n, \cdot-) \cdot Y_{n,i}\}$ is uniformly tight. By Lemma 6.9 and Proposition 6.14, there exists a γ_n such that $\{(U_n \circ \gamma_n, F_n(X_n, \cdot-) \cdot Y_n \circ \gamma_n)\}$ is relatively compact, which in turn implies $\{(X_n \circ \gamma_n, U_n \circ \gamma_n, F_n(X_n, \cdot-) \cdot Y_n \circ \gamma_n)\}$ is relatively compact. Setting $\hat{X}_n = X_n \circ \gamma_n$, etc., Condition C.3a implies

$$\hat{X}_n(t) = \hat{U}_n(t) + G_n(\hat{X}_n, \gamma_n, \cdot-) \cdot \hat{Y}_n(t)$$

and the relative compactness of $\{(\hat{X}_n, \hat{U}_n, \hat{Y}_n)\}$ implies the relative compactness of

$$\{(\hat{X}_n, \hat{U}_n, \hat{Y}_n, G_n(\hat{X}_n, \gamma_n, \cdot), \gamma_n)\}.$$

Recalling that relative compactness in $D_{L^\infty}[0,\infty)$ is equivalent to relative compactness of the first k components in $D_{L^k}[0,\infty)$ for each k, by Lemma 6.16, $\{(X_n, U_n, Y_n)\}$ is relatively compact, and the theorem follows from the analog of Proposition 7.4. □

Theorem 9.4. *Suppose that (U_n, X_n, Y_n) satisfies (9.16), that $(U_n, Y_n) \Rightarrow (U, Y)$, and that $\{F_n\}$ and F satisfy Condition C.3. For each i, assume that $\{Y_{n,i}\}$ is uniformly tight, that* $\sup_n \sup_x \|F_{n,i}(x,\cdot)\|_{\hat{H}} < \infty$, *and that for each sequence of positive numbers $\{\kappa_j\}$, there exists a compact $K_{i,\{\kappa_j\}} \subset \hat{H}$ such that* $\sup_{s\leq t} \|x_j(s)\|_L \leq \kappa_j$ *for all j implies $F_{n,i}(x,s) \in K_{i,\{\kappa_j\}}$ for all $s \leq t$ and n. Then $\{(U_n, X_n, Y_n)\}$ is relatively compact and any limit point satisfies (9.8).*

Proof. Again it is sufficient to verify the relative compactness of $\{(U_n, X_n, Y_n)\}$. The uniform tightness of $\{Y_n\}$ and the boundedness of the F_n imply the stochastic boundedness of $\{\sup_{s\leq t} \|X_{n,i}(s)\|_L\}$ for each i. The compactness condition on the $F_{n,i}$ then ensures that $\{F_{n,i}(X_n, \cdot)\}$ satisfies the compact containment condition. It follows that the sequence $\{F_{n,i}(X_n, \cdot-) \cdot Y_n\}$ satisfies the compact containment condition by Lemma 6.18 which in turn implies $\{X_{n,i}\}$ satisfies the compact containment condition. Lemma 6.15 and Lemma

6.9 then ensure the existence of γ_n as in the proof of Theorem 9.3, and the remainder of the proof is the same. □

The proofs of the following corollaries are the same as for the corresponding results in Section 7..

Corollary 9.5. *a) Let $L = \mathbf{R}^k$. Suppose that F and G satisfy C.3a and C.3c and that for each i, $\sup_x \|F_i(x,\cdot)\|_{H^k} < \infty$ and Y_i is a standard $H^\#$-semimartingale. Then weak existence holds for (9.8).*

b) For general L, suppose that F and G satisfy C.3a and C.3c and that for each i, $\sup_x \|F_i(x,\cdot)\|_{\hat{H}} < \infty$, F_i satisfies the compactness condition of Theorem 9.4, and Y_i is a standard $(L,\hat{H})^\#$-semimartingale. Then weak existence holds for (9.8).

Corollary 9.6. *Suppose, in addition to the conditions of Corollary 9.5, that strong uniqueness holds for (9.8) for any version of (U,Y) for which each Y_i is a standard $H^\#$ (or $(L,\hat{H})^\#$)-semimartingale. Then any solution of (9.8) is a measurable function of (U,Y), that is, if X satisfies (9.8) and the finite linear combinations of $\{\varphi_k\}$ are dense in H, there exists a measurable mapping*

$$g : D_{L^\infty \times \mathbf{R}^\infty}[0,\infty) \to D_{L^\infty}[0,\infty)$$

such that $X = g(U, Y(\varphi_1,\cdot-), Y(\varphi_2,\cdot), \ldots)$. In particular, there exists a strong solution of (9.8).

Corollary 9.7. *Suppose, in addition to the conditions of Theorem 9.3 or Theorem 9.4, that strong uniqueness holds for (9.8) for any version of (U,Y) for which each Y_i is a standard $H^\#$ (or $(L,\hat{H})^\#$)-semimartingale and that $(U_n, Y_n) \to (U,Y)$ in probability. Then $(U_n, Y_n, X_n) \to (U,Y,X)$ in probability.*

10. McKean-Vlasov limits

We now consider an infinite system indexed by $i \in \mathbf{Z}$

$$X_i(t) = U_i(t) + F_i(X, Z, \cdot-) \cdot Y_i(t)$$

which we assume to be shift invariant in the sense that $\{(U_i, Y_i)\}$ is a stationary sequence and $F_i(x,z,t) = F_0(x_{\cdot+i}, z, t)$. We require that $\{(X_i, U_i, Y_i)\}$ also be stationary (which it will be if uniqueness holds) and that Z be the $\mathcal{P}(L)$-valued process given by

$$Z(t) = \lim_{k\to\infty} \frac{1}{2k+1} \sum_{i=-k}^{k} \delta_{X_i(t)}$$

where convergence is in the weak topology. Note that a.s. existence of the limit follows from the ergodic theorem and that $Z(t,\Gamma) = P\{X_i(t) \in \Gamma|\mathcal{I}\}$ a.s., independently of i, where $\mathcal{I}$ is the σ-algebra of invariant sets for the stationary sequence $\{(X_i, U_i, Y_i)\}$. Since $D_L[0,\infty)$ is a complete, separable metric space, there will exist a regular conditional distribution Q on $\mathcal{B}(D_L[0,\infty))$ for X_i given $\mathcal{I}$, and we can take $Z(t,\Gamma)$ to be $Q\{X_i(t) \in \Gamma\}$. Since for any probability measure on $D_L[0,\infty)$, the one dimensional distributions will be cadlag as $\mathcal{P}(L)$-valued functions, Z will be a cadlag process.

Typically, the driving processes in models of this type are assumed to be independent. (See, for example, Graham [20], Kallianpur and Xiong [30, 31], and Méléard [48] for further discussion and references.) In Section 11.1, we will see how a model of Kotelenez [37] can be interpreted as a system of the present type in which the Y_i are identical.

Let $a_i, b > 0$ and $\sum_i a_i < \infty$. We assume

$$\|F_0(x,z,t) - F_0(\tilde{x},\tilde{z},t)\|_{\hat{H}} \leq \sum_i a_i \sup_{s\leq t} \|x_i(s) - \tilde{x}_i(s)\| + b \sup_{s\leq t} \rho_W(z(s), \tilde{z}(s))$$

where ρ_W is the Wasserstein metric on $\mathcal{P}(L)$, that is, letting $B_1 = \{f \in \bar{C}(L) : |f(x)| \leq 1, |f(x) - f(y)| \leq \|x-y\|_L, x, y \in L\}$,

$$\rho_W(\mu,\nu) = \sup_{f\in B_1} \left| \int f d\mu - \int f d\nu \right|.$$

In addition, we assume that Y_i satisfies (9.9) of Theorem 9.1 and that $\sup_{x,z,t} \|F_0(x,z,t)\|_{\hat{H}} < \infty$. Then uniqueness holds as in the proof of Theorem 9.1 with $p = 1$ and $q = \infty$. Observe that

$$\rho_W(Z(t), \tilde{Z}(t)) \leq \lim_{k\to\infty} \frac{1}{k} \sum_{i=1}^{k} \|X_i(t) - \tilde{X}_i(t)\|_L .$$

To see that existence holds, let $Z^0(t) \equiv \mu$ for some fixed $\mu \in \mathcal{P}(L)$, and define X^{n+1} recursively as the solution of

$$X_i^{n+1}(t) = U_i(t) + F_i(X^{n+1}, Z^n, \cdot-) \cdot Y_i(t)$$

where Z^n is defined by

$$Z^n(t,\Gamma) = \lim_{k\to\infty} \frac{1}{k} \sum_{i=1}^{k} \delta_{X_i^n(t)}(\Gamma) \ a.s.$$

Note that X^{n+1} exists and is (strongly) unique, given Z^n, as in Theorem 9.1. By the strong uniqueness, X^n is a functional of (U,Y), and we can take $Z^n(t,\Gamma) = P\{X_i^n(t) \in \Gamma|\mathcal{I}\}$, where $\mathcal{I}$ is the invariant σ-algebra for (U,Y). In particular, as noted above, we can take Z^n to be a cadlag process in $\mathcal{P}(L)$. As in the proof of Theorem 9.1, we have

$$E[\sup_{s\le t}\|X_i^{n+1}(s)-X_i^n(s)\|_L] \tag{10.1}$$
$$\le C_T(t)\left(\sum_j a_j E[\sup_{s\le t}\|X_{i+j}^{n+1}(s)-X_{i+j}^n(s)\|_L]+bE[\sup_{s\le t}\rho_W(Z^n(s),Z^{n-1}(s))]\right)$$

Since by stationarity,

$$E[\sup_{s\le t}\|X_i^{n+1}(s)-X_i^n(s)\|_L]=E[\sup_{s\le t}\|X_j^{n+1}(s)-X_j^n(s)\|_L]\,,$$

we have

$$(1-C_T(t)\sum_j a_j)E[\sup_{s\le t}\|X_i^{n+1}(s)-X_i^n(s)\|_L]\le C_T(t)bE[\sup_{s\le t}\|X_i^n(s)-X_i^{n-1}(s)\|_L]$$

and we have convergence on $[0,t]$ provided

$$\frac{C_T(t)b}{1-C_T(t)\sum_j a_j}<1.$$

Existence for all t then follows.

We now obtain X as the limit of finite particle systems. In particular, we consider the system

$$X_i^n(t)=U_i(t)+F_i(X^n,Z^n,\cdot-)\cdot Y_i(t)$$

for $-n\le i\le n$ with "wrap-around" boundary conditions, that is, for $j\notin\{-n,\ldots,n\}$, we set $X_j^n=X_{j(n)}^n$ where $-n\le j(n)\le n$ and $|j-j(n)|$ is a multiple of $2n+1$. Z^n is the empirical measure

$$Z^n(t)=\frac{1}{2n+1}\sum_{i=-n}^{n}\delta_{X_i^n(t)}.$$

Let $Z^{(n)}$ be defined by

$$Z^{(n)}(t)=\frac{1}{2n+1}\sum_{i=-n}^{n}\delta_{X_i(t)}$$

and note that $\lim_{n\to\infty}\rho_W(Z^{(n)}(t),Z(t))=0$ a.s. We want this convergence to be uniform. In general, for sequences $\{q_n\}\subset\mathcal{P}(D_L[0,\infty))$, $q_n\to q$ does not imply that the marginals converge, let alone uniformly, unless the marginals of the limit q are continuous. Consequently, we assume that Z is continuous which implies that

$$\lim_{n\to\infty}\sup_{s\le t}\rho_W(Z^{(n)}(s),Z(s))=0\ a.s. \tag{10.2}$$

for each $t>0$. We suspect that in the current situation (10.2) will hold without the continuity assumption; however, the assumption is in fact rather mild and will hold, for example, if for all i, j, (U_i, Y_i) and (U_j, Y_j) have a.s. no common discontinuities.

As in (10.1), we have

$$\begin{aligned}
&E[\sup_{s\le t} \|X_i(s) - X_i^n(s)\|_L] \\
&\le C_T(t)\left(\sum_j a_{j-i} E[\sup_{s\le t}\|X_j(s) - X_j^n(s)\|_L] + bE[\sup_{s\le t}\rho_W(Z(s), Z^n(s))]\right) \\
&\le C_T(t)\left(\sum_j a_{j-i} E[\sup_{s\le t}\|X_{j(n)}(s) - X_j^n(s)\|_L] + bE[\sup_{s\le t}\rho_W(Z^{(n)}(s), Z^n(s))]\right. \\
&\qquad \left. + \sum_{|j|>n} a_{j-i} E[\sup_{s\le t}\|X_j(s) - X_{j(n)}(s)\|_L] + bE[\sup_{s\le t}\rho_W(Z(s), Z^{(n)}(s))]\right)
\end{aligned}$$

Noting that

$$E[\sup_{s\le t}\rho_W(Z^{(n)}(s), Z^n(s))] \le \frac{1}{2n+1}\sum_{j=-n}^{n} E[\sup_{s\le t}\|X_j(s) - X_j^n(s)\|_L],$$

if we define the matrix D^n by

$$D_{ij}^n = \sum_{j'(n)=j} a_{j'-i} + \frac{b}{2n+1},$$

and set $R_i^n(t) = E[\sup_{s\le t}\|X_i(s) - X_i^n(s)\|_L]$ and

$$S_i^n(t) = \sum_{|j|>n} a_{j-i} E[\sup_{s\le t}\|X_j(s) - X_{j(n)}(s)\|_L] + bE[\sup_{s\le t}\rho_W(Z(s), Z^{(n)}(s))]$$

for $-n \le i \le n$, then for t sufficiently small, $C_T(t)(\sum_j a_j + b) < 1$, and $(I - C_T(t)D^n)^{-1} = \sum_{k=0}^{\infty}(C_T(t)D^n)^k$ is a matrix with positive entries which implies

$$R^n(t) \le (I - C_T(t)D^n)^{-1} S^n(t)\,,$$

and the convergence follows.

11. Stochastic partial differential equations

Let Y be an $(L, \hat{H})^{\#}$-semimartingale. We consider equations which, formally, can be written as

$$X(t) = X(0) + \int_0^t AX(s)ds + B(X(\cdot-)) \cdot Y(t) \tag{11.1}$$

where A is a linear, in general unbounded, operator on L. Typically, L is a Banach space of functions on a Euclidean space $\mathbf{R}^d$ and A is a differential operator (for example, $A = \Delta$), but for our purposes we will let L be arbitrary and assume that A is the generator of a strongly continuous semigroup on L. In most interesting cases (see Walsh [60] and Da Prato and Zabczyk [11] for systematic developments of the theory), (11.1) cannot hold rigorously in that no solution will exist taking values in the domain $\mathcal{D}(A)$ of the unbounded operator A. Consequently, (11.1) must be interpreted in a weak sense. For example, letting A^* denote the adjoint of A defined on some subspace $\mathcal{D}(A^*)$ of L^*, we can write

$$\langle h, X(t)\rangle = \langle h, X(0)\rangle + \int_0^t \langle A^* h, X(s)\rangle ds + \langle h, B(X(\cdot -)) \cdot Y(t)\rangle \tag{11.2}$$

for $h \in \mathcal{D}(A^*)$. For the right side of (11.2) to be defined, we need only require that X takes values in L and that $B(X)$ takes values in $\hat{H}$. More generally, it would be sufficient for $h \in \mathcal{D}(A^*)$ to be extended to the range of B in such a way that $\langle h, B(X(t))\rangle \in H$. Then (11.2) becomes

$$\langle h, X(t)\rangle = \langle h, X(0)\rangle + \int_0^t \langle A^* h, X(s)\rangle ds + \langle h, B(X(\cdot -))\rangle \cdot Y(t) . \tag{11.3}$$

A process X satisfying (11.2) is usually refered to as a *weak* solution of (11.1); however, note that this functional analytic notion of "weak" should not be confused with the "solution in distribution" notion of "weak" considered in Section 7..

A third notion of solution arises when A is the generator of an operator semigroup $\{S(t\}$ on L that can be extended to a semigroup on $\hat{H}$, starting with the obvious definition $S(t)(\sum a_{ij} f_i \varphi_j) = \sum a_{ij} \varphi_j S(t) f_i$. An infinite-dimensional, stochastic version of the variation of parameters formula leads to

$$X(t) = S(t)X(0) + S(t - \cdot)B(X(\cdot -)) \cdot Y(t) . \tag{11.4}$$

To clarify the meaning of the last term, if Y is given by a semimartingale random measure, then the last term can be written as

$$\int_{U \times [0,t]} S(t-s)B(X(s-), u)Y(du \times ds).$$

See Da Prato and Zabczyk [11, Chapter 6] for a discussion of the relationships among these notions of solution in one setting.

A variety of weak convergence results for stochastic partial differential equations exist in the literature. See, for example, Bhatt and Mandrekar [2], Blount [3, 4], Brzeźniak, Capiński, and Flandoli [5], Fichtner and Manthey [19], Jetschke [28]), Kallianpur and Perez-Abreu [29], Gyõngy [21], and Twardowska [58]. We have not yet had the opportunity to explore the application of the convergence results developed in previous sections to stochastic partial differential equations. In this section, we collect a few of the results that may be useful in making that application.

11.1 Estimates for stochastic convolutions

Let V be an adapted, cadlag, $\hat{H}$-valued process. If for each $t > 0$, the mapping $s \in [0,t] \to S(t-s)V(s) \in \hat{H}$ is cadlag, then

$$Z(t) = S(t - \cdot)V(\cdot -) \cdot Y(t)$$

is well-defined for each t; however, the properties of Z as a process are less clear. In particular, for the solution of (11.4) to be cadlag, the stochastic integral must be cadlag. We begin with a discussion of a result of Kotelenez [35].

Lemma 11.1. *Let Z be a cadlag, L-valued, $\{\mathcal{F}_t\}$-adapted process, and let $\psi : L \to [0,\infty)$. Suppose that*

$$E[\psi(Z(t+h))|\mathcal{F}_t] \le E[\Lambda(t+h) - \Lambda(t)|\mathcal{F}_t] + e^{\beta(t+h)-\beta(t)}\psi(Z(t))$$

for $t, h \ge 0$, where Λ is a cadlag, nondecreasing $\{\mathcal{F}_t\}$-adapted process with $\alpha(t) = E[\Lambda(t)] < \infty$ and β is a nondecreasing function. Then for any stopping time τ and $\delta > 0$,

$$\begin{aligned} P\{\sup_{s \le \tau \wedge t} \psi(Z(s)) > \delta\} & \\ \le \frac{1}{\delta}\left(E[\psi(Z(0))]e^{\beta(t)-\beta(0)} + E\left[\int_0^{t\wedge\tau} e^{\beta(t)-\beta(s)} d\Lambda(s)\right]\right) & \\ \le \frac{1}{\delta}\left(E[\psi(Z(0))]e^{\beta(t)-\beta(0)} + \int_0^t e^{\beta(t)-\beta(s)} d\alpha(s)\right). & \qquad (11.5) \end{aligned}$$

If, in addition, Λ is predictable and $\psi(Z(0) = 0$, then for $0 < p < 1$,

$$E[(\sup_{s \le \tau \wedge t} \psi(Z(s)))^p] \le \left(\frac{e^{\beta(t)}}{1-p} + 1\right) E[(\Lambda(\tau \wedge t))^p]. \qquad (11.6)$$

Proof. Observe that

$$\begin{aligned} E[e^{-\beta(t+h)}\psi(Z(t+h) - \int_0^{t+h} e^{-\beta(s)} d\Lambda(s)|\mathcal{F}_t] & \\ \le e^{-\beta(t+h)} E[\Lambda(t+h) - \Lambda(t)|\mathcal{F}_t] + e^{-\beta(t)}\psi(Z(t) & \\ -E\left[\int_0^{t+h} e^{-\beta(s)} d\Lambda(s)|\mathcal{F}_t\right] & \\ \le e^{-\beta(t)}\psi(Z(t)) - \int_0^t e^{-\beta(s)} d\Lambda(s) & \\ \equiv U(t) & \end{aligned}$$

so that U is a supermartingale. Consequently,

$$\begin{aligned}
P\{\sup_{s\le\tau\wedge t}\psi(Z(s))\ge\delta\} &\le P\{\sup_{s\le\tau\wedge t}U(s)\ge e^{-\beta(t)}\delta\}\\
&\le \frac{1}{\delta}e^{\beta(t)}(E[U(0)]+E[U^-(t)])\\
&\le \frac{1}{\delta}\Big(e^{\beta(t)-\beta(0)}E[\psi(Z(0))]\\
&\qquad +E\Big[\int_0^{t\wedge\tau}e^{\beta(t)-\beta(s)}dA(s)\Big]\Big)
\end{aligned}$$

where $U^-=(-U)\vee 0$ and the last inequality follows from Doob's inequality, and (11.5) follows.

To prove (11.6), we follow an argument from Ichikawa [22]. Let $\sigma_x=\inf\{t: A(t)\ge x\}$. Since A is predictable, σ_x is predictable and can be approximated by an increasing sequence of stopping times σ_x^n satisfying $\sigma_x^n<\sigma_x$. Then

$$\begin{aligned}
P\{\sup_{s\le\tau\wedge t}\psi(Z(s))>x\} &\le \frac{1}{x}E\left[\int_0^{t\wedge\tau\wedge\sigma_x^n}e^{\beta(t)-\beta(s)}dA(s)\right]+P\{\tau\wedge t>\sigma_x^n\}\\
&\le \frac{1}{x}e^{\beta(t)}E[x\wedge A(t\wedge\tau)]+P\{\tau>\sigma_x^n\}.
\end{aligned}$$

Noting that $\lim_{n\to\infty}P\{\tau>\sigma_x^n\}=P\{\tau\wedge t\ge\sigma_x\}=P\{A(t\wedge\tau)\ge x\}$, we have

$$P\{\sup_{s\le\tau\wedge t}\psi(Z(s))>x\}\le\frac{1}{x}e^{\beta(t)}E[x\wedge A(t\wedge\tau)]+P\{A(t\wedge\tau)\ge x\},$$

and multiplying both sides by px^{p-1} and integrating gives (11.6). □

Lemma 11.2. *Let $L=L_2(\nu)$, and let M be a worthy martingale measure with dominating measure K satisfying (2.10). Let $\hat K$ be given by (2.11), and suppose that M determines a standard $(L,\hat H)^\#$-martingale. For $0\le s\le t$, let $\tilde\Gamma(t,s)$ and $\Gamma(t,s)$ be bounded linear operators on L satisfying $\|\tilde\Gamma(t,s)\|\le C(t)$ and $\|\Gamma(t,s)\|\le e^{\beta(t)-\beta(s)}$, where C and β are nondecreasing. Suppose that for $V\in S^0_{\hat H}$, Z, defined by*

$$Z(t,\cdot)=\int_{U\times[0,t]}\tilde\Gamma(t,s)V(s-,\cdot,u)M(du\times ds),\tag{11.7}$$

is cadlag and

$$\begin{aligned}
&\int_{U\times[0,t]}\tilde\Gamma(t+h,s)V(s-,\cdot,u)M(du\times ds)\\
&\qquad=\Gamma(t+h,t)\int_{U\times[0,t]}\tilde\Gamma(t,s)V(s-,\cdot,u)M(du\times ds).
\end{aligned}$$

For $0\le t\le T$, define

$$A(t) = C(T) \int_{U\times[0,t]} \|V(s-,\cdot,u)\|_L^2 \hat{K}(du \times ds).$$

Then for each stopping time τ,

$$P\{\sup_{s\le\tau\wedge T} \|Z(s)\|_L^2 > \delta\} \tag{11.8}$$
$$\le \frac{1}{\delta} E\left[C(T)\int_{U\times[0,T\wedge\tau]} e^{\beta(T)-\beta(s)} \|V(s-,\cdot,u)\|_L^2 \hat{K}(du\times ds)\right]$$

and assuming $\hat{K}$ *is predictable, for* $0 < p < 2$,

$$E[\sup_{s\le\tau\wedge T} \|Z(s)\|_L^p]$$
$$\le \left(\frac{2e^{\beta(t)}}{2-p}+1\right) E[(C(T)\int_{U\times[0,\tau\wedge T]} \|V(s-,\cdot,u)\|_L^2 \hat{K}(du\times ds))^{p/2}]. \tag{11.9}$$

Remark 11.3. A number of closely related results exist in the literature, formulated in terms of Hilbert space-valued martingales. The inequality (11.8) is essentially Theorem 1 of Kotelenez [35]. The restriction to $0 < p < 2$ in (11.9) is not necessary. Kotelenez [36] estimates the second moment of of $\sup_{0\le s<\tau\wedge T} \|Z(s)\|_L$ in the setting of Hilbert space-valued martingales. Ichikawa [22] gives moment estimates similar to those of Kotelenez [36] for $0 < p \le 2$, and under the assumption that the driving martingale is continuous, for $p > 2$. In the particular case of finite dimensional and Hilbert space-valued Wiener processes, Tubaro [57], Da Prato and Zabczyk [12], Da Prato and Zabczyk [11, Theorem 7.3]), and Zabczyk [62] give estimates for $p > 2$. Walsh [60, Theorem 7.13] also considers more general convolution integrals of the form $\int_{U\times[0,t]} g(t,s,u)M(du\times ds)$ under Hölder continuity assumptions on g.

Proof. Observe that

$$E[Z(t+h,\cdot)^2|\mathcal{F}_t]$$
$$= E\left[\left(\int_{U\times(t,t+h]} \tilde{\Gamma}(t+h,s)V(s-,\cdot,u)M(du\times ds)\right)^2\middle|\mathcal{F}_t\right]$$
$$+\left(\int_{U\times[0,t]} \tilde{\Gamma}(t+h,s)V(s-,\cdot,u)M(du\times ds)\right)^2$$
$$\le E\left[\int_{U\times(t,t+h]} \left|\tilde{\Gamma}(t+h,s)V(s-,\cdot,u)\right|^2 \hat{K}(du\times ds)\middle|\mathcal{F}_t\right]$$
$$+\left(\Gamma(t+h,t)\int_{U\times[0,t]} \tilde{\Gamma}(t,s)V(s-,\cdot,u)M(du\times ds)\right)^2 \tag{11.10}$$

Then for $0 \le t < t+h \le T$, integrating (11.10) with respect to ν, we have

$$E[\|Z(t+h)\|_L^2|\mathcal{F}_t] \le E[\Lambda(t+h) - \Lambda(t)|\mathcal{F}_t] + e^{\beta(t+h)-\beta(t)}\|Z(t)\|_L^2$$

which, by Lemma 11.1 with $\psi(z) = \|z\|_L^2$, gives the result. □

For a filtration $\{\mathcal{F}_t\}$, let $\mathcal{T}$ denote the collection of $\{\mathcal{F}_t\}$-stopping times, let $\mathcal{S}_{\hat{H}}^0$ be given in Definition 5.2, let $\mathcal{D}_{\hat{H}}$ be the collection of cadlag, $\hat{H}$-valued, $\{\mathcal{F}_t\}$-adapted processes, and let $\mathcal{D}_L$ be the collection of cadlag, L-valued, $\{\mathcal{F}_t\}$-adapted processes. In the following development, the primary examples of interest are mappings of the form

$$G(X,t) = T(t-\cdot)X(\cdot)\cdot Y(t),$$

or in the case of semimartingale random measures,

$$G(X,t) = \int_{U\times[0,t]} T(t-s)X(s)Y(du\times ds).$$

Proposition 11.4. *Suppose that $G : \mathcal{S}_{\hat{H}}^0 \to \mathcal{D}_L$ has the following properties:*

– *$G(X)$ is linear, that is $G(aX+bY) = aG(X)+bG(Y)$, $X, Y \in \mathcal{S}_{\hat{H}}^0$, $a, b \in \mathbf{R}$.*
– *For each $t > 0$,*

$$\mathcal{H}_t^0 = \{\|G(X,\tau\wedge t)\|_L : X \in \mathcal{S}_{\hat{H}}^0, \sup_{s\le t}\|X(s)\|_{\hat{H}} \le 1, \tau \in \mathcal{T}\}$$

is stochastically bounded.

Then G extends to a mapping $G : \mathcal{D}_{\hat{H}} \to \mathcal{D}_L$.

Remark 11.5. Note that if Y is a standard, $(L,\hat{H})^{\#}$-semimartingale, then G defined by $G(X,t) = X_-\cdot Y(t)$ satisfies the conditions of the proposition. Recall that $X_-\cdot Y(\tau\wedge t) = X_-^{\tau}\cdot Y(t)$, where $X^{\tau} = I_{[0,\tau)}X$.

Proof. The proof is essentially the same as in the definition of the stochastic integral. As before, for each $t, \delta > 0$, there exists a $K(t,\delta)$ such that

$$P\{\|G(X,\tau\wedge t)\|_L \ge K(t,\delta)\} \le \delta$$

for all $X \in \mathcal{S}_{\hat{H}}^0$ with $\|X\|_{\hat{H}} \le 1$. For $X \in \mathcal{D}_{\hat{H}}$, let X^{ϵ} be given by (5.4). Then

$$P\{\sup_{s\le t}\|G(X^{\epsilon_1},s) - G(X^{\epsilon_2},s)\|_L \ge (\epsilon_1+\epsilon_2)K(t,\delta)\} \le \delta$$

and it follows that $\lim_{\epsilon\to 0} G(X^{\epsilon})$ converges to a cadlag process which we define to be $G(X)$. □

Theorem 11.6. *Let G satisfy the conditions of Proposition 11.4. Suppose that for each stopping time τ, bounded by a constant, and $t, \delta > 0$, there exists $K_\tau(t,\delta)$ with $\lim_{t\to 0} K_\tau(t,\delta) = 0$ such that for all stopping times σ satisfying $\tau \le \sigma \le \tau + t$*

$$P\{\|G(X,\sigma) - G(X,\tau)\|_L \ge K(t,\delta)\} \le \delta$$

for all $X \in S^0_{\hat{H}}$ with $\|X\|_{\hat{H}} \le 1$. Let $F : D_L[0,\infty) \to D_{\hat{H}}[0,\infty)$ satisfy $\sup_{s\le t} \|F(x,s) - F(y,s)\|_{\hat{H}} \le M \sup_{s\le t} \|x(s) - y(s)\|_L$. Then for $U \in \mathcal{D}_L$, there exists at most one solution of

$$X(t) = U(t) + G(F(X),t).$$

Proof. The proof is the same as for Theorem 7.1. □

11.2 Eigenvector expansions

Suppose that L is spanned by a sequence $\{f_k\}$ of eigenvectors for A, that is, $Af_k = -\lambda_k f_k$, and that $h_k \in L^*$ satisfies $\langle h_k, f_l\rangle = \delta_{kl}$. Then $A^*_k h_k = -\lambda_k h_k$ and (11.11) implies

$$\langle h_k, X(t)\rangle = \langle h_k, X(0)\rangle - \lambda_k \int_0^t \langle h_k, X(s)\rangle ds + \langle h_k, B(X(\cdot -))\rangle \cdot Y(t)\,. \quad (11.11)$$

At least heuristically, if we define $V_k(t) = \langle h_k, X(t)\rangle$, then

$$X(t) = \sum_{k=1}^{\infty} V_k(t) f_k. \quad (11.12)$$

To study the convergence of (11.12), we need to be able to estimate V_k, and the following lemma of Blount [3, 4] gives a useful approach.

Lemma 11.7. *Let M be a continous, $\mathbf{R}$-valued martingale and Γ a constant with $[M]_t = \int_0^t U(s)ds$ and $|U(s)| \le \Gamma$, let C be an adapted process and C_0 a constant satisfying*

$$\sup_{s\le t} |C(s)| \le C_0,$$

and let $\lambda > 0$. Suppose V satisfies

$$V(t) = V(0) - \lambda \int_0^t V(s)ds + \int_0^t C(s)ds + M(t)\,.$$

Then for each $a > 0$,

$$P\{\sup_{s\le t} |V(s)| \ge a + |V(0)| + \frac{C_0}{\lambda}\} \le \frac{\lambda t}{\exp\{\frac{\lambda a^2}{4\Gamma}\} - 1}$$

Proof. The proof is based on a comparison with the Ornstein-Uhlenbeck process satisfying $d\tilde{V} = -\lambda \tilde{V} dt + \sqrt{\Gamma} dW$ (W a standard Brownian motion). See Lemma 3.19 of Blount [3] and Lemma 1.1 of Blount [4]. □

11.3 Particle representations

With the results of Section 10. in mind, consider the system $\{(X_i, A_i)\}$ with $X_i \in \mathbf{R}^d$ and $A_i \in \mathbf{R}$ satisfying

$$
\begin{aligned}
X_i(t) = X_i(0) \quad &+ \int_0^t \sigma(X_i(s), V(s))dW_i(s) + \int_0^t c(X_i(s), V(s))ds \\
&+ \int_{U\times[0,t]} \alpha(X_i(s), V(s), u)W(du \times ds)
\end{aligned}
$$

and

$$
\begin{aligned}
A_i(t) = A_i(0) \quad &+ \int_0^t A_i(s)\gamma^T(X_i(s), V(s))dW_i(s) + \int_0^t A_i(s)d(X_i(s), V(s))ds \\
&+ \int_{U\times[0,t]} A_i(s)\beta(X_i(s), V(s), u)W(du \times ds)
\end{aligned}
$$

where the W_i are independent, standard $\mathbf{R}^d$-valued Brownian motions and W is Gaussian white noise with $E[W(A,t)W(B,t)] = \mu(A \cap B)t$. Assume that $\{(A_i(0), X_i(0))\}$ are iid and independent of W_i and W. V is the signed measure-valued process obtained by setting

$$
\langle \varphi, V(t)\rangle = \lim_{n\to\infty} \frac{1}{n}\sum_{i=1}^n A_i(t)\varphi(X_i(t)).
$$

For simplicity, assume that σ, c, γ, α, β, and d are bounded. Letting Z be the $\mathcal{P}(\mathbf{R}^{d+1})$-valued process

$$
Z(t) = \lim_{n\to\infty} \frac{1}{n}\sum_{i=1}^n \delta_{(X_i(t),A_i(t))},
$$

we have

$$
\langle \varphi, V(t)\rangle = \int_{\mathbf{R}^{d+1}} a\varphi(x)Z(t, dx \times da),
$$

and the uniqueness result of Section 10. can be translated into a uniqueness result here. Applying Itô's formula to $A_i(t)\varphi(X_i(t))$ we obtain

$$
\begin{aligned}
A_i(t)\varphi(X_i(t)) \quad = \quad & A_i(0)\varphi(X_i(0)) + \int_0^t A_i(s)\varphi(X_i(s))\gamma(X_i(s), V(s))dW_i(s) \\
&+ \int_0^t A_i(s)\varphi(X_i(s))d(X_i(s), V(s))ds \\
&+ \int_{U\times[0,t]} A_i(s)\varphi(X_i(s))\beta(X_i(s), V(s), u)W(du \times ds) \\
&+ \int_0^t A_i(s)L(V(s))\varphi(X_i(s))ds
\end{aligned}
$$

$$+\int_0^t A_i(s)\nabla\varphi(X_i(s))\cdot\sigma(X_i(s),V(s))dW_i(s)$$
$$+\int_{U\times[0,t]} A_i(s)\nabla\varphi(X_i(s))\alpha(X_i(s),V(s),u)W(du\times ds)$$

where $L(v)\varphi(x)=\frac{1}{2}\sum a_{ij}(x,v)\partial_{x_i}\partial_{x_j}\varphi(x)+\sum b_i(x,v)\partial_{x_i}\varphi(x)$ with

$$b(x,v)=c(x,v)+\sigma(x,v)\gamma(x,v)+\int_U \beta(x,v,u)\alpha(x,v,u)\mu(du)$$

and

$$a(x,v)=\sigma(x,v)\sigma^T(x,v)+\int_U \alpha(x,v,u)\alpha^T(x,v,u)\mu(du).$$

Observing that the terms involving W_i and $\tilde{W}_i$ will average to zero, we have

$$\begin{aligned}\langle\varphi,V(t)\rangle &= \langle\varphi,V(0)\rangle+\int_0^t (\langle d(\cdot,V(s))\varphi,V(s)\rangle+\langle L(V(s))\varphi,V(s)\rangle)ds\\ &\quad+\int_{U\times[0,t]}\langle\beta(\cdot,V(s),u)\varphi+\alpha(\cdot,V(s),u)\cdot\nabla\varphi,V(s)\rangle W(du\times ds),\end{aligned}$$

and it follows that V is a weak solution of the stochastic partial differential equation

$$\begin{aligned}dv(x,t) &= (L^*(V(t))v(x,t)+d(x,V(t))v(x,t))\,dt \qquad (11.13)\\ &\quad+(\beta(x,V(t),u)v(x,t)-\mathrm{div}_x[\alpha(x,V(t),u)v(x,t)])\,W(du\times dt)\end{aligned}$$

where $V(t)$ is the signed measure given by $V(A,t)=\int_A v(x,t)dx$.

Remark 11.8. The immediate motivation for the material in this subsection is Kotelenez [37] who considers a model, formulated in a somewhat different way, which is essentially the case $\gamma=d=\beta=\sigma=0$. In particular, the weights A_i are constant. Perkins [54] uses time-varying weights for models based on historical Brownian motion (rather than an infinite system of stochastic differential equations) to obtain weak solutions for stochastic partial differential equations related to superprocesses. Donnelly and Kurtz [14, 15] obtain particle representations similar to those given here for a large class of measure-valued processes including many Fleming-Viot and Dawson-Watanabe processes. We suspect that methods used in Donnelly and Kurtz [15] to prove uniqueness for the martingale problem corresponding to the measure-valued process based on uniqueness for the particle model may extend to the present setting and give uniqueness of the weak solution of (11.13).

12. Examples

No attempt has been made at ultimate generality in the following examples. They are intended to illustrate the variety of models that can be represented as solutions of stochastic differential equations of the type considered here. In regard to technical points, note that for most of the examples there are many possible choices for the space H.

12.1 Averaging

The study of the behavior of stochastic models with a rapidly varying component dates back at least to Khas'minskii [33, 34]. Characterization of the processes as solutions of martingale problems has proved to be an effective approach to proving limit theorems for these models. (See Kurtz [40] for a discussion and additional references.) Formulating such a model as a solution of a stochastic differential equation, let W be a standard Brownian motion in $\mathbf{R}^\beta$, let $X_n(0)$ be independent of W, and let ξ be a stochastic process with state space U, independent of W and $X_n(0)$. Set $\xi_n(t) = \xi(nt)$, and for $\sigma : \mathbf{R}^\beta \times U \to \mathbf{M}^{\alpha\beta}$ and $b : \mathbf{R}^\beta \times U \to \mathbf{R}^\alpha$, let X_n satisfy

$$X_n(t) = X_n(0) + \int_0^t \sigma(X_n(s), \xi_n(s))dW(s) + \int_0^t b(X_n(s), \xi_n(s))ds .$$

We assume that

$$\frac{1}{t}\int_0^t f(\xi(s))ds \to \int_U f(u)\nu(du) \tag{12.1}$$

in probability for each $f \in \bar{C}(U)$. Define a sequence of orthogonal martingale random measures on $U \times \{1, \ldots, \beta\}$ by setting

$$M_n(A \times \{k\}, t) = \int_0^t I_A(\xi_n(s))dW_k(s) . \tag{12.2}$$

Observe that

$$\langle M_n(A \times \{k\}, \cdot), M_n(B \times \{l\}, \cdot)\rangle_t = \int_0^t \delta_{kl} I_{A\cap B}(\xi_n(s))ds$$

and

$$M_n(\varphi, t) = \sum_{k=1}^{\beta} \int_0^t \varphi(\xi_n(s), k)dW_k(s)$$

and that for $\varphi_1, \ldots, \varphi_m \in \bar{C}(U \times \{1, \ldots, \beta\})$, the martingale central limit theorem (see, for example, Ethier and Kurtz [17, Theorem 7.1.4]) implies

$$(M_n(\varphi_1, \cdot), \ldots, M_n(\varphi_\beta, \cdot)) \Rightarrow (M(\varphi_1, \cdot), \ldots, M(\varphi_\beta, \cdot))$$

where M corresponds to Gaussian white noise with

$$\langle M(A \times \{k\}, \cdot), M(B \times \{l\}, \cdot)\rangle_t = \delta_{kl} t \nu(A \cap B).$$

(Approximation of martingale random measures by integrals against scalar Brownian motions, as in (12.2) has been studied by Méléard [47] in the context of relaxed control theory.)

A variety of norms can be used to determine H. We will assume that U is locally compact. Let $\gamma \in C(U)$, $\gamma > 0$, and assume that $\{u : \gamma(u) \leq c\}$ is compact for each $c > 0$. Define

$$||\varphi||_H \equiv \sup_{u,k} |\varphi(u,k)/\gamma(u)|$$

and define H to be the completion of $\bar{C}(U \times \{1, \ldots, \beta\})$ under $||\cdot||_H$. If

$$\sup_{t \geq 0} \frac{1}{t} \int_0^t E\left[\gamma^2(\xi(s))\right] ds < \infty \tag{12.3}$$

then $\{M_n\}$ is uniformly tight, as is the sequence $\{V_n\}$ defined by

$$V_n(A,t) = \int_0^t I_A(\xi_n(s)) ds,$$

which, by (12.1) converges to $\nu(du)ds$. Note, for example, that if ξ is stationary and ergodic, then there exists γ such that the above conditions are satisfied.

Theorem 12.1. *Let ξ and X_n be as above, and assume that (12.1) and (12.3) hold. If σ and b are bounded and continuous and $X_n(0) \Rightarrow X(0)$, then $\{X_n\}$ is relatively compact and any limit point satisfies*

$$X(t) = X(0) + \int_0^t \int_U \sigma(X(s), u) M(du \times ds) + \int_0^t \int_U b(X(s), u) \nu(du) ds\,.$$

Remark 12.2. In order to apply Theorem 7.5 we need the mappings $x \to \sigma(x, \cdot)$ and $x \to b(x, \cdot)$ to be bounded and continous as mappings from $\mathbf{R}^\alpha$ to H. The assumption that σ and b are bounded and continous as mappings from $\mathbf{R}^\alpha \times U$ to $\mathbf{R}$ is a significantly stronger requirement.

Proof. For $A \subset U$ and $B \subset \{0, \ldots, \beta\}$, define

$$Y_n(A \times B, t) = I_B(0) V_n(A,t) + \sum_{k=1}^{\beta} I_B(k) M_n(A \times \{k\}, t)\,.$$

Then $\{Y_n\}$ is uniformly tight for H defined above, and the theorem follows immediately from Theorem 7.5. □

12.2 Diffusion approximations for Markov chains

Any discrete-time Markov chain with stationary transition probabilities can be written as a recursion

$$X_{k+1} = F(X_k, \xi_{k+1})$$

where the $\{\xi_k\}$ are independent and identically distributed. Consider a sequence of such chains with values in $\mathbf{R}^\alpha$ satisfying

$$X_{k+1}^n = X_k^n + \sigma_n(X_k^n, \xi_{k+1})\frac{1}{\sqrt{n}} + b_n(X_k^n, \zeta_{k+1})\frac{1}{n}$$

where $\{(\xi_k, \zeta_k)\}$ is iid in $U_1 \times U_2$. We again assume that U_1 and U_2 are locally compact. Let μ be the distribution of ξ_k and ν the distribution of ζ_k, and suppose $\int_{U_2} G_n(x, u_2)\mu(du_2) = 0$ for all $x \in \mathbf{R}^\alpha$ and $n = 1, 2, \ldots$. Define $X_n(t) = X_{[nt]}^n$, $M_n(A, t) = \frac{1}{\sqrt{n}}\sum_{k=1}^{[nt]}(I_A(\xi_k) - \mu(A))$, and $V_n(B, t) = \frac{1}{n}\sum_{k=1}^{[nt]} I_B(\zeta_k)$. Note that $V_n(A, t) \to t\nu(A)$ and $M_n(A, t) \Rightarrow M(A, t)$ where M is Gaussian with covariance

$$E[M(A, t)M(B, s)] = t \wedge s(\mu(A \cap B) - \mu(A)\mu(B))$$

(see Example 2.3).

Theorem 12.3. *Let X_n be as above, and assume that* $\lim_n \sup_{(x,u_2)\in K} |b_n(x, u_2) - b(x, u_2)| = 0$ *for each compact $K \subset \mathbf{R}^\alpha \times U_2$ and*

$$\lim_{n\to\infty} \sup_{(x,u_1)\in K} |\sigma_n(x, u_1) - \sigma(x, u_1)| = 0$$

for each compact set $K \subset \mathbf{R}^\alpha \times U_1$. Suppose that

$$\sup_{n,x,(u_1,u_2)} (|b_n(x, u_2)| + |\sigma_n(x, u_1)|) < \infty\,,$$

that σ and b are bounded and continuous, and that $X_n(0) \Rightarrow X(0)$. Then $\{X_n\}$ is relatively compact and any limit point satisfies

$$X(t) = X(0) + \int_0^t \int_{U_2} \sigma(X(s), u)M(du \times ds) + \int_0^t \int_{U_1} b(X(s), u)\nu(du)ds\,.$$

Proof. Let $U = U_1 \cup U_2$ and let H be the space of functions on U with norm

$$\|h\|_H = \sqrt{\int_{U_1} h^2(u)\mu(du)} + \int_{U_2} |h(u)|\nu(du).$$

Uniform tightness is again easy to check, and the result follows by Theorem 7.5. □

12.3 Feller diffusion approximation for Wright-Fisher model

The Wright-Fisher model is a discrete generation genetic model for the evolution of a population of fixed size. For simplicity, we only consider the two-type case. Let N denote the size of the population and let X_n^N denote the fraction of the population in the nth generation that is of type I. Given the population in the nth generation, we assume that the population in the $(n+1)$st generation is obtained as follows: For each of the N individuals in generation $n+1$ a "parent" is selected at random (with replacement) from the population in generation n. If the parent is of type I, then with high probability the "offspring" will be of type I, but there is a small probability (which we will write as μ_1/N) of a "mutation" occuring and producing an offspring of type II. Similarly, if the parent is of type II, then the offspring is of type II with probability $(1-\mu_2/N)$ and of type I with probabilty μ_2/N. These "birth-events" are assumed to be independent conditioned on X_n^N. Note that if $X_n^N = x$, then the probabilty that an offspring is of type I is

$$\pi_N(x) = x(1-\mu_1/N) + (1-x)\mu_2/N = x + ((1-x)\mu_2 - x\mu_1)/N.$$

We can construct this model in the following way. Let $\{\xi_k^n, k = 1, \ldots, N, n = 1, 2, \ldots\}$ be iid uniform $[0,1]$ random variables. Then, given X_0^N, we can obtain X_n^N recursively by

$$\begin{aligned} X_{n+1}^N &= \frac{1}{N}\sum_{k=1}^N I_{[0,\pi_N(X_n^N)]}(\xi_k^N) \\ &= \frac{1}{N}\sum_{k=1}^N \left(I_{[0,\pi_N(X_n^N)]}(\xi_k^N) - \pi_N(X_n^N)\right) + \pi_N(X_n^N). \end{aligned}$$

Define a martingale random measure with $U = [0,1]$ by

$$M_N(A,t) = \frac{1}{N}\sum_{n=1}^{[Nt]}\sum_{k=1}^{N}(I_A(\xi_k^n) - m(A))$$

and let $A_N(t) = \frac{[Nt]}{N}$. Then setting $X_N(t) = X_{[Nt]}^N$,

$$\begin{aligned} X_N(t) = X_N(0) &+ \int_{U\times[0,t]} I_{[0,\pi_N(X_N(s-))]}(u) M_N(du \times ds) \\ &+ \int_0^t ((1 - X_N(s-))\mu_2 - X_N(s-)\mu_2)\, dA_N(s). \end{aligned}$$

Noting that

$$E[M_N(A,t)M_N(B,s)] = \frac{[N(t\wedge s)]}{N}(m(A\cap B) - m(A)m(B)),$$

it is easy to check that $M_N \Rightarrow M$ where M is the Gaussian martingale random measure in Example 2.3 (with $\mu = m$). We can take $H = L_2(m) \times \mathbf{R}$, and observing that $x \to F_N(x,\cdot) = (I_{[0,\pi_N(x)]}(\cdot), (1-x)\mu_2 - x\mu_1)$ is a continuous mapping from $[0,1] \to H$ and that $F_N(x,\cdot) \to F(x,\cdot) = (I_{[0,x]}, (1-x)\mu_2 - x\mu_1)$ uniformly in x, we can apply Theorem 7.5 to conclude that $\{X_N\}$ is relatively compact and that any limit point satisfies

$$X(t) = X(0) + \int_{[0,1]\times[0,t]} I_{[0,X(s)]}(u) M(du \times ds) + \int_0^t ((1-X(s))\mu_2 - X(s)\mu_1) ds.$$

12.4 Limit theorems for jump processes

The following results are essentially due to Kasahara and Yamada [32]. They treat some cases with discontinuous coefficients which we do not cover here. We include an additional parameter u that can be used to model different kinds of jump behavior. For simplicity, we assume the u takes values in a compact metric space U.

Let N_n be a point process on $\mathbf{R} \times U \times [0,\infty)$ such that if $A \subset \mathbf{R} - (-\epsilon,\epsilon)$ for some $\epsilon > 0$, $N_n(A \times U \times [0,t]) < \infty$ a.s. We assume that N_n is adapted to $\{\mathcal{F}_t^n\}$ in the sense that $N_n(A \times [0,\cdot])$, $A \in \mathcal{B}(\mathbf{R}) \times \mathcal{B}(U)$, is adapted, and we let Λ_n denote an adapted positive random measure such that $\tilde{N}_n(A \times [0,t]) = N_n(A \times [0,t]) - \Lambda_n(A \times [0,t])$ defines a σ-finite, orthogonal martingale random measure on $\mathbf{R} \times U \times [0,\infty)$ and $\Lambda_n(A \times [0,\cdot])$ is continuous for $A \subset (\mathbf{R} - (-\epsilon,\epsilon)) \times U$. Note that the orthogonality implies $N_n(A \times \{t\}) \le 1$ for all A and t. We assume that

$$\int_{\mathbf{R}\times U} f(z,u)\Lambda_n(dz \times du \times [0,t]) \to t\int_{\mathbf{R}\times U} f(z,u)\nu(dz \times du) \qquad (12.4)$$

in probability for all bounded continuous f that vanish for $|z| \le \epsilon$ for some $\epsilon > 0$, and that

$$\int_{\mathbf{R}\times U} 1 \wedge z^2 \nu(dz \times du) < \infty.$$

Let N denote the Poisson point process on $\mathbf{R} \times U \times [0,\infty)$ with mean measure $\nu(dz \times du)dt$. The convergence in (12.4) implies that

$$\int_{\mathbf{R}\times U\times[0,\cdot]} f(z,u,s) N_n(dz \times du \times ds) \Rightarrow \int_{\mathbf{R}\times U\times[0,\cdot]} f(z,u,s) N(dz \times du \times ds),$$

for all bounded continous f on $\mathbf{R} \times U \times [0,\infty)$ that vanish for $|z| \le \epsilon$ for some $\epsilon > 0$. (See Theorem 2.6).) Let V_n be a positive random measure adapted to $\{\mathcal{F}_t^n\}$, and assume that there exists a finite measure μ such that

$$\int_{\mathbf{R}\times U\times[0,t]} f(z,u,s) V_n(dz \times du \times ds) \to \int_0^t \int_{\mathbf{R}\times U} f(z,u,s)\mu(dz \times du) ds$$

in probability for all bounded continuous f.

Let c satisfy $\nu(\{c\} \times U) = \nu(\{-c\} \times U) = 0$, and suppose X_n satisfies

$$\begin{aligned} X_n(t) = X_n(0) \quad &+ \int_{[-c,c]\times U\times[0,t]} \sigma_n(s, X_n(s), z, u) z \tilde{N}_n(dz \times du \times ds) \\ &+ \int_{(\mathbf{R}-[-c,c])\times U\times[0,t]} b_n^1(s, X_n(s), z, u) N_n(dz \times du \times ds) \\ &+ \int_{\mathbf{R}\times U\times[0,t]} b_n^2(s, X_n(s), z, u) V_n(dz \times du \times ds) \end{aligned}$$

Theorem 12.4. *Let X_n, N_n, and V_n be as above. Suppose that there exists a constant C such that $|\sigma_n| + |b_n^1| + |b_n^2| \le C$ and that there are bounded and continous σ, b^1, and b^2, such that for each compact $K \subset [0,\infty) \times \mathbf{R}^\alpha \times \mathbf{R} \times U$*

$$\begin{aligned} \lim_{n\to\infty} \sup_{(s,x,u)\in K} (|\sigma(s,x,z,u) - \sigma_n(s,x,z,u)| \;&+\; |b^1(s,x,z,u) - b_n^1(s,x,z,,u)| \\ &+\; |b^2(s,x,z,u) - b_n^2(s,x,z,u)|) \\ &=\; 0. \end{aligned}$$

Suppose that

$$\sup_n E[\int_{\mathbf{R}\times[0,t]} (z \wedge c)^2 \Lambda_n(dz \times du \times ds)] < \infty \qquad (12.5)$$

and that there exist a positive measure ρ and positive $\epsilon_n \to 0$ such that $\delta_n \ge \epsilon_n$ and $\delta_n \to 0$ implies

$$\int_{(-\delta_n,\delta_n)\times U\times[0,t]} z^2 g(u) \Lambda_n(dz \times du \times ds) \to t \int_U g(u)\rho(du) \qquad (12.6)$$

in probability for all $g \in \bar{C}(U)$. If $X_n(0) \to X(0)$, then $X_n \Rightarrow X$ satisfying

$$\begin{aligned} X(t) = X(0) \quad &+ \int_{U\times[0,t]} \sigma(s, X(s), 0, u) W(du \times ds) \\ &+ \int_{[-c,c]\times U\times[0,t]} \sigma(s, X(s), z, u) z \tilde{N}(du \times ds) \\ &+ \int_{(\mathbf{R}-[-c,c])\times[0,t]} b^1(s, X(s), z, u) N(dz \times du \times ds) \\ &+ \int_0^t \int_{\mathbf{R}} b^2(s, X(s), z, u) \nu(dz \times du) ds \end{aligned}$$

where W is a Gaussian random measure on $U \times [0,\infty)$ satisfying

$$E[W(A \times [0,t]) W(B \times [0,s])] = (t \wedge s)\rho(A \cap B).$$

Proof. It is enough to verify convergence for each bounded time interval $[0,T]$. Let N_n^c be N_n restricted to $(\mathbf{R}-[c,c])\times U\times[0,\infty)$, and $M_n(A\times B\times[0,t]) = \int_{A\cap[-c,c]\times B\times[0,t]} z\tilde{N}_n(dz\times du\times ds)$. The convergence of $\{N_n^c\}$ and $\{V_n\}$ to finite (random) measures implies, through Prohorov's theorem, a stochastic version of tightness which in turn implies the existence of $\gamma:\mathbf{R}\to[1,\infty)$ with $\lim_{|z|\to\infty}\gamma(z)=\infty$ such that $\{\int\gamma(z)N_n^c(dz\times U\times[0,T])\}$ and $\{\int\gamma(z)V_n(dz\times U\times[0,T])\}$ are stochastically bounded. Let $\|h\|_H=\sup_{z,u}|h(u,z)/\gamma(z)|$. The uniform tightness of $\{V_n\}$ and of $\{N_n^c\}$ as $H^\#$-semimartingales is immediate and that of M_n follows from (12.5). The convergence of M_n to M given by

$$M(A\times B\times[0,t]) = \delta_{\{0\}}(A)W(B\times[0,t]) + \int_{A\cap[-c,c]\times B\times[0,t]} z\tilde{N}(dz\times du\times ds)$$

follows from (12.4) and (12.6) and Theorem 2.7. □

12.5 An Euler scheme

Consider

$$X(t) = X(0) + \int_0^t\int_U \sigma(X(s),u)W(du\times ds) + \int_0^t b(X(s))ds \qquad (12.7)$$

where for definiteness we take $U=[0,1]$ and W to be Gaussian white noise determined by Lebesgue measure m on $U\times[0,\infty)$. The solution of this equation will be a diffusion with generator

$$Lf(x) = \frac{1}{2}\sum a_{ij}(x)\frac{\partial^2}{\partial x_i\partial x_j}f(x) + \sum b_i(x)\frac{\partial}{\partial x_i}b(x)$$

where $a(x)=\int_U\sigma(x,u)\sigma^T(x,u)du$. Of course, this diffusion could be rewritten as the solution of an Itô equation in its usual form and numerical schemes applied to approximate the solution of that equation; however, our interest here is in developing methods for approximating stochastic equations driven by martingale measures, and we use this simple case to explore the possibilities.

A simple simulation scheme for (12.7) might involve discretization in both time and in U to give an iteration of the form

$$\hat{X}(t_{i+1}) = \hat{X}(t_i) + \sum\sigma(\hat{X}(t_i),v_i)W((u_i,u_{i+1}]\times(t_i,t_{i+1}]) + b(\hat{X}(t_i))(t_{i+1}-t_i) \qquad (12.8)$$

where the sum is over the partition $0=u_0<u_1<\cdots<u_m=1$ and $u_i\le v_i\le u_{i+1}$. This iteration gives the simplest Euler-type scheme for (12.7). Consistency for the scheme follows easily from Theorem 7.5. We are interested in analyzing the error of the scheme in a manner similar to that used in Kurtz and Protter [43] to study the Euler scheme for Itô equations. In particular, $\hat{X}$ defined in (12.8) can be extended to a solution of

$$\hat{X}(t) = X(0) + \int_0^t \int_U \sigma(\hat{X} \circ \eta(s), \gamma(u)) W(du \times ds) + \int_0^t b(X \circ \eta(s)) ds$$

where $\eta(s) = t_i$ for $s \in [t_i, t_{i+1})$ and $\gamma(u) = v_i$ for $u \in (u_i, u_{i+1}]$. Then

$$\begin{aligned} X(t) - \hat{X}(t) &= \int_0^t \int_U \left(\sigma(X(s), u) - \sigma(\hat{X}(s), u)\right) W(du \times ds) \\ &\quad + \int_0^t \left(b(X(s)) - b(\hat{X}(s))\right) ds \\ &\quad + \int_0^t \int_U \left(\sigma(\hat{X}(s), u) - \sigma(\hat{X} \circ \eta(s), u)\right) W(du \times ds) \\ &\quad + \int_0^t \int_U \left(\sigma(\hat{X} \circ \eta(s), u) - \sigma(\hat{X} \circ \eta(s), \gamma(u))\right) W(du \times ds) \\ &\quad + \int_0^t \left(b(\hat{X}(s)) - b(\hat{X} \circ \eta(s))\right) ds\,. \end{aligned}$$

We assume that σ and b are bounded and have two bounded continuous derivatives. Observing that

$$\hat{X}(t) - \hat{X} \circ \eta(t) = \int_U \sigma(\hat{X} \circ \eta(t), \gamma(u)) W(du \times (\eta(t), t]) + b(\hat{X} \circ \eta(t))(t - \eta(t)),$$

fix $\alpha > 0$ and define

$$U(t) = \alpha(X(t) - \hat{X}(t))$$

$$Y(A \times [0,t]) = \int_0^t \alpha W(A \times (\eta(s), s]) ds$$

$$Z(A \times B \times [0,t]) = \int_0^t \alpha W(A \times (\eta(s), s]) W(B \times ds)$$

$$V(A \times [0,t]) = \int_0^t I_A(u) \alpha(u - \gamma(u)) W(du \times ds)$$

$$R(A \times [0,t]) = \int_0^t \alpha(s - \eta(s)) W(A \times ds)\,.$$

Assuming $d = 1$ to simplify notation,

$$\begin{aligned} U(t) &= \int_0^t \int_U \left(\frac{\sigma(X(s), u) - \sigma(\hat{X}(s), u)}{X(s) - \hat{X}(s)}\right) U(s) W(du \times ds) \\ &\quad + \int_0^t \left(\frac{b(X(s)) - b(\hat{X}(s))}{X(s) - \hat{X}(s)}\right) U(s) ds \\ &\quad + \int_0^t \int_U \sigma_x(\hat{X} \circ \eta(s), v) \sigma(\hat{X} \circ \eta(s), \gamma(u)) Z(du \times dv \times ds) \end{aligned}$$

$$+\int_0^t\int_U \sigma_x(\hat{X}\circ\eta(s),v)b(\hat{X}\circ\eta(s))R(dv\times ds)$$
$$+\int_0^t\int_U\left(\frac{\sigma(\hat{X}\circ\eta(s),u)-\sigma(\hat{X}\circ\eta(s),\gamma(u))}{u-\gamma(u)}\right)V(du\times ds)$$
$$+\int_0^t b_x(\hat{X}\circ\eta(s))\sigma(\hat{X}\circ\eta(s),u)Y(du\times ds)+Err$$

where Err will be negligible under our asymptotic assumptions on α and η.

Note that Z, V, and R are martingale random measures with

$$\langle Z(C\times\cdot),Z(D\times\cdot)\rangle_t=\int_0^t\int_U \alpha^2 W(C_v\times(\eta(s),s])W(D_v\times(\eta(s),s])dvds,$$

where $C_v=\{u:(u,v)\in C\}$,

$$\langle V(A\times\cdot),V(B\times\cdot)\rangle_t=t\int_{A\cap B}\alpha^2(u-\gamma(u))^2du$$

and

$$\langle R(A\times\cdot),R(B\times\cdot)\rangle_t=m(A\cap B)\int_0^t\alpha^2(s-\eta(s))^2ds.$$

Let Z_n, V_n, R_n, U_n be defined by replacing η by $\eta_n(s)=\frac{[ns]}{n}$, α by $\sqrt{n}$, and $\gamma_n(u)=\frac{\beta k}{\sqrt{n}}$ if $\frac{\beta(k-\frac{1}{2})}{\sqrt{n}}<u\leq\frac{\beta(k+\frac{1}{2})}{\sqrt{n}}$ for $k=0,1,\ldots$. Then as $n\to\infty$ $Z_n\Rightarrow\tilde{Z}$ with

$$\langle\tilde{Z}(C,\cdot),\tilde{Z}(D,\cdot)\rangle_t=\frac{1}{2}m_2(C\cap D)t$$

$V_n\Rightarrow\tilde{V}$ with

$$\langle\tilde{V}(A,\cdot),\tilde{V}(B,\cdot)\rangle_t=\frac{\beta^2}{12}m(A\cap B)t,$$

and $R_n\Rightarrow 0$.

We should have $U_n\Rightarrow\tilde{U}$ satisfying

$$\begin{aligned}\tilde{U}(t)\ &=\ \int_0^t\int_U\sigma_x(X(s),u)\tilde{U}(s)W(du\times ds)+\int_0^t b'(X(s))\tilde{U}(s)ds\\&\quad+\int_0^t\int_U\sigma_x(X(s),v)\sigma(X(s),u)\tilde{Z}(du\times dv\times ds)\\&\quad+\int_0^t\int_U\sigma_u(X(s),u)\tilde{V}(du\times ds);\end{aligned}$$

but the conclusion does not follow from Theorem 7.5 since the sequence $\{Z_n\}$ is not uniformly tight. In particular, Z_n is not worthy.

The desired convergence does, however, hold. Note that the integrand for Z_n is adapted to the filtration $\{\mathcal{F}_t^n\}$ with $\mathcal{F}_t^n=\mathcal{F}_{\eta_n(t)}\subset\mathcal{F}_t$. This observation leads us to define the notion of a "conditionally worthy" martingale random measure. Let M be an $\{\mathcal{F}_t\}$-adapted, martingale random measure, and let $\mathcal{G}_t \subset \mathcal{F}_t$. Then M is $\{\mathcal{G}_t\}$-*conditionally worthy* if there exists a *dominating* random measure K on $U \times U \times [0,\infty)$ such that

$$|E[\langle M(A), M(B)\rangle_{t+r} - \langle M(A), M(B)\rangle_t | \mathcal{G}_t]| \leq E[K(A \times B \times (t, t+r]) | \mathcal{G}_t].$$

In the present setting, we have

$$E[\langle Z_n(C), Z_n(D)\rangle_{t+r} - \langle Z_n(C), Z_n(D)\rangle_t | \mathcal{F}_t^n] = \int_t^{t+r} \int_U nm(C_v \cap D_v)(s - \eta_n(s))dvds$$

so Z_n is "conditionally orthogonal" uniformly in n. The convergence theorems extend to this setting and the convergence of U_n to $\tilde{U}$ follows.

References

1. Araujo, A., Giné, E., *The Central Limit Theorem for Real and Banach Valued Random Variables*, Wiley, New York (1980).
2. Bhatt, A.G., Mandrekar, V., "On weak solution of stochastic PDE's", preprint (1995).
3. Blount, D., "Comparison of stochastic and deterministic models of a linear chemical reaction with diffusion", *Ann. Probab.* 19, 1440-1462 (1991).
4. Blount, D., "A simple inequality with applications to SPDE's", preprint (1995).
5. Brzeźniak, Z., Capiński, M., Flandoli, F., "A convergence result for stochastic partial differential equations", *Stochastics* 24, 423-445 (1988).
6. Brown, T.C., A martingale approach to the Poisson convergence of simple point processes. *Ann. Probab.* 6, 615-628 (1978).
7. Burkholder, D.L., "Distribution function inequalities for martingales", *Ann. Probab.* 1, 19-42, (1973).
8. Cho, N., *Weak convergence of stochastic integrals and stochastic differential equations driven by martingale measure and its applications*, PhD Dissertation, University of Wisconsin - Madison (1994).
9. Cho, N., "Weak convergence of stochastic integrals driven by martingale measure", *Stochastic Process. Appl.* (to appear) (1995).
10. Çinlar, E., Jacod, J., "Representation of semimartingale Markov processes in terms of Wiener processes and Poisson random measures", *Seminar on Stochastic Processes 1981*, (E. Çinlar, K.L. Chung, R.K. Getoor, eds.), Birkhäuser, Boston, 159-242 (1981).
11. Da Prato, G., Zabczyk, J., *Stochastic Equations in Infinite Dimensions*. Cambridge University Press, Cambridge (1992).
12. Da Prato, G., Zabczyk, J., "A note on stochastic convolution", *Stoch. Anal. Appl.* 10, 143-153. (1992).
13. Dellacherie, C., Maisonneuve, B., Meyer, P.A., *Probabilites et Potentiel, Chapitres XVII a XXIV: Processus de Markov (fin); Compléments de Calcul Stochastique*, Hermann, Paris (1992).
14. Donnelly, P.E., Kurtz, T.G., "A countable representation of the Fleming-Viot measure-valued diffusion", *Ann Probab.*, to appear (1996).

15. Donnelly, P.E., Kurtz, T.G., "Particle representations for measure-valued population models", (in preparation (1996).
16. Engelbert, H.J., "On the theorem of T. Yamada and S. Watanabe", *Stochastics* 35, 205-216 (1991).
17. Ethier, S.N., Kurtz, T.G., *Markov Processes: Characterization and Convergence.* Wiley, New York (1986).
18. Feller, W., *An Introduction to Probability Theory and Its Applications II*, 2nd ed., Wiley, New York (1971).
19. Fichtner, K.H,, Manthey, R., "Weak approximation of stochastic equations", *Stochastics Stochastics Rep.* 43, 139-160. (1993).
20. Graham, C., "McKean-Vlasov Ito-Skorohod equations and nonlinear diffusions with discrete jump sets", *Stochastic Process. Appl.* 40, 69-82 (1992).
21. Gyöngy, I., "On the approximation of stochastic partial differential equations I, II", *Stochastics* 25, 59-85, 26, 129-164 (1988,1989).
22. Ichikawa, A., "Some inequalities for martingales and stochastic convolutions", *Stoch. Anal. Appl.* 4, 329-339 (1986).
23. Ikeda, N., Watanabe, S., *Stochastic Differential Equations and Diffusion Processes*, North Holland, Amsterdam (1981).
24. Itô, K., "On stochastic differential equations", *Mem. Amer. Math. Soc.* 4 (1951).
25. Jakubowski, A., "Continuity of the Ito stochastic integral in Hilbert spaces", preprint (1995).
26. Jakubowski, A., "A non-Skorohod topology on the Skorohod space" (1995).
27. Jakubowski, A. Mémin, J.,Pagès, G., "Convergence in loi des suites d'intégrales stochastique sur l'espace D^1 de Skorohod", *Probab. Theory Related Fields* 81, 111-137 (1989).
28. Jetschke, G., "Lattice approximation of a nonlinear stochastic partial differential equation with white noise", *Random Partial Differential Equations, Oberwolfach (1989)*, Birkhäuser Verlag, Basel, 107-126 (1991).
29. Kallianpur, G., Pérez-Abreu, V.,"Weak convergence of solutions of stochastic evolution equations on nuclear spaces", *Stochastic Partial Differential Equations and Applications II*, LNM 1390, 119-131 (1989).
30. Kallianpur, G., Xiong, J., "Stochastic models of environmental pollution", *J. Appl. Probab.* (1994).
31. Kallianpur, G., Xiong, J., "Asymptotic behavior of a system of interacting nuclear space valued stochastic differential equations driven by Poisson random measures", *Appl. Math. Optim.* 30, 175-201 (1994).
32. Kasahara, Y., Yamada, K., "Stability theorem for stochastic differential equations with jumps", *Stochastic Process. Appl.* 38, 13-32 (1991).
33. Khas'minskii, R.Z., "On stochastic processes defined by differential equations with a small parameter", *Theory Probab. Appl.* 11, 211-228 (1966).
34. Khas'minskii, R.Z., "A limit theorem for the solutions of differential equations with random right-hand sides", *Theory Probab. Appl.* 11, 390-406 (1966).
35. Kotelenez, P., "A submartingale type inequality with applications to stochastic evolution equations", *Stochastics.* 8, 139-151 (1982).
36. Kotelenez, P., "A stopped Doob inequality for stochastic convolution integrals and stochastic evolution equations", *Stoch. Anal. Appl.* 2, 245-265 (1984).
37. Kotelenez, P., "A class of quasilinear stochastic partial differential equations of McKean-Vlasov type with mass conservation", *Probab. Theory Relat. Fields* 102, 159-188 (1995).
38. Kurtz, T.G., "Representations of Markov processes as multiparameter time changes", *Ann. Probab.* 8, 682-715 (1980).

39. Kurtz, T.G., "Random time changes and convergence in distribution under the Meyer-Zheng conditions", *Ann. Probab.* 19, 1010-1034 (1991).
40. Kurtz, T.G., "Averaging for martingale problems and stochastic approximation", *Applied Stochastic Analysis. Proceedings of the US-French Workshop. Lect. Notes. Control. Inf. Sci.* 177, 186-209 (1992).
41. Kurtz, T.G., Marchetti, F., "Averaging stochastically perturbed Hamiltonian systems", *Proceedings of Symposia in Pure Mathematics*, 57, 93-114 (1995).
42. Kurtz, T.G., Protter, P., "Weak limit theorems for stochastic integrals and stochastic differential quations", *Ann. Probab.* 19, 1035-1070 (1991).
43. Kurtz, T.G., Protter, P., "Wong-Zakai corrections, random evolutions, and simulation schemes for sde's", *Stochastic Analysis: Liber Amicorum for Moshe Zakai*, Academic Press, San Diego. 331-346 (1991).
44. Lenglart, E., Lepingle, D., Pratelli, M., "Presentation unifiée de certaines inegalités des martingales", *Séminaires de Probabilités XIV*, LNM, Springer, Berlin, 26-61 (1980).
45. Liggett, T.M., "Existence theorems for infinite particle systems", *Trans. Amer. Math. Soc.* 165, 471-481 (1972).
46. Liggett, T.M., *Interacting Particle Systems*, Springer-Verlag, New York (1985).
47. Méléard, S., "Representation and approximation of martingale measures", *Stochastic PDE's and their Applications. Lect. Notes in Control and Information Sci.* 176, Springer, Berlin-New York, 188-199 (1992).
48. Méléard, S., "Asymptotic behaviour of some interacting particle systems; McKean-Vlasov and Boltzmann models", this volume (1996).
49. Merzbach, E., Zakai, M., "Worthy martingales and integrators", *Stat. Prob. Letters.* 16, 391-395 (1993).
50. Métivier, M., Pellaumail, J., *Stochastic Integration*, Academic Press, New York (1980).
51. Meyer, P.A., Zheng, W.A., "Tightness criteria for laws of semimartingales", *Ann. Inst. H. Poincaré, Probab. Statist.* 20, 353-372 (1984).
52. Mikulevicius, R., Rozovskii, B.L., "On stochastic integrals in topological vector spaces", *Stochastic Analysis. Proceedings of Symposia in Pure Mathematics.* 57, 593-602 (1994).
53. Protter, P., *Stochastic Integration and Differential Equations*, Springer-Verlag, New York (1990).
54. Perkins, E., "On the martingale problem for interactive measure-valued branching diffusions", *Mem. Amer. Math. Soc.* 549 (1995).
55. Shiga, T., Shimizu, A., "Infinite-dimensional stochastic differential equations and their applications", *J. Mat. Kyoto Univ.* 20, 395-416 (1980).
56. Stricker, C., "Lois de semimartingales et critères de compacité", *Séminaires de Probabilités XIX*, LNM 1123 (1985).
57. Tubaro, L., "An estimate of Burkholder type for stochastic processes defined by the stochastic integral", *Stochastic Analysis and Appl.*, 187-192 (1984).
58. Twardowska, K., "Approximation theorems of Wong-Zakai type for stochastic differential equations in infinite dimensions", *Dissertationes Mathematicae CCXXV* (1993).
59. Ustunel, S., "Stochastic integration on nuclear spaces and its applications", *Ann. Inst. Henri Poincaré* 18, 165-200 (1982).
60. Walsh, J., "An introduction to stochastic partial differential equations", *Lect. Notes in Math.* 1180, 265-439 (1986).
61. Yamada, T., Watanabe, S., "On the uniqueness of solutions of stochastics differential equations" *J. Math. Kyota Univ.* 11, 155-167 (1971).

62. Zabczyk, J., "The fractional calculus and stochastic evolution equations", *Barcelona Seminar on Stochastic Analysis (St. Feliu de Goixols, 1991). Progr. Probab.* 32, 222-234. Birkhäuser, Basel (1993).

C.I.M.E. Session on "Probabilistic Models for Nonlinear PDE's and Numerical Applications"

List of participants

O. ARCUDI, Dipartimento di Matematica, Via Belzoni 7, 35131 Padova, Italy

M. BEN ALAYA, CERMICS-ENPC, La Courtine, 93167 Noisy le Grand cedex, France

P. BERNARD, Laboratoire de Math. Appl., Les Cezeaux, 63177 Aubière cedex, France

M. BOSSY, INRIA, 2004 route de Lucioles, PB 63, 06902 Sophia Antipolis cedex, France

PH. BRIAND, Laboratoire de Math. Appl., Univ. Blaise Pascal, 63177 Aubière cedex, France

B. CADRE, IRMAR, Univ. de Rennes I, Campus de Beaulieu, 35042 Rennes cedex, France

L. CARAMELLINO, Dipartimento di Matematica, Univ. di Tor Vergata, Via della Ricerca Scientifica, 00133 Roma, Italy

D. CHEVANCE, INRIA, 2004 route des Lucliose, BP 93, 06902 Sophia Antipolis cedex, France

J.M.C. CLARK, Dept. of Electrical and Electronic Eng., Imperial College, Exhibition Road, London SW7 2BT, UK

F. COQUET, IRMAR, Univ. de Rennes I, Campus de Beaulieu, 35042 Rennes cedex, France

R. DI LISIO, Dipartimento di Matematica, Univ. La Sapienza, P.le Aldo Moro 2, 00185 Roma, Italy

T. FUJIWARA, Dept. of Math., Hyogo Univ. of Teacher Education, Yashiro, Hyogo, 673-14 Japan

G. GIACOMIN, Inst. f. Ang. Math., Univ. Zurich, Winterhurer Str. 190, CH-8057 Zurich 7

A. GRORUD, C.M.I. Univ. de Provence, 39 rue Joliot-Curie, 13453 Marseille cedex 13, France

E.I. HAUSENBLAS ERIKA, Institute of Mathematics, Hellbrunnerstr. 34, A-5020 Salzburg

A. JAKUBOWSKI, Faculty of Math. and Info., Nicholas Copernicus Univ., ul. Chopina 12/18, 87-100 Torun, Poland

V. KATSOUROS, Dept. of Electrical and Electronic Eng., Imperial College, Exhibition Road, London SW7 2BT, UK

D. LEPINGLE, Mathématiques, Univ. d'Orléans, B.P. 6759, 45067 Orléans cedex, France

Z. LOZANOV CRVENKOVIC, Inst. of Math., Univ. of Novi Sad, Trg D. Obradovica 4, 21000 Novi Sad, Yugoslavia

A. MORO, Dipartimento Statistico, Viale Morgagni 59, 50134 Firenze, Italy

B. PACCHIAROTTI, Dip.to di Matematica, Univ. La Sapienza, P.le Aldo Moro 2, 00185 Roma, Italy

L. PARESCHI, Dipartimento di Matematica, Piazza di Porta S. Donato 5, 40127 Bologna, Italy

A. RAMPONI, Dipartimento di Matematica, Univ. di Tor Vergata, Via della Ricerca Scientifica, 00133 Roma, Italy

H. REGNIER, INRIA, 2004 route des Lucioles, BP 63, 06902 Sophia Antipolis cedex, France

V. RICCI, Dipartimento di Matematica, Univ. La Sapienza, Piazzale A. Moro 2, 00185 Roma, Italy

A. ROZKOSZ, Faculty of Math. and Info., Nicholas Copernicus Univ., ul. Chopina 12/18, 87-100 Torun, Poland

P. SEUMEN TONOU, INRIA, 2004 route des Lucioles, BP 63, 06902 Sophia Antipolis cedex, France

L. SLOMINSKI, Faculty of Math. and Info., Nicholas Copernicus Univ., ul. Chopina 12/18, 87-100 Torun, Poland

M.B. ZAVELANI ROSSI, Dipartimento di Matematica, Univ. La Sapienza, P.le A. Moro 2, 00185 Roma, Italy

M. ZERVOS, Dept. of Electrical and Electronic Eng., Imperial College, Exhibition Road, London SW7 2BT, UK

FONDAZIONE C.I.M.E.
CENTRO INTERNAZIONALE MATEMATICO ESTIVO
INTERNATIONAL MATHEMATICAL SUMMER CENTER

"Integral Geometry, Radon Transforms and Complex Analysis"

is the subject of the first 1996 C.I.M.E. Session.

The session, sponsored by the Consiglio Nazionale delle Ricerche (C.N.R.) and the Ministero dell'Università e della Ricerca Scientifica e Tecnologica (M.U.R.S.T.), will take place, under the scientific direction of Professor ENRICO CASADIO TARABUSI (Università di Roma "La Sapienza"), Professor MASSIMO PICARDELLO (Università di Roma "Tor Vergata") and Professor GIUSEPPE ZAMPIERI (Università di Padova), in Venezia, **from 3 to 12 June 1996.**

Courses

a) **Radon Transforms, Wavelets and Applications.** (6 lectures in English).

Prof. C. A. BERENSTEIN (University of Maryland, College Park, USA)

1) The use of wavelets to localize the 2-dimensional Radon Transform. Application to Computerized Axial Tomography. Possible extensions to Nuclear Magnetic Resonance and comparison with Λ–tomography. Open problems for the exterior Radon Transforms.
2) Edge detections and singularities of functions in terms of their Radon transforms. Applications to the work of Boman-Quinto (Theorem of Hörmander- Kashiwara-Kawai).
3) The inverse conductivity problem and the hyperbolic Radon transform. Inversion, range, relation with the spherical transform.

b) **Holomorphic mappings between real analytic submanifolds in** C^N

Prof. P. EBENFELT (University of California, San Diego)

Summary: Let M and M' be real analytic submanifolds in C^N, and let $p_0 \in M$ and $p'_0 \in M'$. We shall study the class C of holomorphic mappings H from the neighborhood of p_0 (in C^N) into C^N such that $H(p_0) = p'_0$, $H(M) \subset M'$ and such that Jac(H)≠0 on M. Properties of this class depend intimately on the geometric properties of M and M' near the points p_0 and p'_0 respectively. We shall define the notion of CR (Cauchy-Riemann), finite type and minimality, holomorphic nondegeneracy, etc. We shall also introduce a new invariant sequence of sets, called *the Segre sets*, attached to M at p_0 that is useful in analyzing the class C and that provides a new characterization of finite type. One of the main results presented here gives essentially necessary and sufficient conditions on M, assuming that M and M' are *real algebraic*, such that every holomorphic mapping in C is *algebraic*. We shall also give results and state open problems in the general case (i.e. M and M' are real analytic) as well as consider some variations of the situation described above.

c) **Complex Integral Geometry, d-bar Cohomology, Representations.** (6 lectures in English)

Prof. S. GINDIKIN (Rutgers University, New Brunswick, USA)

1) Admissible complexes of lines in C^3 and SL(2,C).
2) The Gelfand-Graev-Shapiro operator k.
3) Integral Geometry on SL(n,C).
4) Rational Curves and twistors.

5) Integral geometry and Plancherel formula on complex semi-simple Lie groups.
6) Real integral geometry and d-bar cohomology on tubes. Connection with hyperfunctions.
7) The Penrose transform and holomorphic language for d-bar cohomology.

The course will be concentrated around two fundamental concepts of integral geometry:
- universal structure of local inversion formulas and the operator \kappa;
- integral geometrical language for d-bar cohomology and hyperfunctions.
Both these concepts have essential connections with the theory of representations.

d) Analytic and Group-theoretic Methods in Integral Geometry. (6 lectures in English)

Prof. S. HELGASON (MIT, Cambridge, USA)

Radon transforms for double fibrations. Inversion and range problems. Radon transform on symmetric spaces. Various applications to differential equations, solvability results, wave equations on symmetric spaces, including systems of multi-temporal wave equations. The Poisson integral as a Radon transform with integral geometric interpretation of the Hua equations on bounded symmetric domains. Bundle-valued Radon transforms.

e) Singular Radon Transforms (6 lectures in English)

Prof. E. M. STEIN (Princeton University, Princeton, USA)

1) The background: Euclidean case.
2) Singular Radon transforms and the Heisenberg group.
3) Relations with the d-bar Neumann problem.
4) Other directions: nihilpotent groups and discrete analogues.

f) Analytic discs and the theory of CR functions (6 lectures in English)

Prof. A. TUMANOV (University of Illinois, Urbana-Champaign, USA)

1) Introduction to the theory of CR functions: the classical theorems on the holomorphic extendibility of CR functions on hypersurfaces in complex space by Hartogs and Bochner and H. Lewy. The approximation theorem by Baouendi and Treves. Constructions of analytic discs with boundaries in a prescribed generic manifold in complex space. Bishop's equation. Extending CR functions to analytic discs.
2) Geometry of analytic discs with boundaries in a generic manifold in complex space. The defect of a disc. Minimal CR manifolds. Extending CR functions to wedges. Extending CR functions from manifolds with boundaries and edges.
3) Propagation of extendibility of CR functions: the conormal bundle of a generic manifold. The CR structure in the conormal bundle. Propagation of extendibility of CR functions along CR orbits. Application: Trepreau's example of a CR function that cannot be represented as a sum of boundary values of holomorphic functions.
4) Other applications of analytic discs: Extending CR function from globally minimal manifolds. A Morera theorem for CR functions. Regularity of CR mappings.

FONDAZIONE C.I.M.E.
CENTRO INTERNAZIONALE MATEMATICO ESTIVO
INTERNATIONAL MATHEMATICAL SUMMER CENTER

"Calculus of Variations and Geometric Evolution Problems"

is the subject of the second 1996 C.I.M.E. Session.

The session, sponsored by the Consiglio Nazionale delle Ricerche (C.N.R.) and the Ministero dell'Università e della Ricerca Scientifica e Tecnologica (M.U.R.S.T.), will take place, under the scientific direction of Professors STEFAN HILDEBRANDT (Universität Bonn) and MICHAEL STRUWE (ETH-Zentrum, Zürich) at Grand Hotel San Michele, Cetraro (Cosenza), **from 15 to 22 June, 1996.**

Courses

a) **Variational methods for Ginzburg-Landau equations** (6 lectures in English)
Prof. Fabrice BETHUEL (E.N.S., Cachan)

Ginzburg-Landau models arise in various areas of physics. In their simplest form, they involve a functional of the type

$$E_\varepsilon(u) = \frac{1}{2}\int_\Omega |\nabla u|^2 + \frac{1}{2\varepsilon^2}(1-|u|^2)^2$$

where ε is some positive parameter, Ω a domain in R^2 and u a complex valued map from Ω to C. We will be interested in the case ε is small, so that in the limit $\varepsilon \to 0$, we have to deal with S^1 valued maps, and corresponding topological defects. The aim of the course is to present recent developments concerning stationary maps for E_ε, as well as some dynamical properties. The emphasis will be put on the notion of renormalized energies, defined on a space of configuration of particles of positive and negative charges.

1. Different models from physics: superconductivity, abelian-Higgs models, Gross-Pitacyskii equations
2. The Dirichlet problem: asymptotic behavior of minimizers, renormalized energies
3. Non minimizing solutions: Morse theory and topological methods
4. Problems with symmetries: use of an S^1 -index
5. Dynamics

References

[1] For 1) one may usefully consult
D. Saint James, C. Sarma and E. J. Thomas, Type II Superconductivity, Pergamon Press, New York and Oxford (1969).
A. Jaffe and C. Taubes, Vortices and Monopoles, Birkhäuser, Boston, (1980).

[2] 2) relies most on results in
F. Béthuel, H. Brézis et F. Hélein, Ginzburg-Landau vortices, Birkhauser 1994. See also
M. Struwe, On the asymptotic behavior of minimizers of the Ginzburg-Landau model in 2-dimensions, J. Differential Integral Equations 7 (1994) Erratum 8 (1995)

[3] For 3) a good introduction to Morse theory can be found in
M. Struwe, Variational methods, Springer (1990) or
P. Rabinowitz, Minimax methods in critical point theory with applications to differential equations, CBMS Regional Conference Series Math. 65, Amer. Math. Soc., Providence 1986.

The results presented are taken mainly from
L. Almeida and F. Béthuel, Topological methods for the Ginzburg-Landau equation, C.R. Acad. Sci. Paris 320 (1995), 935-938 and detailed paper to appear.

[4] For 4) Morse theory with symmetries is presented for instance in
T. Bartsch, Topological methods for variational problems with symmetries, Springer (1993).
The results presented are taken from
L. Almeida and F. Béthuel, Multiplicity for the Ginzburg-Landau equation in presence of symmetries, to appear in Houston J. of Math.

[5] Finally for 5) we will present results of
L. Pismen and J. Rubinstein, Notion of vertex lines in the Ginzburg-Landau model, Physica D47 (1991), 353-360 and
F. H. Lin, Some dynamical properties of Ginzburg-Landau vortices, to appear in CPAM

b) **Diffusion in science and geometry** (6 lectures in English)
Prof. R. HAMILTON (University of California, San Diego)

The course will discuss various diffusion processes which arise in science and geometry, and the parabolic partial differential equations describing them, including the Heat Equation, Diffusion Reaction Equations, the Porous Medium Equation and the Motion of a Surface by its Mean Curvature or by its Gauss Curvature.

The course will also cover the interesting phenomena which arise in these processes, including speed of propagation, travelling wave solutions, formation of singularities and continuation through singularities.

c) **Variational models for microstructure and phase transitions** (6 lectures in English)
Prof. Stefan MÜLLER (ETH-Zentrum, Zürich)

Elastic crystals and other physical systems that undergo phase transitions often exhibit a complex microgeometry of different phases. In the last ten years important progress has been made to understand such microstructures as minimizers or almost minimizers of suitable energy functionals. The analysis of these variational problems leads to easily stated but rather deep mathematical problems that involve an interplay of different mathematical areas such as the calculus of variations, nonlinear partial differential equations, differential geometry and functional analysis.

In the lectures I will first describe the setting and the basic questions and then discuss some recent progress in the area. The prerequisites are some familiarity with Sobolev spaces and weak convergence. No knowledge of continuum mechanics is assumed.

Since the subject is rather young there is no textbook presentation available. References [1] and [2] below are two research papers which fundamentally influenced the subject. Their reading is tough at times (and not required as prerequisite) but certainly very rewarding. Sverak's ICM article is a very compact survey of current results and interconnections between questions in different areas of mathematics. References [4] and [5] address, in the context of simple model problems, the special topic of regularizing perturbations and their effect on the geometry of the microstructure. All of the above contain plenty of references for further study.

References

[1] J. M. Ball and R. D. James, Fine phase mixtures as minimizers of energy, Arch. Rat. Mech. Anal. 100 (1987), pp. 13-52.
[2] J. M. Ball and R. D. James, Proposed tests of a theory of fine microstructure and the two-well problem, Phil. Trans. R. Soc. London A 338 (1992), pp. 389-450.
[3] V. Sverak, Lower semicontinuity of variational integrals and compensated compactness, to appear in: Proc. Int. Congress Math. (Zürich 1994), Birkhäuser.
[4] S. Müller, Singular perturbations as a selection criterion for periodic minimizing sequences, Calculus of Variations 1 (1993), pp. 169-204.
[5] R. V. Kohn and S. Müller, Surface energy and microstructure in coherent phase transitions, Comm. Pure Appl. Math. 47 (1994), pp. 405-435.

d) **Parametric surfaces of prescribed mean curvature** (6 lectures in English)
Prof. Klaus STEFFEN (Universität Düsseldorf)

In these lectures we treat the following existence problem. For a given function $H : R^{n+1} \to R^n$ and a given (n-1)-dimensional boundary contour Γ in R^{n+1} to find an oriented n-dimensional surface in R^{n+1} bordered by Γ with prescribed mean curvature *H*, i.e. the mean curvature of the surface at each point *x* of its support is *H(x)*. The problem zill be considered first in the setting of 2-dimensional parametric surfaces $x: R^2 \supset D \rightarrow R^3$. For $H \equiv 0$ this is the classical Plateau problem. For $H \neq 0$ it is known that the problem does not always have a solution and various sufficient conditions on H and Γ have been given in the literature. For example if sup $|H|$ is sufficiently small in terms of geometric quantitites associated with Γ, then a solution exists.

Using the direct method of the Calculus of Variations one finds that a sufficient condition for existence (and also a necessary one for the method) is an isoperimetric condition of the type $\left|\int_A H dx\right| \leq cP(A)$ with c< 1/n for all sets $A \subset R^{n+1}$ with finite perimeter *P(A)*. Such isoperimetric conditions can be deduced in various ways from the isoperimetric inequality and they can be employed to give a unified treatment of the existence theory.

The method of isoperimetric conditions is also suitable to treat the existence problem in the setting of Geometric Measure Theory. One then obtains hypersurfaces of prescribed mean curvature in R^{n+1} with certain possible singularities if $n \geq 7$. Moreover, one can study the problem in an ambient Riemannian manifold *M* instead of R^{n+1}. Here various isoperimetric inequalities enter and new phenomena then arise because there are boundaries $\Gamma \subset M$ wich do not border a hypersurface of constant mean curvature $H \neq 0$ in *M*. It is thus interesting to exhibit conditions on which ensure the solvability of the problem at least for sufficiently small $|H|$ and then to find reasonable geometric bounds on the range of the functions *H* which can be admitted. Some recent results in this direction will be discussed and applied also to the setting of 2-dimensional parametric surfaces in a 3-dimensional Riemannian manifold.

References

Dierkes, U.,Hildebrandt, S., Kuster, A., Wohlrab, O.: Minimal Surfaces Vol. 1, Vol. 2, Grundlehren math. Wiss. 295, 296. Springer-Verlag, Berlin-Heidelberg-New York, 1992.
Duzaar, F.: On the existence of surfaces with prescribed mean curvature and boundary in higher dimensions. Ann. Inst. Henri Poincaré (Anal. Non Lin.) **10** (1993), 191-214.
Duzaar, F., Steffen, K.: λ minimizing currents. Manuscr. Math. **80** (1993), 403-447.
Federer, H.: Geometric measure theory. Springer Verlag, Berlin-Heidelberg-New York, 1969.
Steffen, K.: Isoperimetric inequalities and the problem of Plateau. Math. Ann. **222** (1976), 97-144.
Struwe, M.: Plateau's problem and the calculus of variations. Mathematical Notes 35, Princeton University Press, Princeton 1988.

FONDAZIONE C.I.M.E.
CENTRO INTERNAZIONALE MATEMATICO ESTIVO
INTERNATIONAL MATHEMATICAL SUMMER CENTER

"Financial Mathematics"

is the subject of the third 1996 C.I.M.E. Session.

The session, sponsored by the Consiglio Nazionale delle Ricerche (C.N.R.)and the Ministero dell'Università e della Ricerca Scientifica e Tecnologica (M.U.R.S.T.), will take place, under the scientific direction of Professor WOLFGANG RUNGGALDIER (Università di Padova) in Bressanone, **from 8 to 13 July, 1996.**

Courses

a) **Market microstructure and adverse selection** (5 lectures in English)
Prof. Bruno BIAIS and Prof. Jean Charles ROCHET (Université de Toulouse)

1. Outline

The course will study how private information is incorporated into asset prices. We will start with competitive agents and exogenous market structure. Then we will relax the competitiveness assumption. In the third part we will relax the exogenous market structure assumption and look for optimal trading mechanisms.

1.1 Information revelation with competitive agents
Grossman and Stiglitz, 1980
1.2 Information revelation with strategic agents
Kyle, 1985; Kyle, 1989; Biais, Hillion, 1994
1.3 Optimal mechanisms
Rochet, Vila, 1993; Glosten, 1989; Biais, Bossaerts, Rochet, 1995

2. References

Biais, B., 1994, Insider and liquidity trading in stock and options markets, *Review of Financial Studies*, 743--781
Biais, B., P. Bossaerts, and J.C. Rochet, 1995 An optimal mechanism to sell unseasoned equity, mimeo, Université de Toulouse.
Glosten, L., 1989, Insider trading, liquidity and the role of the monopolist specialist, *Journal of Business*, 211--235
Grossman, S., and J. Stiglitz, 1980, On the impossibility of informationally efficient markets, *American Economic Review*, 393--408
Kyle, A. S, 1985, Continuous auctions and insider trading, *Econometrica*, 1315--1335
Kyle, A. S. 1989, Informed speculation with imperfect competition, *Review of Economic Studies*, 317--356
Rochet, J.C., and J.L. Vila, Insider trading without normality, *Review of Financial Studies*, 131--152

b) **An introduction to the theory of interest rate derivatives** (5 lectures in English)
Prof. Tomas BJÖRK (Stockholm School of Economics)

1. The term structure of interest rates. Classical approach to interest rate derivatives via locally riskless portfolios. Incomplete markets. The basic PDE. Market price of risk. Martingale formulation.
2. Interest rate models under Q-measure. Affine term structures. Fitting the yield curve. Hull & White. Heath-Jarrow-Morton.
3. Forward measures. Computation of interest rate options. Forwards,futures and swaps.
4. Extensions to models including jumps. (i.e. Shirakawa and BKR.)

References

[1] Artzner, P. & Delbaen, F. (1989) Term Structure of Interest Rates: the Martingale Approach. *Advances in Appl. Math.* 10, 95-129. Björk, T. (1995) *in* The term structure of discontinuous interest rates. Working Paper, Royal Institute of Technology, Stockholm. Forthcoming.

[2] Björk, T. & Kabanov, Y. & Runggaldier, W. (1995) Bond market structure in the presence of marked point processes. Working Paper, Stockholm School of Economics.
[3] Cox, J. & Ingersoll, J. & Ross, S. (1985) A Theory of the Term Structure of Interest Rates. *Econometrica* 53, 385-408.
[4] Geman, H. & El Karoui, N. & Rochet, J-C. (1995) Changes of numeraire, changes of probability measure and option pricing. *J. Appl. Prob.* 32, 443-458.
[5] Heath, D. & Jarrow, R. & Morton, A. (1987) Bond Pricing and the Term Structure of Interest Rates. *Econometrica* 60:1, 77-106.
[6] Ho, T. & Lee, S. (1986) Term Structure Movements and Pricing Interest Rate Contingent Claims. *Journal of Finance* 41, 1011-1029.
[7] Hull, J. & White, A. (1990) Pricing Interest-Rate Derivative Securities. *The Review of Financial Studies* 3:4, 573-592.
[8] Jamshidian, F. (1989) An Exact Bond Option Formula. *Journal of Finance*. 44, 205-209.
[9] Shirakawa, H. (1991) Interest rate options pricing with Poisson-Gaussian forward rate curve processes. *Mathematical finance*. 1, 77-94.
[10] Vasicec, O. (1977) An Equilibrium Characterization of the Term Structure. *Journal of Financial Economics*. 5, 177-188.

c) **Optimal trading under constraints** (6 lectures in English)
Prof. Jaksa CVITANIC (Columbia University, New York)

- continuous-time, Ito diffusions model for the financial market: linear Stochastic Differential Equations; the problem of hedging (super-replication) of contingent claims; the fair price of a claim in a complete market; martingale approach to Black-Scholes: martingale representation theorem, Feynman-Kac Partial Differential Equation.

- incomplete markets and markets with convex portfolio constraints: no short-selling, no borrowing, etc.; no-arbitrage upper and lower bounds for the price; convex duality, support functions, dynamic programming, variational inequalities.

- the problem of utility maximization from consumption and terminal wealth in a complete market and markets with constraints; a dual problem; existence of optimal portfolios, examples with explicit solutions; Hamilton-Jacobi-Bellman Partial Differential Equations.

- the problem of "drawdown constraints": how to invest optimally while always having more than a given fraction of your wealth's running maximum.

- hedging for a "large investor": nonlinear price equations, prices influenced by investor's strategy; Forward-Backward Stochastic Differential Equations, nonlinear Feynman-Kac Partial Differential Equations.

- markets with transaction costs: hedging and utility maximization; non-existence of nontrivial hedging portfolios.

Bibliography

- CVITANIC, J. & KARATZAS, I. (1992) Convex duality in constrained portfolio optimization. *Ann Appl Probab* 2, 767-818.
- CVITANIC, J. & KARATZAS, I. (1993) Hedging contingent claims with constrained portfolios. *Ann Appl Probab* 3, 652-681.
- CVITANIC, J. & KARATZAS I. (1995) On portfolio optimization under drawdown constraints. IMA Volumes in Mathematics and its Applications 65, 35-46.
- DAVIS, M.H.A & NORMAN, A. (1990) Portfolio selection with transaction costs. *Math Operations Research* 15, 676-713.
- GROSSMAN, S.J. & ZHOU, Z. (1993) Optimal investment strategies for controlling drawdowns. *Math Finance* 3, 241-276.
- KARATZAS, I. (1989) Optimization problems in the theory of continuous trading. *SIAM J. Control Optim.* 27, 1221-1259.
- KARATZAS, I. & SHREVE, S. (1991) Brownian Motion and Stochastic Calculus (2nd edition). Springer-Verlag, New York.
- MA, J., PROTTER, P. & YONG, J. (1994) Solving Forward-Backward Stochastic Differential Equations explicitly - a four step scheme. *Probab Theory and Related Fields* 98, 339-359.
- SONER, H.M., SHREVE, S. & CVITANIC, J. (1995) There is no nontrivial hedging portfolio for option pricing with transaction costs. *Ann. Appl. Probab.* 5, 327-355.

d) **Backwards stochastic differential equations. Finance and optimization.** (6 lectures in English)
Prof. Nicole EL KAROUI (Université de Paris VI))

1) Backwards stochastic differential equations (BSDE)
   - Existence and uniqueness, a priori estimates, linear BSDE
   - A comparison theorem
   - Dependence upon parameters; continuity, differentiability
2) Concave BSDE's and associated control problem
   - BSDE as minimum or minimax
   - Concave BSDE's as value function of a control problem
   - European option pricing with convex bounded portfolio constraints
   - Recursive utility : classical properties and variational

- Consol rate, swap rates, and credit quality

3) The Markovian case
   - Markov properties of solutions of BSDE's
   - BSDE's and non-linear partial differential equations
   - Applications to non-linear pricing
   - Simulations of BSDE's
4) The problem of recursive utility maximization from consumption and terminal wealth
   - a maximum principle
5) Forward-Backward Stochastic Differential Equations
   - Existence and uniqueness
   - nonlinear Feynman-Kac partial differential equations
   - optimal dynamics and dual process
   - hedging for a "large investor" (Cvitanic)
6) Reflected BSDE's and non-linear pricing of American options.

e) Market imperfections and non-linear pricing rules (6 lectures in English)
Prof. Elyés JOUINI (ENSAE, Malakoff, France)

1) Martingales, arbitrage and equilibrium in securities markets with imperfections (sub-linear pricing rules).
2) Arbitrage prices in securities markets with shortsales constraints and different borrowing and lending rate (the discrete and continuous cases, numerical results).
3) Stochastic dominance, portfolio choices and efficient strategies with market frictions.
4) Stationarity and arbitrage, characterization of the martingale measures.
5) Incomplete markets and transaction costs : optimal portfolio choice.
6) Fixed costs and absence of arbitrage opportunities.

Reference

- Bensaid,B., H. Pages, J.-P. Lesne and J. Scheinkman, "Derivative asset pricing with transaction costs" *Mathematical Finance*, 1992, 2, 63-86.
- Carassus, L. and E. Jouini, "Investment opportunities, short-sales constraints and arbitrage opportunities", Working Paper, CREST, 1995.
- Cvitanic, J. and I. Karatzas, "Hedging contingent claims with constraint portfolios", *Ann. Appl. Probab.* 1993, 3, 652-681.
- Harrison, J. and D. Kreps, "Martingales and arbitrage in multi-period securities markets", *J. Economic Th.*, 1979, 20, 381-408.
- Jouini, E. "Imperfections de marché et arbitrage", to appear in Y.Simon (Ed.), *Enciclopédie des marchés financiers*, Economica.
- Jouini,E. and H. Kallal, "Martingales and arbitrage in securities markets with transaction costs", *J. Economic Th.*, 1995, 66(1), 178-197.
- Jouini,E. and H. Kallal, "Arbitrage in securities markets with short-sales constraints", *Mathematical Finance*, 1995, 5(3), 197-232.
- Jouini,E. and H. Kallal, "Efficient trading strategies in the presence of market frictions", Working Paper, CREST, 1995.
- Jouini,E. and H. Kallal, "Equilibrium in securities markets with bid-ask speads", Working Paper, CREST, 1995.
- Jouini,E., P.-F. Koehl and N. Touzi, "Incomplete markets, transaction costs and liquidity effects", Working Paper, CREST, 1995.

LIST OF C.I.M.E. SEMINARS

		Publisher
1954 -	1. Analisi funzionale	C.I.M.E.
	2. Quadratura delle superficie e questioni connesse	"
	3. Equazioni differenziali non lineari	"
1955 -	4. Teorema di Riemann-Roch e questioni connesse	"
	5. Teoria dei numeri	"
	6. Topologia	"
	7. Teorie non linearizzate in elasticità, idrodinamica, aerodinamica	"
	8. Geometria proiettivo-differenziale	"
1956 -	9. Equazioni alle derivate parziali a caratteristiche reali	"
	10. Propagazione delle onde elettromagnetiche	"
	11. Teoria della funzioni di più variabili complesse e delle funzioni automorfe	"
1957 -	12. Geometria aritmetica e algebrica (2 vol.)	"
	13. Integrali singolari e questioni connesse	"
	14. Teoria della turbolenza (2 vol.)	"
1958 -	15. Vedute e problemi attuali in relatività generale	"
	16. Problemi di geometria differenziale in grande	"
	17. Il principio di minimo e le sue applicazioni alle equazioni funzionali	"
1959 -	18. Induzione e statistica	"
	19. Teoria algebrica dei meccanismi automatici (2 vol.)	"
	20. Gruppi, anelli di Lie e teoria della coomologia	"
1960 -	21. Sistemi dinamici e teoremi ergodici	"
	22. Forme differenziali e loro integrali	"
1961 -	23. Geometria del calcolo delle variazioni (2 vol.)	"
	24. Teoria delle distribuzioni	"
	25. Onde superficiali	"
1962 -	26. Topologia differenziale	"
	27. Autovalori e autosoluzioni	"
	28. Magnetofluidodinamica	"

1963 - 29. Equazioni differenziali astratte "
30. Funzioni e varietà complesse "
31. Proprietà di media e teoremi di confronto in Fisica Matematica "

1964 - 32. Relatività generale "
33. Dinamica dei gas rarefatti "
34. Alcune questioni di analisi numerica "
35. Equazioni differenziali non lineari "

1965 - 36. Non-linear continuum theories "
37. Some aspects of ring theory "
38. Mathematical optimization in economics "

1966 - 39. Calculus of variations Ed. Cremonese, Firenze
40. Economia matematica "
41. Classi caratteristiche e questioni connesse "
42. Some aspects of diffusion theory "

1967 - 43. Modern questions of celestial mechanics "
44. Numerical analysis of partial differential equations "
45. Geometry of homogeneous bounded domains "

1968 - 46. Controllability and observability "
47. Pseudo-differential operators "
48. Aspects of mathematical logic "

1969 - 49. Potential theory "
50. Non-linear continuum theories in mechanics and physics and their applications "
51. Questions of algebraic varieties "

1970 - 52. Relativistic fluid dynamics "
53. Theory of group representations and Fourier analysis "
54. Functional equations and inequalities "
55. Problems in non-linear analysis "

1971 - 56. Stereodynamics "
57. Constructive aspects of functional analysis (2 vol.) "
58. Categories and commutative algebra "

1972 - 59.	Non-linear mechanics	"
60.	Finite geometric structures and their applications	"
61.	Geometric measure theory and minimal surfaces	"
1973 - 62.	Complex analysis	"
63.	New variational techniques in mathematical physics	"
64.	Spectral analysis	"
1974 - 65.	Stability problems	"
66.	Singularities of analytic spaces	"
67.	Eigenvalues of non linear problems	"
1975 - 68.	Theoretical computer sciences	"
69.	Model theory and applications	"
70.	Differential operators and manifolds	"
1976 - 71.	Statistical Mechanics	Ed Liguori, Napoli
72.	Hyperbolicity	"
73.	Differential topology	"
1977 - 74.	Materials with memory	"
75.	Pseudodifferential operators with applications	"
76.	Algebraic surfaces	"
1978 - 77.	Stochastic differential equations	"
78.	Dynamical systems	Ed Liguori, Napoli and Birhäuser Verlag
1979 - 79.	Recursion theory and computational complexity	"
80.	Mathematics of biology	"
1980 - 81.	Wave propagation	"
82.	Harmonic analysis and group representations	"
83.	Matroid theory and its applications	"
1981 - 84.	Kinetic Theories and the Boltzmann Equation (LNM 1048)	Springer-Verlag
85.	Algebraic Threefolds (LNM 947)	"
86.	Nonlinear Filtering and Stochastic Control (LNM 972)	"
1982 - 87.	Invariant Theory (LNM 996)	"
88.	Thermodynamics and Constitutive Equations (LN Physics 228)	"
89.	Fluid Dynamics (LNM 1047)	"

1983 -	90. Complete Intersections	(LNM 1092)	Springer-Verlag
	91. Bifurcation Theory and Applications	(LNM 1057)	"
	92. Numerical Methods in Fluid Dynamics	(LNM 1127)	"
1984 -	93. Harmonic Mappings and Minimal Immersions	(LNM 1161)	"
	94. Schrödinger Operators	(LNM 1159)	"
	95. Buildings and the Geometry of Diagrams	(LNM 1181)	"
1985 -	96. Probability and Analysis	(LNM 1206)	"
	97. Some Problems in Nonlinear Diffusion	(LNM 1224)	"
	98. Theory of Moduli	(LNM 1337)	"
1986 -	99. Inverse Problems	(LNM 1225)	"
	100. Mathematical Economics	(LNM 1330)	"
	101. Combinatorial Optimization	(LNM 1403)	"
1987 -	102. Relativistic Fluid Dynamics	(LNM 1385)	"
	103. Topics in Calculus of Variations	(LNM 1365)	"
1988 -	104. Logic and Computer Science	(LNM 1429)	"
	105. Global Geometry and Mathematical Physics	(LNM 1451)	"
1989 -	106. Methods of nonconvex analysis	(LNM 1446)	"
	107. Microlocal Analysis and Applications	(LNM 1495)	"
1990 -	108. Geoemtric Topology: Recent Developments	(LNM 1504)	"
	109. H_∞ Control Theory	(LNM 1496)	"
	110. Mathematical Modelling of Industrical Processes	(LNM 1521)	"
1991 -	111. Topological Methods for Ordinary Differential Equations	(LNM 1537)	"
	112. Arithmetic Algebraic Geometry	(LNM 1553)	"
	113. Transition to Chaos in Classical and Quantum Mechanics	(LNM 1589)	"
1992 -	114. Dirichlet Forms	(LNM 1563)	"
	115. D-Modules, Representation Theory, and Quantum Groups	(LNM 1565)	"
	116. Nonequilibrium Problems in Many-Particle Systems	(LNM 1551)	"

1993 -	117.	Integrable Systems and Quantum Groups	(LNM 1620)	Springer-Verlag
	118.	Algebraic Cycles and Hodge Theory	(LNM 1594)	
	119.	Phase Transitions and Hysteresis	(LNM 1584)	"
1994 -	120.	Recent Mathematical Methods in Nonlinear Wave Propagation	to appear	"
	121.	Dynamical Systems	(LNM 1609)	"
	122.	Transcendental Methods in Algebraic Geometry	to appear	"
1995 -	123.	Probabilistic Models for Nonlinear PDE's	(LNM 1627)	"
	124.	Viscosity Solutions and Applications	to appear	"
	125.	Vector Bundles on Curves. New Directions	to appear	"

Printing: Weihert-Druck GmbH, Darmstadt
Binding: Theo Gansert Buchbinderei GmbH, Weinheim